T0073042

Sustainable Agricultural Chemistry in the 21st Century

Agriculture is one of the oldest and most global human enterprises, and as the world struggles with sustainable practices and policies, agricultural chemistry has a clear role to play. This book highlights the ways in which science in agriculture is helping to achieve global sustainability in the twenty-first century, and demonstrates that this science can and should be a leading contributor in discussions on environmental science and chemistry. The four drivers of this subject are presented, those being economic, environmental, regulatory and scientific, and help showcase agricultural chemistry as a dynamic subject that is contributing to this necessity of global sustainability in the twenty-first century.

Sustainable Agricultural Chemistry in the 21st Century

Green Chemistry Nexus

William M. Nelson

CRC Press
Taylor & Francis Group
Boca Raton London New York

CRC Press is an imprint of the
Taylor & Francis Group, an **informa** business

Designed cover image: Shutterstock

First edition published 2023
by CRC Press
6000 Broken Sound Parkway NW, Suite 300, Boca Raton, FL 33487-2742

and by CRC Press
4 Park Square, Milton Park, Abingdon, Oxon, OX14 4RN

CRC Press is an imprint of Taylor & Francis Group, LLC

© 2023 William M. Nelson

Library of Congress Cataloging-in-Publication Data
Names: Nelson, William M., author.
Title: Sustainable agricultural chemistry in the 21st Century : green chemistry nexus / William M. Nelson.
Other titles: Green chemistry nexus
Description: Boca Raton, FL : Taylor and Francis, 2023. |
Includes bibliographical references and index.
Identifiers: LCCN 2022057821 (print) | LCCN 2022057822 (ebook) |
ISBN 9780367741211 (hardback) | ISBN 9780367744588 (paperback) |
ISBN 9781003157991 (ebook)
Subjects: LCSH: Agricultural chemistry. |
Green chemistry. | Environmental chemistry.
Classification: LCC S585 .N39 2023 (print) |
LCC S585 (ebook) | DDC 630.2/4–dc23/eng/20230215
LC record available at https://lccn.loc.gov/2022057821
LC ebook record available at https://lccn.loc.gov/2022057822

ISBN: 9780367741211 (hbk)
ISBN: 9780367744588 (pbk)
ISBN: 9781003157991 (ebk)

DOI: 10.1201/9781003157991

Typeset in Times
by Newgen Publishing UK

Contents

Preface

Agriculture and agricultural chemistry face a multidimensional challenge that is unprecedented in the history of the world. To meet the goals for global sustainability, it must produce more food, produce less waste, and have minimal environmental impact. The success of this work in agriculture is critical to the survival of our world as we know it now. In the face of an increasing global population, decreasing availability of arable land, shrinking water supplies, and climate change, this challenge is even more daunting.

Historical agricultural practices have enabled us to increase food production to produce enough to feed the entire world. However, the use of crop protection chemicals and intensive farming practices has left in its wake an unsustainable amount of pollution and greenhouse gas emissions. These issues were not envisioned in the Industrial Revolution nor the Green Revolution, but they stand as massive obstacles to the Earth's survival.

Winds of change have begun to clear the air enough to provide a vision of hope. The call for environmental protection and pollution prevention caused us to re-examine the ways we conducted business, and even the ways we provide for the necessities of life like food (and employment.) We owe a debt of gratitude to those who heard the cry of "Silent Spring."[1] Green chemistry continues to challenge us to re-think how we do chemistry and this has rippled throughout the natural and engineering sciences, as well as chemistry. This has led to actions being taken by individuals, by universities, by industries, by cities, and by entire countries! We are stewards of our planet, and we must work with the environment and with nature to secure its (our) survival.

Green and sustainable chemistries provide insights, knowledge, and tools. These find applications in the practice of agriculture. This practice is changing and evolving to meet the demands of UN Sustainable Development Goals.[2] In the twenty-first century agriculture and agricultural chemistry are using Smart and Intensive agricultural practices to foster Smart Intensive Agriculture. In the process of doing this, they are working to achieve circularity, the hallmark of true sustainability. In this book the reader will see how these efforts are building upon past and current research endeavors. More importantly, the book provides a lens to focus our vision on the needed paths to take to achieve sustainable agricultural chemistry.

The transformation of agriculture and agricultural chemistry from being the major contributor to environmental problems to a potential avenue to sustainability is still a work in progress. Utilizing a vision of circularity that can lead to the valorization of agricultural waste, this field is not only interesting, but it is exciting in how it is assuming a position of leadership. The epitome of the expression of sustainability in agriculture and agricultural chemistry will be found in the biorefinery, which takes full advantage of all the advances occurring within agricultural chemistry, closes the loop of food production, and demonstrates a way this can be accomplished.

If you are concerned about the environmental and ecological direction of our world, this book lays out a path to consider and then focuses it on the areas of agriculture and agricultural chemistry. The book weaves the information into a logical and cohesive story. It is very readable and will challenge the reader to at least consider that we are moving in a positive direction toward sustainability. Are we "there yet" (as we always ask during a car ride)? NO. But we are moving forward!

What will you learn from this book? Fundamentally, every development in green chemistry and engineering can educate the future of agriculture. Additionally, any current advance in the internet of things (IoT), robotics, nanochemistry, and artificial intelligence (AI) will find expression in food production. And maybe most interesting, and certainly desirable, food waste can be valorized into energy and useful products for society.

When you work on something that is so important to you, the time spent is its own reward. The importance of environmental and global sustainability motivates me. I live with family and friends who support me in this passion. My wife Millie and my children (Maria, Milee, Liam, and Maddie) have supported me throughout, and have given me time and loving understanding. Additionally, my friends have also encouraged me to continue in this worthwhile endeavor. Finally, my faith and belief in God give me reason to continue to work for our Earth.

This book is ultimately for the future generations of the world who will benefit from the success of the work we have started toward sustainability....

William M Nelson
Environmental Laboratory
ERDC, US Army Corps of Engineers
(217) 369–2335
mahernelson@gmail.com

REFERENCES

1. Boaler, S.B. and G.D. Holmes, *Silent Spring by Rachel Carson.* Forestry, 1964. **37**(1): pp. 115–116.
2. O'Riordan, T.J.C., *UN Sustainable Development Goals: How can sustainable/green chemistry contribute? The view from the agrochemical industry.* Current Opinion in Green and Sustainable Chemistry, 2018. **13**: pp. 158–163.

Author Bio

William M. Nelson is Research Physical Scientist at the Engineering Research and Development Center for the US Army Corps of Engineers in Vicksburg, MS. He received a BS in chemistry from the University of North Carolina in Chapel Hill in 1972, an MA in chemistry from Washington University in St. Louis in 1982, and a PhD in organic chemistry from the Johns Hopkins University in Baltimore in 1989, during which time he studied the synthesis and photobiology of analogs of the environmental carcinogen benzo[a]pyrene. After that he did postdoctoral studies in photochemistry at the University of Illinois in Champaign–Urbana. After leading R&D for a small pharmaceutical company, he worked for the State of Illinois, helping companies minimize waste and prevent pollution. Since 1995 he has been involved with Green and Environmental chemistries while working in the US Environmental Protection Agency and the US Food and Drug Administration, and teaching at colleges and universities. During all this time he has also served as Supervisor, Branch Chief, and a mentor to many aspiring scientists. The emphases in his research have been on organic reaction mechanisms, preparation of alternative cleaning technologies, green solvents, photocatalysis, and synthesis of energetic compounds. He has jointly authored 2 patent applications and has published/presented more than 40 papers.

1 Criteria for sustainable agricultural chemistry

The photo of our earth from space is an incredibly beautiful sight (See Figure 1.1). Even more amazing are the many natural systems sustaining and supporting life on this beautiful planet. To begin the task of presenting sustainable agricultural chemistry, we first need to envision global sustainability. To understand agricultural chemistry, we must begin with the broadest depiction of the systems that describe our earth.

1.1 ENVIRONMENTAL SCIENCE

Environmental science describes how all areas of the environment are related and how they are mutually dependent. This field includes physical, biological and information sciences (including ecology, biology, physics, chemistry, plant science, zoology, mineralogy, oceanography, limnology, soil science, geology and physical geography, and atmospheric science) in order to analyze and study the environment, and in the process lay the foundation for sustainability. From this effort we obtain an integrated, quantitative, and interdisciplinary approach to the study of environmental systems.[1]

The results of environmental science lead to the understanding earth processes, evaluating alternative energy systems, pollution control and mitigation, natural resource management, and the effects of global climate change. Fundamentally, these become a systems approach to addressing the challenges of maintaining environmental equilibrium. We have learned that the systems are interrelated.

More importantly, these are active fields of scientific investigation, and they are driven by (a) the need for a multi-disciplinary approach to analyze complex environmental problems, (b) the development of substantive environmental laws requiring specific environmental protocols of investigation, and (c) the need to expand public awareness of a need for action in addressing environmental problems. Let us begin our discussion by examining the four environmental spheres.[2]

DOI: 10.1201/9781003157991-1

FIGURE 1.1   Earth from Space. (Use www.shutterstock.com/image-photo/planet-earth-starry-sky-solar-system-1666159696)

1.1.1   FOUR SPHERES

We begin by envisioning the earth consisting of four major, interconnected systems. (see Figure 1.2) Everything in Earth's ecosystem can be placed into one of four major subsystems: land, water, living things, and air. These four subsystems are often referred to as "spheres." They are like four dimensions of the earth which work simultaneously, independently, but also interdependently. In this book we will focus mostly on the "lithosphere" (land), but the actions of agriculture will influence, and be influenced by the "hydrosphere" (water), "biosphere" (living things), and "atmosphere" (air).

Scientifically, the ways these spheres interact is fascinating. Through chemical/physical processes, the different spheres are able to exchange mass and/or energy amongst each other. The exchange of these moderate the balance of the total ecology of the earth. However, the total mass/energy remains constant; it is only the relative proportions that change. An example is how the soil draws resources from water/air and promotes the growth of plants. In terms of size, the physical dimensions of the lithosphere (geosphere or solid Earth) only refers to the uppermost layers of the solid Earth (oceanic and continental crustal rocks and uppermost mantle).[4]

The most relevant actions for the sustainability of the lithosphere occur in the biosphere. This is also called the zone of life on Earth. The biosphere is virtually a closed system with regards to matter, with minimal inputs and outputs. With regards to energy, however, it is an open system, with photosynthesis capturing

FIGURE 1.2   The Four Spheres.[3] (Use www.shutterstock.com/image-vector/carbon-cycle-diagram-nature-farm-landscape-2040984617)

solar energy. It is a self-regulating system close to energetic equilibrium.[5] The biosphere is the global ecological system integrating all living beings with their relationships, and interaction with the elements of the other spheres. The biosphere is heavily reliant on the lithosphere for survivability and actions within the biosphere affect sustainability.

The atmosphere of Earth is the layer of gases, retained by Earth's gravity, surrounding the planet Earth and forming its planetary atmosphere. The atmosphere of Earth protects life on Earth by creating pressure allowing for liquid water to exist on the Earth's surface, absorbing ultraviolet solar radiation, warming the surface through heat retention (greenhouse effect), and reducing temperature extremes between day and night (the diurnal temperature variation). The quality of gases and relative amounts of chemicals will affect the lithosphere. This dynamism directly influences the sustainability of the lithosphere.[6] Air composition, temperature, and atmospheric pressure vary with altitude, and air suitable for use in photosynthesis by terrestrial plants and breathing of terrestrial animals is found only in Earth's troposphere and in artificial atmospheres.

Lastly, the hydrosphere establishes the system of waters as a conduit of nutrients needed for growth in the lithosphere. The hydrosphere is the combined mass of water found on, under, and above the surface of the earth. Earth's hydrosphere has been around for about 4 billion years and it continues to change in shape. This is caused by seafloor spreading and continental drift, which rearranges the land and ocean. Saltwater accounts for 97.5% of this amount, whereas fresh water accounts for only 2.5%. Of this fresh water, 68.9% is in the form of ice and permanent snow cover in the Arctic, the Antarctic and mountain glaciers; 30.8% is in the form of fresh groundwater; and only 0.3% of the fresh water on Earth is in easily accessible lakes, reservoirs and river systems. From this 0.3% of the water, nutrients and minerals flow through the lithosphere. [7, 8]

1.1.2 DYNAMIC INTERACTION AMONG SPHERES

The dynamic interaction among the spheres defines the boundary conditions for the growth of the environment. A detailed description of this complex harmony is beyond the scope of this book but suffice it to note that a change in one area can cause a change in another. These dynamic processes, as we will see, will force changes and choices in the realm of agricultural chemistry.[9]

Throughout this book, as we discuss agricultural chemistry, we will identify the interdependence of the spheres and their role in influencing agricultural activity which affects the sustainability of our planet. The most important flows in the earth system are those concerned with the transfer of energy and the cycling of key materials in biogeochemical cycles.

1.1.3 FOUNDATION OF SUSTAINABILITY

Fundamentally, the dynamic beneficial interactions among the spheres lead to sustainability. (See Figure 1.2) This interdependence is the mechanism for the planet to maintain a healthy environment through an integrated harmony. As scientists, as humans, we must recognize and support this interdependence. Underlying these systems are the foundational three pillars of sustainability – environment (planet), social (people) and economic (profit). It will become clear through this book how they can promote sustainability.

Sustainability is linked to the livelihood of future generations, and this provides the proper perspective. (See Figure 1.3) For a particular process to be sustainable, it should not cause irreversible change to the environment, it should be economically viable, and it should benefit society. In this view, sustainability becomes the synergy between society, economics, and environment. The scientific and economic factors are drivers for growth, profit, reducing costs, and investments into research and development. They also meet the demands of global food requirements, which is the domain of agricultural chemistry. The combination of economic and environmental interests demand global stewardship for sustainable use of natural resources. This is also the domain of agricultural chemistry.

1.2 AGRICULTURAL CHEMISTRY

Agricultural chemistry is the study of chemistry, especially organic chemistry and biochemistry, as they relate to agriculture – agricultural production, the processing

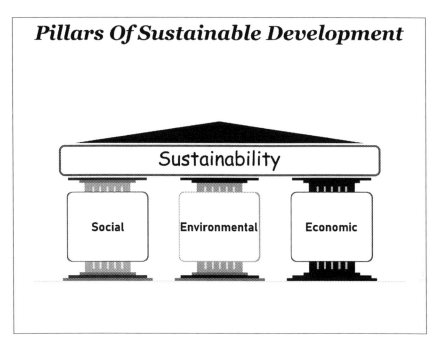

FIGURE 1.3   Pillars of Sustainability.[10] (See www.shutterstock.com/image-vector/
three-pillars-sustainable-development-icons-vector-2154172365)

of raw products into foods and beverages, and environmental monitoring and
remediation.[11] The growth, development, and nourishment of society relies on
agriculture and agricultural chemistry. Agricultural chemistry studies the chemical
compositions and reactions involved in the production, protection, and use of
crops and livestock; all of these are germane to life in the twenty-first century. This
chemistry includes all the life processes through which humans obtain food and
fiber for themselves and feed for their animals. Applied science and technology are
directed toward control of those processes to increase yields, improve quality, and
reduce costs.

An essential component of agricultural chemistry is the soil, which forms the
interface between the hydrosphere, lithosphere, atmosphere, and biosphere. Soils
are complex assemblies of solids, liquids, and gases. For example, in a typical
silt loam soil that is ideal for plant growth the solid component in the surface
represents about 50% of the volume (about 45% mineral and 5% organic matter),
gases (air) comprise about 20–30%, and water typically makes up the remaining
20–30%. The inorganic components of soils represent more than 90% of the solid
components. Its properties such as size, surface area, and charge, greatly affect
many important equilibrium and kinetic reactions and processes in soils. This
influences the movement of mass and energy throughout the four spheres.

Agriculture often places significant pressure on natural resources and the environment. Through the exchange of mass and energy, the soil in dynamic interconnectivity with all four spheres. Sustainable agricultural practices are necessary to protect the environment, expand the Earth's natural resource base, and maintain and improve soil fertility. This contributes to Earth's sustainability. Agriculture's role in sustainability must be to rejuvenate the soil and environment, so as to maintain its fecundity and productivity.

Sustainable agricultural chemistry development is critical to global sustainability. Sustainable farming (agriculture) employs practices that should contribute to the equilibrium. Sustainable farming draws upon all environmental methods and practices that are economically viable, environmentally sound, and protect public health. This contributes to the growth of nutritious and healthy food as well as brings up the standard of living of the farmer.

Agricultural chemistry promotes sustainability by expanding our knowledge of the causes and effects of biochemical reactions related to plant and animal growth, to reveal opportunities for controlling those reactions, and to develop chemical/agricultural products that will facilitate that control. We can measure the growth and maturity of agricultural chemistry through an evaluation of the chemistry involved in the practice.

Agricultural chemistries include the chemical materials developed to assist in the production of food, feed, and fiber, including herbicides, insecticides, fungicides, and other pesticides, plant growth regulators, fertilizers, and animal feed supplements. Chief among these groups from the commercial point of view are manufactured fertilizers, synthetic pesticides (including herbicides), and supplements for feeds. In the chapters that follow, it will be important to assure these are effective, efficient, and environmentally sound.

When agriculture is considered with ecology, chemistry becomes an effective tool to measure sustainability. The wise use of natural resources is one of the significant ways that agriculture, through agricultural chemistry contributes to sustainability. Responsible use of natural resources includes consideration of social and scientific responsibilities in production and land use. Responsibility for land and natural resources involve maintaining or enhancing the quality of these resources and using them in ways that allow them to be regenerated for the future, while producing enough food for current needs.

An agroecosystems and food systems perspective can help evaluate the contributions of agricultural chemistry to sustainability. Agroecosystems in the broadest sense will include all chemistry involved from field to farm. A systems approach also gives us the tools to assess the impact of human society and its institutions on farming and its environmental sustainability.

The connection to sustainability occurs when the production of food and agricultural chemical activities affects basic natural resources. This is a nexus point: sustainability demands, agricultural chemistry processes, and green chemistry intersect. At this juncture, the ability of future generations to produce and flourish is paramount. A sustainable agriculture approach seeks to utilize natural resources in such a way that they achieve the necessary immediate goals, while regenerating their productive capacity, and minimizing harmful impacts. By incorporating green chemistry at this point agroecosystems can utilize sustainable chemistry and environmental harmony, so that it is possible to maintain an economically viable production system with fewer potentially toxic effects.

1.3 COMPONENTS OF SUSTAINABLE AGRICULTURAL CHEMISTRY

Current agricultural science builds upon centuries of previous practices. These practices were guided by different cultural, societal, and physical demands. Heightened awareness of the need to develop more responsible and sustainable chemistries is a compelling factor in this century. This new imperative will become stronger as the world's population approaches 9 billion and the planet's natural resources become more limited. This book presents a multi-dimensional approach to the ways in which agricultural chemistry promotes and drives sustainability.

1.3.1 FOUR DRIVERS

The central importance of this book is built on the Four Drivers of Sustainability:

* Environmental Driver
* Economic Driver
* Regulatory Driver
* Scientific Driver

The four drivers and their effects on sustainability in the twenty-first century are presented in the first section of the book. The discussions showcase agricultural chemistry as a dynamic subject that is contributing to this necessity of global sustainability. Primarily it does this in four ways:

* Using earth's resources in an environmentally responsible manner to produce, preserve, and deliver food;
* Providing food to feed a growing planet, as economically as possible;
* Being a strong player in Environmental Science and Chemistry, to work within national and international guidelines; and
* Participating in the synergism of scientific research and development through Green Chemistry.

As the world struggles to incorporate and use sustainable practices and policies regarding environmental protection, agricultural chemistry has a clear role to play. Agricultural chemistry must lead by example. The chemistries involving food production must be efficient, but also must evolve and change in order to meet new restrictions and footprint guidelines. In this process, agricultural chemistries can provide insights, new chemical understanding, and green processes. Importantly, this leads to a better appreciation of the four spheres discussed earlier and how we as a global population must maintain their equilibrium.

The agricultural chemistries presented in this book illustrate that this science can and should be a leading contributor in discussions on environmental science and chemistry. Science involving agriculture is essential to human survivability and it is important to situate agricultural chemistry in the context of global sustainability. The chapter topics reveal that the chemistries currently practiced in agriculture are both fundamental yet exquisite in their efficiency and simplicity. Incorporating recent developments in Green Chemistry can offer new directions in the science undergirding agriculture. This can, in turn, establish a dynamic reciprocity where agricultural and green chemistries are mutual beneficiaries.

This leads to the role that existing and developing agricultural chemistries will play in sustainability in the twenty-first century. By describing new products, processes, applications, and development in this book, we will demonstrate that the dynamic synergism in agriculture shows how this field is adapting in the new age, while helping to ensure its sustainability.

This book will show the ways in which chemistry in agriculture is incorporating the four drivers while helping to achieve global sustainability in the twenty-first century. The drivers are elaborated later, as we discuss what is being done, where there are gaps, and where the world needs to go. After presenting criteria for Sustainable Agricultural Chemistry, we will explore how each of the four drivers, both individually and in tandem, influence and at times determine how agriculture is practiced. Finally we will look to the future, trying to trace the trajectories of Agricultural and Sustainable Chemistries.

1.4 CONTRIBUTING TO THE SUCCESS OF SUSTAINABILITY THROUGH A GREEN CHEMISTRY NEXUS

1.4.1 GUIDANCE FROM GREEN CHEMISTRY

Green chemistry principles (GCP) are recommended in research, chemical industries, in industrial management, governmental policy, educational practice, and technology development.[12] These principles must balance economic growth, resource sustainability, and environmental protection. GCP encompass (i) pollution and accident prevention, (ii) safety and resource sustainability, and (iii) energy and resource sustainability. To integrate GCP into sustainability, agricultural chemistry needs to include five strategies:

- establishment of cross-departmental collaboration,
- development of cleaner production and green products,
- provision of integrated chemical management system,
- implementation of green chemistry education program, and
- construction of a business model.

Throughout the book these will be illustrated with specific examples and relevant references.

1.4.2 RESPONSE OF AGRICULTURAL CHEMISTRY TO THE WORLD

Agriculture is facing an enormous challenge: It must ensure that enough high-quality food is available to meet the needs of a continually growing population. It must do this in an environmentally responsible (aka, sustainable) manner. Current and future agronomic production of food, feed, fuel, and fibre requires innovative solutions for existing and future challenges, such as climate change, resistance to pests, increased regulatory demands, renewable raw materials or requirements resulting from food chain partnerships. Agrochemical industry will have the opportunity to shape the future of agriculture by delivering innovative integrated solutions.

1.4.3 LITHOSPHERE

Let us return to our initial discussion in this chapter: the four spheres. Soil science must play a crucial role in meeting present and emerging societal needs of the twenty-first century and beyond for a population expected to stabilize around 9 billion while having increased aspirations for a healthy diet and a rise in the standards of living. Soil resources must be managed regarding numerous global needs through interdisciplinary collaborations. Challenges facing the lithosphere include mitigation of global warming; improvement of quantity and quality of freshwater resources; minimizing desertification; serving as a repository of waste; meeting growing energy demands; and rapidly urbanizing and industrializing societies. These diverse and complex demands on soil resources necessitate a shift in strategic thinking and conceptualizing sustainable management of soil resources. In ways that will become clear in this book, agricultural chemists must stay on the cutting edge of science, while responding to the demands of society.

Pursuit of sustainability, being a moral, ethical, and political challenge, must be addressed in cooperation with economists and political scientists. Agricultural chemists must work in cooperation with industrial ecologists and urban planners toward sustainable development and management of aspects of food production in urban and industrial ecosystems. More than half of the world's population (3.3 billion) live in towns and cities, and the number of urban dwellers is expected to increase to 5 billion by 2030. Thus, the study of urban soils for industrial use, human habitation, recreation, infrastructure forestry, and urban agriculture is a high priority.

REFERENCES

1. Bringezu, S. and R. Bleischwitz, *Sustainable resource management global trends, visions and policies*. Sustainable Resource Management: Global Trends, Visions and Policies. 2017. Taylor and Francis. 1–338.

2. Larson, P., *The four spheres of Earth.* 2016. Huntington Beach, CA: Teacher Created Materials.

3. Navarro, S.J. *Four spheres.* For Earth and Life Science Senior High School. 2017 [cited 2021, May 23, 2021].

4. Mbachu, O., et al., *The rise of artificial soil carbon inputs: Reviewing microplastic pollution effects in the soil environment.* Science of the Total Environment, 2021. **780**: pp. 1–12.

5. Nielsen, S.N., *Reductions in ecology and thermodynamics. On the problems arising when shifting the concept of exergy to other hierarchical levels and domains.* Ecological Indicators, 2019. **100**: pp. 118–134.

6. Kleppel, G.S., *Do differences in livestock management practices influence environmental impacts?* Frontiers in Sustainable Food Systems, 2020. **4**: pp. 1–15.

7. Sapozhnikov, Y.A., *Isotopic properties of earth hydrosphere.* Water Resources, 2011. **38**(6): pp. 837–839.

8. Manahan, S.E., *Water chemistry: Green science and technology of nature's most renewable resource.* Water Chemistry: Green Science and Technology of Nature's Most Renewable Resource. 2010. CRC Press. 1–387.

9. Fenelon, J. and J. Alford, *Envisioning indigenous models for social and ecological change in the anthropocene.* Journal of World-Systems Research, 2020. **26**(2): pp. 372–399.

10. Boussemart, J.P., et al., *Performance analysis for three pillars of sustainability.* Journal of Productivity Analysis, 2020. **53**(3): pp. 305–320.

11. *Agricultural chemistry.* [cited 2021, May 24, 2021].

12. Anastas, P. and N. Eghbali, *Green Chemistry: Principles and Practice.* Chemical Society Reviews, 2010. **39**(1): pp. 301–312.

2 Agricultural chemistry in global sustainability

2.1 SUSTAINABILITY AND SUSTAINABLE DEVELOPMENT

According to the Worldometer, the current world population was 7.8 billion people as of August 2020, and it is projected to be 10 billion people in 2057.[1] The high growth of world population growth is adding increasing pressures on global issues pertaining to environmental problems and food security. This will adversely affect the Sustainable Development Goals (SDGs) proposed by the United Nations. (See Figure 2.1) The concepts of sustainability and sustainable development are relevant not only in geophysical, but also in scientific research about environmental issues and industrial and agricultural production.[2] While the concept of sustainable development is often associated with that of sustainability and thus both terms are used as synonyms, their interaction is key to the success of agriculture and agricultural chemistry (AC).[3] Though it is not totally accurate, the definition of sustainability that is most commonly used comes from the World Commission on Environment and Development in 1987, which defined the concept of sustainable development as "...development that meets the needs of the present without compromising the ability of future generations to meet their own needs..."[4] Throughout this book, soil lies at the nexus of crucial societal and environmental challenges, offering a means to address food security and sustainability, addressing poverty, access to water, biodiversity loss, and climate change.[5] AC plays an essential and indispensable role in agricultural processes.

The United Nations 2030 agenda produced what are known as the Sustainable Development Goals (SDGs). These goals present the milestones needed to achieve sustainability. These areas include hunger eradication, poverty reduction, and sustainable farming.[6] Sustainable agriculture is positioned as the most critical sector of the SDGs, and its success affects most of the 17 SDGs, directly and indirectly. Both the SDG2 (Zero Hunger) and SDG1 (No Poverty) will require the greatest assistance from agriculture, while SDG9 (innovative technology and infrastructure) and SDG7 (affordable energy) have less reliance on agriculture. The goals of eight SDGs are centered on sustainable production and utilization, reliance on biodegradable feedstock, and environmental protection, which are to some extent the goals of green chemistry (GC) principles.[7]

DOI: 10.1201/9781003157991-2

FIGURE 2.1 The UN sustainable development goals. (www.shutterstock.com/image-vec tor/sdgs-17-development-goals-environment-2033729978)

As we explained in the Introduction, the earth's physical environment is in a constant process of maintaining equilibrium. It is important to deepen our knowledge about the planet scientifically and holistically and acknowledge the complexity of the processes taking place on it. The four spheres become our way of tracking the movement of mass and energy to understand how this occurs. Through this analysis the ways that AC contributes to sustainability become clear. There is, however, no unique way this contribution is accomplished, but the science and the art of AC attempt to assist nature in maintaining homeostasis, while ensuring food security in the context of global sustainability. This is a complex effort with many dimensions. An example that would illustrate this might be companies that are developing technologies that offer routes from natural gas to olefins and aromatics via methane to methanol. A potential feedstock option is plant-based bioethanol, made predominantly from corn, sugarcane, or biomass from agricultural and wood waste. Using wastes or residues or nonfood biomass for biofuels and biofeedstock generation will reduce pressure on land use, have improved CO_2 performance, improve fuel properties, and may be noncompetitive with the food chain.[8] Does science and economic capital exist to make this a reality?

2.2 AGRICULTURAL CHEMISTRY (AC) THROUGH THE LENS OF THE FOUR SPHERES

As explained in the previous chapter, the natural environment can be regarded as being composed of four spheres (lithosphere, hydrosphere, atmosphere, and biosphere). The activities within each one and among them forms an interlocking puzzle that influences the drive to sustainability. (See Figure 2.2)

2.2.1 LITHOSPHERE

In addition to meeting increasing demands on agricultural production to support more people with adequate food, the land, or lithosphere, plays an

FIGURE 2.2 The four spheres of the earth's environment: Lithosphere, hydrosphere, atmosphere, and biosphere. (www.shutterstock.com/image-vector/continuous-line-draw ing-hands-holding-terrestrial-2170048301)

increasingly important role in providing a wider variety of roles such as flood control, water purification, and cultural and esthetic values; in lowering global warming by capturing carbon emissions in vegetation and soils; and in the support and preservation of biodiversity.[5] Soil can be difficult to envision as environmentally significant partly because it often appears as "just dirt", seemingly featureless and inert.[9] However, soil is a vital environmental resource and plays ecological, social, and economic functions which are fundamental for life. Its availability to future generations is essential to sustainability. AC must identify soil degradation processes. These will include erosion, decline in organic matter, local and diffuse contamination, sealing, compaction, decline in biodiversity, salinization, floods, and landslides. Using a better understanding of the mass transport, the residence times of elements, nutrients, or contaminants in soil are known to depend on the mobility of its predominant chemical forms. Continuous agriculture practices have led to sustainability problems linked with production decreases, environmental contamination, crop failure, and soil degradation in many regions of the world.[10] In the past AC has contributed to the present problem, and now it can provide leadership to its solution. There are management alternatives, such as crop-pasture rotations that contribute to environmental sustainability and productive diversification, but a better understanding of the dynamic lithosphere is needed. Long-term experiments and research in AC will play a major role in evaluations of the impact of different management practices on soil quality.

The integrity of soil is a dynamic and direct result of the activity of many different organisms (including humans). Soil structure is a complex assemblage of substances and biological relations, where the main processes and relations occur outside and between sections of the environment, and which may be disrupted at any time.[9] Soil degradation is often associated with a decrease in the organic matter content, mainly caused by soil use change and global warming. Improving the quality of organic matter in soil or preventing its reduction has positive effects on soil and water quality, crop yields, biodiversity, and climate leading to a reduction of greenhouse gas emissions from soil to the atmosphere. To accomplish this there needs to be more known about the main processes which govern organic matter stabilization. This requires approaches at both molecular and multidisciplinary levels.[10]

In soils it will be important to assess not simply the types and quantities of chemicals present, but also the risk/toxicity of its concentration. Included in this will be the role of remediation approaches to toxic compounds and elements due to their broad sustainability and applicability. The four spheres help us to understand how the released hazardous materials enter into the food chain and then are biomagnified into living beings via food and vegetable consumption and originate potentially health-threatening effects. A cellular, molecular, and nano-level understanding of the pathways and reactions reveals the reasons for accumulation of potentially toxic compounds and elements. These approaches can enable the development of crop varieties with highly reduced concentrations of these substances in them.[11]

One approach to sustainability is in the production of energy crops, but this also requires a maximization of the current use of soil and the achievement of as much production as possible. As we will see later in this chapter, this can be met by the increasing emphasis on biorefinery crops like sugarcane, one of the most important energy crops. Inevitably, its production will produce side effects: more varieties, pests and diseases, nutrients requirements, mechanical harvest, residue management, and water use. Sugarcane products and coproducts are valuable, and this is highlighted by an energy coming from sugarcane ethanol. The sugarcane sector is a model in the gain of production and in the employment of technologies that reduce the use of chemical defensives, as, for instance, the biological control of pests. Combining these facts, sugarcane expansion provides a possible way that is less harmful to the environment, with land use changes that allow for a smaller emission of greenhouse gases and greater sustainability of the adopted production systems.[12]

If we look at the world, the necessary volume of food production is increasingly dependent on resource-poor rural communities; depleted soils must increasingly provide more crops and yet are further diminished by the increasing threat of pollution and climate change. These areas experience large burdens of environmental pollution due to the expansion of commodity traffic and industrial development. While the major challenges to sustainability come from agricultural production, significant threats also come from mining and extractive industries, urban growth, waste dumping, and infrastructure and energy development.[4] AC has shown that there is a positive and significant correlation among organic carbon, nitrate- and ammonia-nitrogen, phosphorus, soil respiration, and dehydrogenase activity that clearly results in increased soil nutrients with the increases in microbial and other biotic activity. The solution for soil longevity and productivity lies in strategies for nurturing fertility-building soil fauna and managing degraded systems by adopting suitable agricultural practices and in solutions involving smart management.[5]

2.2.2 ROLE OF WATER (HYDROSPHERE)

The earth's environmental system is facing multiple challenges to its supply of fresh water (e.g. climate change, water shortages, ecosystem degradation), while the need remains to ensure food and water security for a growing world population. The desired climate goal to achieve the 1.5 °C target (more about this later) without jeopardizing sustainable development goals (SDGs) such as achieving water security.[13] Freshwater availability and stress is particularly challenging for agriculture. Water shortages, affecting about 1.4–4 billion people already depending on the chosen metric, may strongly increase in the future due to population growth and other impacts of global climate change. From a sustainability perspective, it is important to understand how additional water use for bioenergy production and the emerging biorefinery will change the balance needed to maintain a sustainable supply of potable water.

Water and energy cycles are typically poorly integrated into regional, national, continental, and global planning on climate change reduction, conservation, land use, and water management. The use of science can help protect our planet's climate and life-sustaining functions. Models and studies show that forest, water, and energy interactions provide the foundations for carbon storage, for cooling terrestrial surfaces and for distributing water resources. Forests, trees, and agricultural activities must be recognized as prime regulators within the water, energy, and carbon cycles. Our understanding of how foliage influences water, energy, and carbon cycles has important implications as for how it might be used to improve sustainability, adaptation and mitigation efforts.[14] This will lead to novel low-pressure irrigation technologies which allow both reduced water and energy use. More studies are needed about how to transition from legacy technologies affecting water and energy use at the aquifer scale to more sustainable practices. Water usage and measures to ensure its availability are critical to sustainable AC in the twenty-first century.[15]

Agricultural water management is a critical issue in many parts of the world. Cost-effective water policies, recovery, and re-use are required to stabilize water use. The system currently used to manage groundwater abstraction and allocation depends upon the local geography. As the availability of water becomes strained, measures might include a quota system combined with a tax. Opposition to this use of economics must be measured so that it limits costs for farming.[12] The interplay among science, government, and economics will be important to balance. As an example, irrigation, drinking and using, industry, tourism, ecology related to the use and distribution of water sources compete for the dwindling supply of water. Thus, the determination of the role and importance of water usage and management require an integrated and interdisciplinary approach.[16]

If water, energy, and fuel are equally vital for sustainability, the challenge in prioritizing their use is daunting. As the global population grows in size and increasingly concentrates in cities with lifestyles based on greater material consumption, more attention must be given to an integrated system that will satisfy the needs for energy, water, and food. This system will have reciprocal effects on climate change and the ecosystem. New frameworks and tools are needed that will integrate the societal and technical dimensions. Transformative social and political change is needed to create new structures, markets, and governance to meet sustainable development goals.[17]

The future expected global water scarcity requires that scientists work on sustainable solutions for agriculture needs. Soils can aid in the resolution of some of these issues. Soils in aqueous systems are known to be filters for metals, phosphorus, and particulate matter transported in urban runoff.[18] Contamination of groundwater by compounds from natural soil sources or from anthropogenic sources is a concern to the public health. Remediation of contaminated groundwater if done by soils is valuable, since billions of people all over the world use it for drinking purposes. Addressing the sustainability issues, the technologies encompassing natural chemistry, bioremediation, and biosorption can be adopted in appropriate cases.[19] These, in turn, will advance sustainability. A major effort of AC is in the improvement of soil stability and increasing its water retention ability.[20]

2.2.3   ROLE OF AIR (ATMOSPHERE)

The quality of air is necessary for high productivity of agriculture. The health emergency linked to the spread of COVID-19 provided insights into the twenty-first century lifestyle and has led to reduction in industrial and logistics activities, as well as to a drastic changes in citizens' behaviors and habits. The restrictions on working activities, traveling, and movement imposed by the lockdown have had important consequences, including improvement in environmental quality. The results indicate that the lockdown during the first few months of 2020 reduced air pollution levels, compared to previous periods. The concentrations of particulate matter, nitrogen dioxide, sulfur dioxide, and carbon monoxide also decreased during this period. The blockage of production activities, the limitation of logistical activities and the closure of traffic are the factors that have strongly influenced the rapid improvement in air quality, thanks to the significant reduction of polluting substances.[21] These results are lessons that can improve decisions on future policies and strategies in industrial and logistics activities (including the mobility sector) aimed at their environmental sustainability.[21]

2.2.4   ROLE OF HUMANS (BIOSPHERE)

The effects from human activities can be felt directly in agriculture. A key process of global environmental change is change in land use. The purposes and activities through which people interact with land and terrestrial ecosystems cause many sustainability challenges.[22] Human land use directly influences the chemistry of the air, the diversity of plant and animal species, and the balance of global ecosystems. [5] Land science has grown to become an 'interdiscipline' with progress being achieved in several key areas. AC and researchers in physical and natural sciences are working to address critical knowledge gaps in our understanding of land system change with respect to human behaviors (cognitions, culture, and decision-making) for example; to better incorporate feedbacks between environmental change and human activities; land use-intensity and management; and globalization, world economics, and supply chain dynamics.[5]

The destruction of the ozone layer, together with global warming, is one of the major environmental topics today. The effect of human activities on atmospheric ozone, namely the increase of tropospheric ozone and the general diminution of stratospheric ozone and the production of the Antarctic ozone hole directly affects agricultural practices. Global warming, due to anthropogenic greenhouse gases emissions, reduces the possibility of achieving sustainability on Earth. However, many other threats are potentially just as serious and the degradation of agricultural land is a problem that will need to be addressed to ensure continued food production.[23]

2.3   FACTORS CHALLENGING SUSTAINABLE AGRICULTURE

There are many factors that are challenging the attainment of sustainability, and in particular sustainable agriculture. AC is directly involved in the potential solutions. The major factors are presented below, and they are not in order of importance.

2.3.1   SOIL INTEGRITY

Soil is more than just dirt; it sustains many dimensions of life on earth. In addition to increasing demands on agricultural production to support the increasing population, landscapes serve an important function in supporting services such as flood control, water purification, and cultural and esthetic activities; in stabilizing the environment by sequestration of carbon emissions in vegetation and soils; and in protection of biodiversity.[24] AC seeks to understand the near and long term impacts of how using farmland affects soil bacterial community and how it helps in maintaining soil health and sustainability.[25] Since the mid-1990s research has been done on the critical role of soils in climate change mitigation and adaptation. [26] It is hard to predict the effects of changes on climate, atmospheric composition, and land use on terrestrial ecosystems, including agriculture, forestry and soils, and to consider the terrestrial carbon cycle with an emphasis on underlying drivers and processes of contemporary and future carbon quantities (fluxes and pools).

Changing soil nutrient levels, especially total nutrients, can explain most of the variation in agricultural yield measured over time. While bacteria indirectly boost annual food output through enzyme activities and nutrient levels,[27] farmers have continuously resorted to the use of chemical fertilizers to maximize yields. The excessive use of chemical fertilizers plus the addition of nitrogen to reduce the soil pH, reduces the availability of the nutrients present in fertilizer and thus work against sustainable agricultural development. Using organic fertilizers that are rich in organic matter, humus and beneficial microorganisms is a means to reduce the use of chemical fertilizers. As we will explore later, the combined application of organic and inorganic fertilizers are important measures for sustainable agricultural development.[27]

2.3.2 WATER

Water stress is just one aspect of the wide-range of potential impacts of climate change. Similarly, also every technology designed to enhance the availability of water will entail (potentially not yet known) side effects, which can even be beneficial in some regions but detrimental elsewhere. Ensuring the water supply will require a more holistic analyses of the consequences of mitigation approaches required that take into account all dimensions of the complex earth system as it pertains to agriculture and AC.[13]

The availability of an adequate supply of clean water is one of the *sine qua non* of agriculture. Mitigation of climate change will be imperative to increase the supply of freshwater resources (among other benefits).[28] The agricultural need for water will extend beyond this in order to satisfy requirements for sustainability. For example, large-scale, biomass production and growing agricultural needs will increase the pressure in multiple environmental dimensions, resulting in increased competition for scarce freshwater resources.[17] This tension requires choices to be made in potential available water supply among the competing elements (agricultural production, lifestyle necessities, industrial applications, and human consumption.)

2.3.3 CARBON FOOTPRINT

How does the concept of carbon footprint impact the practice of AC in the twenty-first century? Carbon footprints and other categories of environmental footprints directly measure the effects of human activities in the planetary boundaries and quantify threats to human security. The footprint family of indicators usually consists of ecological, carbon or more precisely greenhouse gas and water footprints and also sometimes the energy footprint. The footprints that are important for ecosystem health in regard to water, health, food, and land and species security are nitrogen, phosphorus, biodiversity, and land footprints. These footprints measure the impact levels that the elements and activities have on the environment. The environmental footprints are tools and warnings for present and future action plans.[12]

The accelerated growth of human civilization over many centuries has left an increasing amount of problems. The expansion of manufacturing and commercial agriculture alongside continuing globalization has resulted in the widespread contamination of freshwater supplies with chemical toxins including persistent organic pollutants. Effective mitigation of such pollution is paramount to the safeguarding of human health, animal and aquatic life, and the environment all directly affected by agriculture. These are the measured targets of footprint indices.[12]

Agricultural chemistry in the twenty-first century is attempting to sustainably address these footprint targets. The strategy of the biorefinery integrated with circular bioeconomy in the perspectives of unravelling the global issues can help to tackle carbon management and greenhouse gas emissions. Biorefining is one of the most important facilitating strategies of the bio-based circular economy that closes the loop of fresh or raw resources, water, minerals, and carbon. They are sustainable bioprocesses that efficiently utilize biomass resources for the production of various marketable products and metabolites (e.g., carbohydrates, proteins, lipids, bioactive compounds, and biomaterials). [15] A waste biorefinery–circular bioeconomy strategy represents a low carbon economy by reducing the greenhouse gases footprint.[20] Bioprocesses utilizing waste resources to produce biomaterials and biofuels can greatly reduce fossil resources as the production feedstock and this prevents the natural resources from complete depletion. This approach sustains the energy–environment nexus and also protects the environment by mitigating the carbon footprints (i.e., Green House Gas (GHG) emission from burning fossil resources).[29] This benefits from the application of the full life cycle of a product through different pathways in cultivation stage which demonstrates the importance of decision-making process for further investment. [30]

2.3.4 ECOLOGY/ECONOMY

The issue of waste is huge, omnipresent, and still growing! Globally, 1.3 billion tons of food are wasted annually, with few uses other than landfilling, anaerobic digestion, or composting. In the twenty-first century food waste repurposing will provide an alternative waste management strategy in the efforts to achieve sustainability. As an example, an integrated biorefinery technology, which recovers and processes potato peel waste, has yielded multiple value-added products.[31] The efficient repurposing of these components for producing bioproducts, opens up a platform for circular economy by keeping resources in use for as long as possible, and deriving the maximum value from them. Interestingly, the lignocellulosic plant material and the unique compositions of some of the other food waste offer opportunities for profitable integrated biorefineries while addressing a key environmental challenge on the road to global sustainability.[31]

In a later chapter we will address how a sustainable and circular approach to waste disposal is critical to preserve the environment and human health. The waste biorefinery has great potential. Valorization of waste or side streams into bioprocessing to produce value-added bioproducts like biopolymers and bio-lipids remarkably contributes to a sustainable circular bioeconomy. A circular bioeconomy can reduce the GHG footprint which helps to resolve other global issues such as environmental protection and food security.[20]

2.3.5 NUTRIENTS AND FOOD SAFETY

Understanding the importance and role of nutrients in the global diet must be based on solid science. Likewise, ensuring their safety and amounts will be critical for meeting the health needs of the growing population. Objective dietary guidelines must lead to nutrient adequacy and positive health outcomes. The parameters are measured using consumption, rather than sustainable provisioning of food.[32] During the twenty-first century, in addition to adequate supplies, understanding and meeting standards will be essential. Currently 88% of the countries in the world for which data are available are struggling with the problem of poor nutrition. This indicates that the world is not on track to achieve the United Nations Sustainable Development Goals.[33]

Food security is closely tied to nutrition. The threats result from population growth, land availability for growing crops, a changing climate, the risks related to the use of agrichemicals, and a reliance on depleting fossil fuel reserves for food production. Government oversight challenges the safety and availability agrichemicals to be available in the future for the control of crop pests and pathogens. The situation heightens the need for the implementation of a more sustainable agricultural system globally, incorporating an integrated approach to disease management.[34]

Nutrients and safety are linked, as shown by several random examples:

- Selenium (Se) and other micronutrients are important for living organisms, since they are involved in several physiological and metabolic processes. Se intake in humans is often low and very seldom excessive, and its bioavailability depends also on its chemical form, with organic Se as the most available after ingestion. The main dietary source of Se for humans is represented by plants.[13]
- Folates are also essential micronutrients for human health. To ensure their supply and availability, it is important to evaluate the effects of storage, processing, and cooking methods on folate content and identify factors with great influence on folate retention in wheat grains and wheat-based foods. [19]
- A necessary component of food sustainability is preservation and storage. The eating and cooking quality of stored rice grains is significantly affected by ageing, but the molecular mechanisms for this are not well understood. Storage resulted in molecular degradation of starch. A more improved molecular-level understanding of the effects of ageing process on rice cooking and eating qualities is needed.[14]

- Lastly, diverse toxins in foods and agriculture will threaten the sustainability of food supplies. Among the array of diverse mycotoxins, aflatoxins have high toxicity and incidence in foods and feeds. Despite the progress made in controlling aflatoxins, they are still a major issue. The changing climate is producing conditions increasingly suitable for aflatoxigenic mold growth and toxin production. Accordingly, it is difficult to harmonize the regulatory standards of aflatoxins worldwide, which prevents agri-foods of developing countries from accessing the markets of industrialized countries.[19]

2.3.6 CLIMATE CHANGE

Previously, it was stated that a global warming of 1.5 °C could trigger the negative impacts of climate change in terms of sea level rise and unsecure food production. [35] Climate activity, more than any other issue, highlights the dynamics and inter-connectivity of the Four Spheres. Chemistry can help to explain and understand effects on the climate, so it represents an important focus of global change research. Factors relating atmospheric chemistry and climate demonstrate effects of emerging issues of climate, stratospheric ozone depletion, and air quality on the practice of agriculture. Understanding how the chemistry and composition of the atmosphere may change over the twenty-first century is essential in preparing adaptive responses or developing mitigation strategies. A changing climate directly impacts human health, agricultural productivity, and natural ecosystems. It is essential to have high-quality, policy-relevant information on the current state of climate and its possible future states, as well as options for mitigation, control, change, and adaptation, in order to take full advantage of AC in the twenty-first century.[11]

An unprecedented climate change (i.e., ever-changing weather patterns) can disrupt the continuous production of food crops which would cause shortages in food sources. Increasing industrialization and urbanization are the major contributing factors on climate change.[20] Climate change impacts the chemical risks associated with food consumption related to growth, development, and microbial contamination. Chemical hazards in food often come from either those present in food due to agricultural practices, environmental contamination or fungal growth or those coming from preparation and processing. The first group includes mycotoxins, while the second include chemicals used in processing and production. Usually, the effective control of chemical hazards is achieved by either minimizing the chemicals or removing/substituting with alternatives. Interestingly, it is possible to develop specific guides to avoid excess of these compounds, but more research is needed to integrate the climate challenges with consumer demands.[12]

Finally, over the last decades, global warming has increasingly stimulated the expansion of cyanobacterial blooms in freshwater ecosystems worldwide, in which toxic cyanobacteria produce mainly microcystins (MCs). MCs can enter into agricultural soils, producing negative effects on the germination, growth and development of plants and their associated microbiota. It is known that MCs can pass through the root membrane barrier, translocate within plant tissues and accumulate into different organs, including edible ones. Also, MCs reduce the microbial activity in soil and can persist in agricultural soils by adsorption to clay-humic acid particles.[13]

2.4 EMERGING AREAS

In response to the need to maintain a sufficient supply of food to feed the growing population, yet doing so in an environmentally responsible manner, AC is utilizing new approaches. These will be briefly presented in this section, and developed in later chapters.

2.4.1 GREEN CHEMISTRY

Green chemistry (GC) was developed with the practice of chemistry as a scientific way to reduce/eliminate the wasteful chemicals and/or hazard in chemistry. Strategies that enable organizations to succeed as businesses and as responsible corporate citizens are necessary for sustainable development. GC has developed proven practical strategies in chemistry that allow researchers and businesses to develop, maintain, or extend their chemical endeavors without causing harm to the environment or its population. In a continually maturing manner GC attempts to set goals and objectives and then monitor, measure, and report on how the results align with the aims of responsible environmental sustainability. GC accomplishes these goals, while improving bottom-line profitability.[11]

GC presents a central concept of materials design, and it aims to achieve sustainability at the molecular level.[36] GC covers a wide range of fields including agriculture and AC takes advantage of the GC tools. The ultimate success of sustainable agriculture to help achieve the SDGs is dependent on the actual practices of agriculture, wise use, and conservation of critical resources, use of green agro-products, sustainable technology, and green energy. The use of green agro-products can influence SDGs through conservation principles, environmental sustainability, and judicious use of critical resources. A direct positive effect of GC is its goal to replace harmful chemicals from production through application, process safety, optimal energy consumption, and improved use of materials (atom economy.)[37]

2.4.2 WORLD HEALTH

The amount, the quality, and the types of food are directly related to world health. The connection is not aways clear, but AC has an influential role. As the earth is under severe stress from several inter-linked factors mainly associated with rising global population, resource consumption, security of resources, and waste generation, the importance of agriculture is clear. Within agriculture there are multi-disciplinary approaches utilizing chemistry, process engineering, sustainability science, and sustainable solutions. Three critical arenas of food (sustainable diet, valorization of unavoidable food supply chain wastes, and circularity of food value chain systems) support the United Nations' seventeen Sustainable Development Goals. The SDGs will lead to sustainability and stability with the help of AC. Food supply chain wastes can be addressed by integrated innovative biorefinery systems which transform food waste into functional and platform chemical products. Circularity of food value chain systems promote novel materials and methods for plant-based protein functionalization for food/nutraceutical applications. Focusing future research on knowledge needed by agriculture and policy makers can move the population at large toward achieving sustainability in the food supply.[38]

Nutrition, which goes beyond mere supply, is a cornerstone of sustainable development and one of the biggest global development challenges. To develop the knowledge of diets from science-based recommendations to develop practical guidelines, food-based dietary guidelines are needed. However, equally important, consumers should have access to appropriate food products.[39] Ensuring the supply and access to nutritious food is a function of agriculture and how markets function at local level. 'What to eat' is an individual decision, but it is the responsibility of the government to ensure a nutritionally adequate diet.[33]

2.4.3 MODELING SUSTAINABILITY

Knowing a definition for sustainability has little value if that definition cannot be actualized. Agriculture, through AC, can be used to develop models of how it can be implemented. The challenge of modeling sustainability is enormous. The components of a sustainable production system should include the reduction of emissions, the efficient use of resources, and the transition to renewable energy. The approach of AC to the modeling of sustainability will include the bioeconomy, the practice of GC, the biorefinery, smart agriculture, and circularity. These present ways to increase efficiency and reduce impacts and risks associated with the use of non-renewable resources. Guidance and evaluation of these practices will come from life-cycle assessment, life-cycle costing, and externality assessment in order to assess the impacts along the whole chain.[14]

Sustainable development evolved from growing awareness of the interdependence of social and economic progress with the limits of the supporting natural environment. The challenge has become how to accomplish this effort. Water, food, energy, climate change, and land subsystems are closely intertwined, and therefore modification of one will affect all others. Quantification of connections, synergies, and tradeoffs across these subsystems is challenging due to the complexities and uncertainties.[40] The practices developed in chemistry and agriculture provide actual tools and actions that will guide measurable results. These can be used to strengthen the model. The model should have the ability to

- Help decision makers identify optimal policy alternatives.
- Provide decision makers insights into in-depth analysis of optimal natural resources allocation strategies associated with different levels of supply and climate change.
- Deal with uncertainties which occur in objective functions and constraints as well as tracking fluctuating attitudes of decision-makers.[40]

If the results from new AC practices can be woven into a more comprehensive model, then greater strides toward sustainability can be made. One approach could be sustainable resources management from a novel water-food-energy-climate change-land nexus perspective. This model would explore management strategies that might unify the practices mentioned previously. Included should be water-food-energy-climate change-land nexus associated with renewable energy and fossil fuel, such as fossil fuel for agricultural machinery and bioenergy produced from crop straw.[14] These will greatly contribute to progress toward the SDGs. Even now the models are showing the soil system is important for the sustainability and economic viability of agriculture, as well as proper management of our natural resources. Soils strongly influence the structure and function of ecosystems and will act as buffers to global environmental change. Soils information is critical for modeling ecological processes and vegetation dynamics, forecasting agricultural potential, predicting climate changes.

2.4.4 Bioeconomy/biorefinery

Some of the biggest lessons learned come from studying and working with nature. Creating and developing processes inspired by nature are becoming major efforts in AC for the twenty-first century. Today most energy and material products come from fossil fuel resources. The cost of fossil resources, their uncertain availability, and environmental concerns resulting from their use, cast uncertainty on oil exploitation. Alternative solutions that mitigate climate change and reduce the consumption of fossil fuels should be promoted. The replacement of oil with biomass as raw material for fuel and chemical production is a viable option and is spurring the development of biorefinery and the bioeconomy. Central to these movements, biomass feedstocks are converted to different classes of biofuels and biochemicals.[41] The goal will be to replace fuel and chemical production derived from non-renewable resources with those in the biorefinery.

Plant-based raw materials (i.e. biomass) have the potential to replace a large fraction of fossil resources as feedstocks for industrial productions, meeting needs in both the energy and non-energy sectors.[42] There are three main drivers for using biomass for production of bioenergy, biofuels, and biochemicals: climate change, energy security, and rural development. This leads to the logical conclusion that the sustainable biomass production is a crucial issue. In the future questions regarding fertile land competition between chemical production and the food and feed industries will need to be resolved.

As will be seen later, the biorefinery employs a wide range of technologies able to separate biomass resources into their building blocks, which can then be converted to value added products, biofuels, and chemicals. The biorefinery is a facility (or network of facilities) that accomplishes biomass conversion to produce needed products from biomass. It is analogous to today's petroleum refinery, which produces multiple fuels and products from petroleum.[41]

The major difference between the petroleum and the biorefinery is the feedstock. The term "feedstock" refers to raw materials used in either process. Renewable carbon-based raw materials for biorefinery are provided from four different sectors: agriculture (dedicated crops and residues); forestry; commercial industries (process residues and leftovers), and household goods (municipal solid waste and wastewaters); and aquaculture (algae and seaweeds). In order for the biorefinery to be a sustainable technology, its feedstock must also be sustainable and not interfere with any of the other SDGs.

The need for biorefineries is spurred by the growth in demand for energy, fuels, and chemicals. The aim of research will be in developing new technologies and creating novel processes, products, and capabilities to ensure the growth is sustainable. When developing chemistry for future biorefineries, it is important that the methods and techniques used minimize impact to the environment and the final products are truly green and sustainable. The use of sustainable feedstock is not enough to ensure a prosperous future for a later generation; protection of the environment using greener methodologies is also required.[41]

2.4.5 BIOCHAR

To meet the demands of a growing population, agriculture has developed in many ways. Many of these areas of growth are now widely seen to be unsustainable, environmentally damaging, and unlikely to keep up with demand. The new needs are: increased yields, reduced negative impacts, better sustainability, and a better standard of living for all farmers. One emerging area where solutions are being identified is in the area of biochar. This substance could be a key input for raising and sustaining production while simultaneously reducing pollution and dependence on fertilizers, it could also improve soil moisture availability and sequester carbon.[43]

Biochar is the carbon-rich product when biomass, such as wood, manure or leaves, is heated in a closed container with little or no available air.[44] It then is the charred organic matter that can be applied to soil in a deliberate manner, with the intent to improve soil properties. As more types of this substance are discovered, its application and uses will only increase. For instance, it is a potential means for sequestering carbon and as a potentially valuable input for agriculture to improve soil fertility, aid sustainable production, and reduce contamination of streams and groundwater. Optimistically, it can contribute to agricultural and environmental sustainability by increasing soil fertility and reactivity.[45]

It has also been recommended as a method for the *in situ* remediation of heavy metals contaminated soils due to its high recalcitrance, stability, specific surface area and retention capacity.[21]

Biochar and the growing efforts to valorize agricultural waste could be a key components agriculture and AC in the twenty-first century. These are some of the best practical ways to move toward sustainability, and it might simultaneously rehabilitate degraded land and counter pollution of streams and groundwater. Adequate research needs to establish the qualities of biochar and to develop the best usage and ways to promote optimum innovation. If biochar is found to have the potential its supporters claim, its use must be carefully controlled to reduce problems.[46]

2.5 CONCLUSION AND A PATH FORWARD

Land, soil, and agriculture are a meeting ground, a boundary object, a nexus in which multiple issues, functions, uses, values, and goals for sustainability interact. With multiple functions for land, for biodiversity, carbon, livelihoods, food production, it is critical to understand how the practice of agriculture and AC can be balanced with complex trade-offs and synergies across multiple functions and across different scales.[5] The challenge will be to see how agriculture and AC will help the world of the twenty-first century meet its sustainability goals.

REFERENCES

1. *Worldometer. World Population Clock.* [10 Aug 2020]; Available from: www.worldometers.info/world-population/.
2. Ruggerio, C.A., *Sustainability and sustainable development: A review of principles and definitions.* The Science of the total environment, 2021. **786**: p. 147481.
3. Olawumi, T.O. and D.W.M. Chan, *A scientometric review of global research on sustainability and sustainable development.* Journal of Cleaner Production, 2018. **183**: pp. 231–250.
4. Frank, B., *Our common future: The "Brundtland Commission" Report.* Ambio, 1987. **16**(4): pp. 217–218.
5. de Bremond, A., *The emergence of land systems as the nexus for sustainability transformations: This article belongs to Ambio's 50th Anniversary Collection. Theme: Agriculture land use.* Ambio, 2021. **50**(7): pp. 1299–1303.
6. Lindstaedt, N., *Human security in disease and disaster.* Human Security in Disease and Disaster. 2021: Taylor and Francis. 1–376.
7. De Luca Peña, L.V., et al., *Towards a comprehensive sustainability methodology to assess anthropogenic impacts on ecosystems: Review of the integration of Life Cycle Assessment, Environmental Risk Assessment and Ecosystem Services Assessment.* Science of the Total Environment, 2022. **808**.
8. Chouffot, R., *Meeting the global sustainability challenge through chemistry.* INFORM – International News on Fats, Oils and Related Materials, 2011. **22**(7): pp. 432–436.

9. O'Brien, A.T., *Ethical acknowledgment of soil ecosystem integrity amid agricultural production in Australia.* Environmental Humanities, 2020. **12**(1): pp. 267–284.
10. Adamo, P. and L. Celi, *Knowledge, conservation and sustainable use of soil: Agricultural chemistry aspects.* Italian Journal of Agronomy, 2009. **4**(3 SUPPL.): pp. 137–149.
11. Borroni, V.N., et al., *Food quality, effects on health and sustainability today: a model case report.* International Journal of Food Sciences and Nutrition, 2017. **68**(1): pp. 117–120.
12. Graveline, N., *Combining flexible regulatory and economic instruments for agriculture water demand control under climate change in Beauce.* Water Resources and Economics, 2020. **29**.
13. Stenzel, F., et al., *Irrigation of biomass plantations may globally increase water stress more than climate change.* Nature Communications, 2021. **12**(1): p. 1512.
14. Ellison, D., et al., *Trees, forests and water: Cool insights for a hot world.* Global Environmental Change, 2017. **43**: pp. 51–61.
15. *Biorefinery.* IEA Bioenergy Task 42 [cited 2019 27 Sept 2019]; Available from: www.iea-bioenergy.task42-biorefineries.com/en/ieabiorefinery/Activities.htm.
16. Koç, C., *A study on the role and importance of irrigation management in integrated river basin management.* Environmental Monitoring and Assessment, 2015. **187**(8).
17. Yamagata, Y., et al., *Estimating water–food–ecosystem trade-offs for the global negative emission scenario (IPCC-RCP2.6).* Sustainability Science, 2018. **13**(2): pp. 301–313.
18. Sansalone, J., et al. *Adsorptive-filtration of runoff subject to controlled and uncontrolled hydraulic and water chemistry loadings.* In *World Environmental and Water Resources Congress 2013: Showcasing the Future.* 2013. Cincinnati, OH: American Society of Civil Engineers (ASCE).
19. Hashim, M.A., et al., *Remediation technologies for heavy metal contaminated groundwater.* Journal of Environmental Management, 2011. **92**(10): pp. 2355–2388.
20. Leong, H.Y., et al., *Waste biorefinery towards a sustainable circular bioeconomy: a solution to global issues.* Biotechnology for Biofuels, 2021. **14**(1).
21. Marinello, S., M.A. Butturi, and R. Gamberini, *How changes in human activities during the lockdown impacted air quality parameters: A review.* Environmental progress & sustainable energy, 2021: p. e13672.
22. Meyfroidt, P., et al., *Middle-range theories of land system change.* Global Environmental Change, 2018. **53**: pp. 52–67.
23. Cracknell, A.P. and C.A. Varotsos, *Remote sensing and atmospheric ozone: Human activities versus natural variability.* Remote Sensing and Atmospheric Ozone: Human Activities Versus Natural Variability. 2012: Springer Berlin Heidelberg. 1–662.
24. Gao, L., et al., *Recent progress of agro-biodiversity conservation and implications for agricultural development in China.* Biodiversity Science, 2021. **29**(2): pp. 177–183.
25. Cheng, Z., Y. Chen, and F. Zhang, *Effect of reclamation of abandoned salinized farmland on soil bacterial communities in arid northwest China.* Science of the Total Environment, 2018. **630**: pp. 799–808.
26. Walker, B.H. and W.L. Steffen, *Global change and terrestrial ecosystems.* 1996, Cambridge; New York: Cambridge University Press.
27. Liu, J., et al., *Long-term organic fertilizer substitution increases rice yield by improving soil properties and regulating soil bacteria.* Geoderma, 2021. **404**: 1–10.

28. Rockström, J., et al., *The world's biggest gamble*. Earth's Future, 2016. **4**(10): pp. 465–470.

29. Li, S.-Y., et al., *Biorefining of protein waste for production of sustainable fuels and chemicals*. Biotechnology for Biofuels, 2018. **11**: p. 256.

30. Azari, A., A.R. Noorpoor, and O. Bozorg-Haddad, *Carbon footprint analyses of microalgae cultivation systems under autotrophic and heterotrophic conditions*. International Journal of Environmental Science and Technology, 2019. **16**(11): pp. 6671–6684.

31. Ebikade, E., et al., *The future is garbage: repurposing of food waste to an integrated biorefinery*. ACS Sustainable Chemistry and Engineering, 2020. **8**(22): pp. 8124–8136.

32. Luckett, B.G., et al., *Application of the Nutrition Functional Diversity indicator to assess food system contributions to dietary diversity and sustainable diets of Malawian households*. Public Health Nutrition, 2015. **18**(13): pp. 2479–2487.

33. Janowska-Miasik, E., et al., *Diet quality in the population of Norway and Poland: differences in the availability and consumption of food considering national nutrition guidelines and food market*. BMC Public Health, 2021. **21**(1): p. 319.

34. Velivelli, S.L.S., et al., *Identification of mVOCs from Andean rhizobacteria and field evaluation of bacterial and mycorrhizal inoculants on growth of potato in its center of origin*. Microbial Ecology, 2015. **69**(3): pp. 652–667.

35. Carus, M. and L. Dammer. *The "Circular Bioeconomy"—concepts, opportunities and limitations*. Bio-based economy 2018–01 [cited 2018 10 Sept 2019]; Available from: www.bio-based.eu/nova-papers.

36. Anastas, P. and N. Eghbali, *Green Chemistry: Principles and Practice*. Chemical Society Reviews, 2010. **39**(1): pp. 301–312.

37. Tickner, J.A., et al., *The nexus between alternatives assessment and green chemistry: supporting the development and adoption of safer chemicals*. Green Chemistry Letters and Reviews, 2021. **14**(1): pp. 21–42.

38. Sadhukhan, J., et al., *Perspectives on "game changer" global challenges for sustainable 21st century: Plant-based diet, unavoidable food waste biorefining, and circular economy*. Sustainability (Switzerland), 2020. **12**(5).

39. Munoz-Plaza, C.E., et al., *Navigating the urban food environment: Challenges and resilience of community-dwelling older adults*. Journal of Nutrition Education and Behavior, 2013. **45**(4): pp. 322–331.

40. Yue, Q., et al., *Fuzzy multi-objective modelling for managing water-food-energy-climate change-land nexus towards sustainability*. Journal of Hydrology, 2021. **596**: 1–13.

41. Cherubini, F., *The biorefinery concept: Using biomass instead of oil for producing energy and chemicals*. Energy Conversion and Management, 2010. **51**(7): pp. 1412–1421.

42. Bajpai, P., *Biotechnology in the chemical industry: Towards a green and sustainable future*. Biotechnology in the Chemical Industry: Towards a Green and Sustainable Future. 2019: Elsevier. 1–246.

43. Sohi, S.P., et al., *A review of biochar and its use and function in soil*, in *Advances in Agronomy*. 2010: Academic Press Inc. 47–82.

44. Lehmann, J. and S. Joseph, *Biochar for environmental management: Science and technology*. Biochar for Environmental Management: Science and Technology. 2012: Taylor and Francis. 1–416.

45. Mia, S., F.A. Dijkstra, and B. Singh, *Aging induced changes in biochar's functionality and adsorption behavior for phosphate and ammonium.* Environmental Science and Technology, 2017. **51**(15): pp. 8359–8367.

46. Barrow, C.J., *Biochar: Potential for countering land degradation and for improving agriculture.* Applied Geography, 2012. **34**: pp. 21–28.

3 Forces in agricultural chemistry and the need for circularity

In the last chapter we discussed the goals, concerns, and challenges facing Agricultural Chemistry (AC) and agriculture in meeting sustainability. The image of the environment consisting of four spheres, as described in the previous chapter, allows the visualization of the flows and demands of energy, resources, and chemicals. There are four forces of anthropological origin that greatly influence how agriculture is practiced (See Figure 3.1). These forces shape and determine the overall effort to achieve sustainability. As we will see in this chapter, the various forces are unifying the overall effort toward sustainability. When shaped by circularity, these forces are unified and complete the global resource utilization loop, which in turn promotes the achievement of sustainability.

AC is mostly confined to terrestrial activities. However, it is influenced by and influences the other dimensions of the environment. These interactions must be guided in the twenty-first century by the principles of sustainability.[1] Due to the expanding nature of agriculture and AC, there will be different approaches and methods used to achieve the goals of sustainability within each individual sector. As we will see, the emerging guidance and unification comes from principles of circularity. As is the case with many areas of the physical sciences, the field of agricultural chemistry draws upon the methods and developments of Green Chemistry (GC); GC, too, will contribute to circularity. The products and processes that will emerge from the synergy of these three areas (agriculture, AC, and GC) in this field will work in harmony with other global efforts in the total sustainability effort.

The four anthropogenic forces mentioned above contain energies and applications that are becoming stronger as they move toward circularity and are captured in life cycle definitions. It is important to state from the outset that new chemistries are not being invented, rather the chemistries inherent in global sustainability problems are being addressed with new approaches.

DOI: 10.1201/9781003157991-3

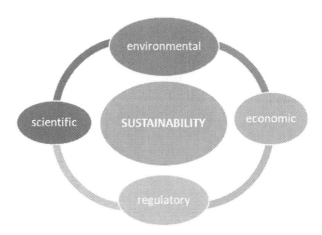

FIGURE 3.1   Four forces shaping sustainability in agriculture

3.1   AGRICULTURE AND THE CASE FOR CIRCULARITY

3.1.1   FOUR FIELDS OF STRATEGIC IMPORTANCE

Agriculture and AC require energy resources and energy management, water management and wastewater treatment, environmental engineering and management. The concepts of sustainability provide a guiding strategy.[2] When the individual activities are "improved," some of them could potentially have unfavorable impacts on the environment, food security, and land use in other spheres of the environment. Sustainability needs to be applied in balanced ways; this should include aspects like the social and economic development, land use, agricultural practices, competition with food, feed, air quality, water resources, agricultural practices, labor conditions, energy efficiency, and Green House Gas (GHG) emissions.

As concepts of sustainability gradually permeate most science and engineering disciplines the steps to achieve it will vary. Previously, the influence of science and engineering followed a linear process, like most industrial processes. This linear process followed a pattern of "make-use-dispose." It was successful, resulting in societal growth, but these activities are also largely responsible for many of the environmental problems facing our world: pollution of water, soil, and air.[2] The sustainability imperatives steer the development of useful policies, education, products, technologies, management procedures, and ethical principles in new directions. These in turn will protect the human health, well-being, the environment, and also protect the future generations. The limits and dimensions of the sustainability in general when previously applied to AC were restricted by the linearity of the fossil energy systems. In the new paradigm, the work will be guided by principles of circularity. The four fields listed above will become more sustainable and stronger as they are guided by a sense of circularity. This circularity, like sustainability itself, is inspired by nature.

3.1.2   EARTH AS A SYSTEM

If the sustainability of our planet is to be achieved, then our world must be treated as a system. This system continuously recycles and recirculates all the materials, substances, and energies. The health of the entire system is ultimately a response to and function of nature. This is no more strongly illustrated than by looking at our water supply and usage, since this is of vital concern to agriculture and agricultural chemistry. Water is an agent in the production and maintenance of crops. Research shows that the consequences of physical or economic water scarcity are becoming more and more common in various geographical locations, and that will impact all sustainability efforts.[3]

The symbiosis of the natural environmental system warrants our attention and scrutiny. In this system of global interconnectedness, flexible approaches are needed for all issues that arise within the system. It is necessary to anticipate and to research how new challenges such as the presence of emerging pollutants in the wastewater streams and land impact agricultural activities, and cause effects of climate change. Both basic research and development of novel process models are urgently needed to enable implementation of environmentally safe, sustainable system solutions. To accomplish this requires effective system approaches for the analysis, design, and evaluation of operation of large- or regional-scale applications. We need to observe how nature accomplishes circularity with high efficiency.

3.1.3   IDENTIFICATION OF ISSUES

All the challenges facing sustainability depend heavily on chemistry (use of materials, cleaning of raw materials and products, and heat removal.) Treating the system as a chemical system, it will be possible to identify key elements, highlight critical reactions, and describe important transfer processes. This task is magnified and made more difficult when human activities in industrial manufacturing both involve and interfere with agriculture. Regardless of the industrial sector, these activities will contribute chemicals, pharmaceuticals, fertilizers, foodstuffs or biofuels into the environment. Industrial activities are largely responsible for the pollution of our water, soil, and air, all of which pose human health and ecological threats. The challenge for sustainability in general and agricultural chemistry in particular will be transforming waste and pollutants into valuable starting materials.

This situation has been exacerbated by the awareness of the finite nature of many resources, as well as the limited environmental tolerance towards human activities. The linear route of production, in which scarce resources are consumed with the concomitant production of waste, is the major contributing factor to global crises such as climate change, diminished biodiversity, as well as food, water, and energy shortages.[4] In order to address this issue, a paradigm shift is needed.

In response to this, AC (in conjunction with GC) is identifying alternative raw materials and environmentally benign processes.

3.1.4 CLOSED LOOP TOWARD CIRCULAR AGRICULTURAL CHEMISTRY

Closing the loop on any process is the way to purposefully design a system to be restorative and regenerative. The goal becomes to keep products, components, and materials at their highest utility and value at all times.[5] This type of circular chemistry provides behaviors and practices in chemistry that seek to replace today's linear 'take–make–dispose' approach in chemical processes with ones where materials are continuously cycled back through the value chain for reuse. By doing this, the goal becomes optimizing resource efficiency and preserving finite feedstocks, and even diversifying the raw materials it requires.[6] An issue remains as long as these resources are often created in a linear production process without sustainable end-of-life options. Resource renewability alone is not a measure of sustainability. What is equally important is the use of waste as a resource.

Chemistry, in particular Green Chemistry, is crucial for achieving this transformation to circular processes.[7] Chemists understand how to design and develop indispensable materials and technologies, but now they also simultaneously recognize the potentially detrimental effects that this may have on their practice; they are therefore becoming increasingly aware that each step they employ must be designed or reassessed with sustainability in mind. The change in materials and applications are similarly directly applicable for AC and agriculture.

The opinion that oil, gas, and coal are harmful, whereas renewables are clean, removes attention from the true sustainability problem: material circulation. What is most important is the inclusion of circularity in the definition of sustainability. This includes the use of waste as a resource.[4] Environmental assessments, typified by the life cycle assessment (LCA), which assesses the impact on the environment of the entire life cycle of a chemical product are valuable. It should provide information on the environmental impact of a chemical from its design to its disposal. This information will help to identify opportunities for innovations in processes and can also identify which feedstocks are most sustainable to use and help to determine applicability of waste.[8]

3.2 MOVING AGRICULTURAL CHEMISTRY TO SUSTAINABILITY THROUGH CHEMISTRY

3.2.1 FOOD SYSTEMS

Food production is essential for survival. Food systems are at the center of various global challenges such as climate change, resource scarcity, and ecosystem degradation[9]. The manner of food production (including handling, processing, distribution, and preparation) creates problems. The failures and vulnerabilities of the current global agri-food system have been highlighted during the ongoing

COVID-19 pandemic, which affected food supply chains, food environments, and consumption patterns[10]. Future food systems must achieve food and nutrition security for the population while addressing various sustainability challenges. The transition towards sustainable agri-food systems utilizing lessons in GC and circularity is urgent for AC in the twenty-first century.[11]

A food system includes all the elements and activities that relate to the production, processing, distribution, preparation, and consumption of food. By its very nature a system must be complete and account for all its mass and energy.[12] This shows that the concept of food systems goes beyond activities (e.g., production, processing, distribution, preparation, and consumption). It encompasses other elements as well as the outputs and outcomes of food-related activities. Food systems must include food supply chains, food environments, and consumer behavior[9]. Food systems overlap with agricultural systems in food production, but they also include the institutions, technologies, and practices that govern the way food is marketed, processed, transported, accessed, and consumed[10]. Food systems, as part of agriculture, can and should be developed in such a way that they contribute to sustainability. In the twenty-first century this sustainability nexus must assume the form of circularity.

Food systems are also of interest in agricultural chemistry.[9] AC provides the connection and the tools by which agriculture and food systems reach sustainability. Future food systems achieve food and nutrition security by addressing various sustainability challenges, highlighting the urgency of fostering transition towards sustainable agri-food systems.[13] Food sustainability must include different aspects, such as sustainable agriculture, and sustainable diets. Food system transformation which includes sustainability will ensure the world will achieve its sustainability goals.[14]

3.2.2 PROBLEMS FACING AGRICULTURE

The food system including agriculture is the biggest user and polluter of land and water, the biggest driver of habitat and biodiversity loss, and one of the largest emission sources of greenhouse gases[15]. This means that agri-food systems have affected not only genetic diversity (of plant and animal species alike) but also landscape diversity. It will be critical for agriculture to achieve sustainability in order that global sustainability is possible.

The current structure of agriculture contains many challenges to meeting its goals of sustainability. Importantly, the entire food system is increasingly acknowledged as the single largest area of concern for the survival of global population. It is also a critical component for the ability of the world to achieve a sustainable future.[16] The way resources are used in agriculture is central in determining the environmental sustainability of agri-food systems. These resources include water[17], land [18], nutrients[19], and energy[20]. AC will continue to play a key role in providing the science and engineering to incorporate the natural resources into a sustainable solution.

3.2.3 ECONOMY, SOCIETY, AND CULTURE

If agriculture is a keystone in the global efforts toward sustainability, its value must be reflected in global economies, societal activities, and the numerous activities that identify many global cultures. The economic dimensions of sustainability of the agri-food system is measured by food accessibility and affordability.[21] These are related to the level of prices of agri-food products as well as their volatility. Price is a misleading measure of the economics of agri-food systems, as the "full cost" or "real cost" is considerably higher[22]. The current, globalized food system supplies relatively inexpensive food to a large proportion of the world's population, but with significant consequential social, environmental and health costs that are poorly understood. This cost is in addition to the actual economics of agricultural food production.

To put this in globally relevant terms, the survival and growth of societies require the entire food system to ensure its sustainability. This requires five areas of research and action: economic and structural costs; political economy; diversity of cultural norms; equity and social justice; and governance and decision support tools [23]. Critical to this are healthy diets from a sustainable food system. The more these diets rely upon and incorporate sound scientific dietary recommendations, the more likely nutritional sustainability will result.

Achieving agri-food system sustainability is one of the most pressing challenges for AC in this century. Addressing this challenge requires drawing upon knowledge and expertise from diverse disciplines and intellectual traditions. The challenges and critical threats to food system sustainability demand a definition of appropriate research, policy, and action. In fact, it is crucial to develop a common understanding, framing, and vision on aspects of the agri-food system that accepts the necessity of agriculture and mandates the resources that are required to practice it and to deliver the necessary products.

3.3 CIRCULAR THINKING IN AGRICULTURAL CHEMISTRY

Agricultural chemistry in the twenty-first century will be driven by circular thinking. It is both logical and necessary. Today's linear economic model is increasingly problematic. Creating a closed-loop model in production and consumption is a preferred alternative. There have been environmental and social damages associated with the linear economy. Applying the unifying principle of the circular economy to both agriculture and AC opens new fields of sustainable applications of sourced biomass, crop residues, industrial side-streams, and wastes. While novel, these applications will become more common, as they are shown to be more beneficial to the environment. The goal will now be transforming them into streams of value-added products, and improving the environmental metrics to measure them. The results could include a wide spectrum of products which address societal and consumer needs.

In order to accomplish this new paradigm novel structures such as biomass supply, biorefineries, value chain clusters, intense farming, plus additional science and engineering funding will all be necessary. These will also need to incorporate climate change mitigation, adaptation, and reducing biodiversity loss.[24] These structures and constraints are fundamental factors, which will prove decisive in the success, failure, movement, or direction of any action toward sustainability in agriculture and AC in the twenty-first century. These novel structures and their movement toward circularity have their basis in the natural environmental processes.

3.3.1 Circular economy

The inspiration for circularity in chemistry and agriculture came from models of circular economies. A circular economy (CE) is "a model of production and consumption, which involves sharing, leasing, reusing, repairing, refurbishing and recycling existing materials and products as long as possible"[25]. This view is not merely recycling; it is an approach to valorizing all phases of an economic system. It is distinctly different from the traditional linear economy. In a linear economy, natural resources are turned into products which are ultimately destined to become waste because of the way they have been designed and made. Recycling was added as a means to recover some value from waste and to manage its effects on the environment. By contrast, a circular economy employs reuse, sharing, repair, refurbishment, remanufacturing, and recycling to create a closed-loop system. This minimizes the use of new resource inputs and lowers the creation of waste, pollution, and carbon emissions.[25] This is not simple, as it demands the design of products and process that will minimize resource consumption and valorize reuse of waste and by-products.

Green chemistry (GC) has been a major driver of sustainable development and has diffused throughout many scientific disciplines in recent years. It is important to look at the ways in which GC, sustainability, and CE concepts are related to each other and how researchers are addressing and utilizing this relation.[26] Agriculture and AC can become the appropriate field where these efforts develop and prosper. The chemical industry is able to contribute to a successful transition towards a greater economic, environmental, and social sustainability. While the main focus of GC is the environment, GC is implementing Sustainable Chemistry (SC) systems and helping to develop chemistries that will help the transition towards sustainability and CE.

3.3.2 Agriculture and chemistry

The work of AC provides essential tools for the tasks needed to achieve sustainability in agriculture. Achieving a circular chemistry and agriculture requires redesign and re-engineering of industrial processes and products, with the effective capture and use of waste streams. This must also include imaginative new approaches to chemical transformations, and careful evaluation of impacts, uses, and costs. All of these require developments in fundamental science and a greater appreciation for the overall natural system.[27] It is possible to use Green Chemistry and its principles as a tool to drive the transition to circularity. In the process, there must be improved opportunities for exchange of information between different individuals from academia, governments and regulatory agencies, business and industrial sectors. This novel path will also be disruptive, requiring the establishment of new supply chains.[28]

In 1998, the 12 principles of green chemistry were published by Paul Anastas and John C. Warner[29] in order to set up guidelines for the development of sustainable chemical processes. These principles marked the beginning of a re-evaluation of accepted chemical practices and development of new approaches in the relationship of our technological society to the environment. It has led to the reinventing of the synthesis, preparation, and use of materials.[30] The goal is to continuously move chemical processes and products towards more environmentally friendly ones, matching the goals of a circular economy. This aligns well with the global sustainability goals described in the previous chapter, and provides practical tools to accomplish them.

Agriculture and food chemistry lie at the center of sustainability. AC is driving efforts in such advancing fields as biorational pest control, green or renewable energy sources, biofuel and biobased product development. This will encourage minimal or recycled water use, no-till conservation farming, development, and use of new cover crops or even new food crops. Green chemistry and engineering will also help food and feed processing that result in improved safety, quality, and efficiency. Producing enough food of sufficient quality to feed an ever-increasing population aided by agricultural chemistry will help to achieve this goal. The attainment of a sustainable food supply will require both higher yields and wiser use of chemicals involved in the agricultural processes. Crop protection chemicals, developed by the agrochemical industry, to ensure that consistently high yields are obtained, provide ease and reliability of harvest and to maintain excellent quality of produce from the crops grown.[31]

Combining agriculture with chemistry to ensure sustainability is central to the tenets of this book. Sustainability will involve a myriad of methods of harvesting or using a resource so that the resource is not depleted or permanently damaged. These sustainable agricultural systems do not deplete the soil, water, or biotic resources. Important to sustainable food supplies will be the considerations of health, nutrition, and proximity to markets so that population segments or whole populations are sustainably served.[32] The path to sustainability in agriculture and food production using chemistry is replete with new actions and potential. Since sustainability is frequently associated with minimal use of limiting resources (for example, tillable land, fresh water, and fuel and energy usage with associated requirements of preservation, packaging, and distribution, carbon footprint, and fuel usage), these become the very same areas where agricultural chemistry contributes to the goals of sustainability.

3.3.3 THE BIOREFINERY AND BIOCHAR AS WAYS TO PROMOTE AC

The linear economy produces waste streams. An ideal solution, mimicking   ,
would be to use waste for subsequent reactions and/or as starting materials for
new processes. The various strategies for the valorization of waste biomass to
synthesize platform chemicals, and the underlying developments in chemical and
biological catalysis which make this possible, are critical for the future sustainable
practice of agriculture. This is a key component for AC in the twenty-first century,
and this path will be shown later to be a viable alternative to production of the
many commodity chemicals upon which society depends.[33] This is illustrated
by the biorefinery and it is shown as a key way that agriculture and agricultural
chemistry will contribute to the global drive to sustainability.

The recovery and regeneration of the soil is essential for agriculture and is
therefore critical to sustainability. Biochar is relevant to the renewability and
sustainability of agriculture. This also will be developed in later chapters. It is
illustrative of the direction and impact of AC. Biochar, a carbonaceous porous
material produced from the pyrolysis of agricultural residues and solid wastes,
and has been widely used as a soil amendment. Its use and application do result in
climate improvement, contaminant immobilization, soil improvement, nutrient
recovery, engineered material production, and waste-water treatment.[34]
Biochar's nutrient value helps in improving plant growth and in fertilizer use
efficiency. The renewability, low-cost, high porosity, and high and customizable
surface chemistry of biochar is an area that epitomizes the state of the art in
agricultural chemistry. Research is showing that biochar can be modified in
new ways and these can expand the applicability in agricultural sustainability
beyond soil.

3.3.4 LIFE CYCLE ANALYSES

Environmental assessments, typified by the life cycle assessment (LCA), which
assess the impact on the environment of the entire life cycle of a chemical product,
are invaluable, as we show in the next chapter. They are useful in identifying
inefficiencies in current chemical processes[4]. Such sustainability metrics, which
provide information on the environmental impact of a chemical from its design to
its disposal can help to identify opportunities for innovation in a process and can
also pinpoint which feedstock is most sustainable to use as resource.[35] In the
work in AC, LCA will be both a retroactive analysis tool and a pro-active gauge
guiding this work in the twenty-first century.

Life cycle thinking when coupled with circularity will reinvent chemistry[36],
and should provide the basic principles for developing novel chemical products
and processes that use waste as resource. In turn, this will contribute to realizing
the circular economy and securing our sustainable future by addressing the United
Nations Sustainable Development Goals[37].

3.4   DEVELOPING A NEW PARADIGM

The traditional linear model of economics and chemistry is problematic. To realistically achieve sustainability, the new circular model must replace it (See Figure 3.2). Agriculture is pivotal to the success of sustainability and the inherent reactivity of chemicals enables their conversion into value-added compounds and productive processes invaluable to the attainment of sustainable goals.

In a similar manner to green chemistry, circular chemistry will illuminate ways to reduce the harmful impact of chemical compounds on the environment, while addressing the growing demands of our world. Although the use of substances

CIRCULAR ECONOMY

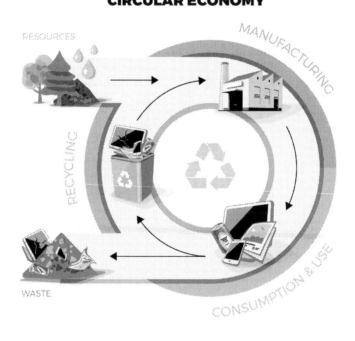

LINEAR ECONOMY

FIGURE 3.2   Circular economy vs the linear economy. (See www.shutterstock.com/ image-vector/comparing-circular-linear-economy-showing-product-379256980)

of concern may be unavoidable in some circumstances[38], these should not be released to the environment, but be recycled or captured. This approach is especially true in agricultural chemistry and requires the continued growth in knowledge and engineering developments.

The most important way in which agricultural chemistry can impact the drive to sustainability is by asking (and helping answer) the following questions:

- "Do we need this material to achieve our goal?,"
- "Do we need to make something new or can available material be reused or repaired?," and
- "Does the used product really need to be disposed of?"[4]

The actions that become answers to these questions provide guidance for optimal product and process design for sustainability. Following the green chemistry principles does not unequivocally lead to an increase in sustainability. However, combining those principles with current circular thinking will produce a powerful force in the progress to achieve the ultimate goal of global sustainability.

REFERENCES

1. Ploeg, P., J. Revald Dorph, and N. Harvey, *Planetary Boundaries and Sustainability Principles: An integrated approach in the context of agriculture.* 2016.
2. Duić, N., K. Urbaniec, and D. Huisingh, *Components and structures of the pillars of sustainability.* Journal of Cleaner Production, 2015. **88**: pp. 1–12.
3. Belmonte-Ureña, L.J., et al., *Circular economy, degrowth and green growth as pathways for research on sustainable development goals: A global analysis and future agenda.* Ecological Economics, 2021. **185**: 1–17.
4. Keijer, T., V. Bakker, and J.C. Slootweg, *Circular chemistry to enable a circular economy.* Nature Chemistry, 2019. **11**(3): pp. 190–195.
5. Stahel, W.R., *The circular economy.* Nature, 2016. **531**(7595): pp. 435–438.
6. Bender, T.A., J.A. Dabrowski, and M.R. Gagné, *Homogeneous catalysis for the production of low-volume, high-value chemicals from biomass.* Nature Reviews Chemistry, 2018. **2**(5): pp. 35–46.
7. Anastas, P.T. and J.B. Zimmerman, *The molecular basis of sustainability.* Chem, 2016. **1**(1): pp. 10–12.
8. Jiménez-González, C., D.J.C. Constable, and C.S. Ponder, *Evaluating the "Greenness" of chemical processes and products in the pharmaceutical industry—a green metrics primer.* Chemical Society Reviews, 2012. **41**(4): pp. 1485–1498.
9. Garnett, T., *Three perspectives on sustainable food security: Efficiency, demand restraint, food system transformation. What role for life cycle assessment?* Journal of Cleaner Production, 2014. **73**: pp. 10–18.
10. Savary, S., et al., *Mapping disruption and resilience mechanisms in food systems.* Food Security, 2020.
11. El Bilali, H., et al., *Food and nutrition security and sustainability transitions in food systems.* Food and Energy Security, 2019. **8**(2).

12. Koester, U., *Food loss and waste as an economic and policy problem.* Intereconomics, 2014. **49**(6): pp. 348–354.
13. El Bilali, H., C. Strassner, and T. Ben Hassen, *Sustainable agri-food systems: Environment, economy, society, and policy.* Sustainability (Switzerland), 2021. **13**(11): pp. 1–67.
14. Rockström, J., et al., *Planet-proofing the global food system.* Nature Food, 2020. **1**(1): pp. 3–5.
15. Campbell, A., *Australian rangelands science-a strategic national asset.* Rangeland Journal, 2020. **42**(5): pp. 261–264.
16. Galli, A., et al., *Sustainable food transition in Portugal: Assessing the Footprint of dietary choices and gaps in national and local food policies.* Science of the Total Environment, 2020. **749**: pp. 1–15.
17. Vanham, D. and A. Leip, *Sustainable food system policies need to address environmental pressures and impacts: The example of water use and water stress.* Science of the Total Environment, 2020. **730**.
18. Villoria, N., *Consequences of agricultural total factor productivity growth for the sustainability of global farming: Accounting for direct and indirect land use effects.* Environmental Research Letters, 2019. **14**(12).
19. Vaccari, D.A., S.M. Powers, and X. Liu, *Demand-driven model for global phosphate rock suggests paths for phosphorus sustainability.* Environmental Science and Technology, 2019. **53**(17): pp. 10417–10425.
20. Schramski, J.R., C.B. Woodson, and J.H. Brown, *Energy use and the sustainability of intensifying food production.* Nature Sustainability, 2020. **3**(4): pp. 257–259.
21. Zurek, M., et al., *Assessing sustainable food and nutrition security of the EU food system-an integrated approach.* Sustainability (Switzerland), 2018. **10**(11): pp. 1–16.
22. O'Kane, G., *What is the real cost of our food? Implications for the environment, society and public health nutrition.* Public Health Nutrition, 2012. **15**(2): pp. 268–276.
23. Béné, C., et al., *Five priorities to operationalize the EAT–Lancet Commission report.* Nature Food, 2020. **1**(8): pp. 457–459.
24. Lange, L., et al., *Developing a Sustainable and Circular Bio-Based Economy in EU: By Partnering Across Sectors, Upscaling and Using New Knowledge Faster, and For the Benefit of Climate, Environment & Biodiversity, and People & Business.* Frontiers in Bioengineering and Biotechnology, 2021. **8**: pp. 1–16.
25. Geissdoerfer, M., et al., *The Circular Economy – A new sustainability paradigm?* Journal of Cleaner Production, 2017. **143**: pp. 757–768.
26. Silvestri, C., et al., *Green chemistry contribution towards more equitable global sustainability and greater circular economy: A systematic literature review.* Journal of Cleaner Production, 2021. **294**: pp. 1–24.
27. Catlow, C.R., et al., *Science to enable the circular economy: Science to enable the circular economy.* Philosophical Transactions of the Royal Society A: Mathematical, Physical and Engineering Sciences, 2020. **378**(2176): pp. 1–3.
28. Loste, N., E. Roldán, and B. Giner, *Is Green Chemistry a feasible tool for the implementation of a circular economy?* Environmental Science and Pollution Research, 2020. **27**(6): pp. 6215–6227.
29. Anastas, P.T. and J.C. Warner, *Green Chemistry: Theory and practice.* 1998: Oxford: Oxford University Press.

30. Anastas, P.T., *Green chemistry next: Moving from evolutionary to revolutionary: A view from the co-author of the 12 principles of green chemistry*. Aldrichimica Acta, 2015. **48**(1): pp. 3–4.

31. O'Riordan, T.J.C., *UN Sustainable Development Goals: How can sustainable/green chemistry contribute? The view from the agrochemical industry*. Current Opinion in Green and Sustainable Chemistry, 2018. **13**: pp. 158–163.

32. Seiber, J.N., *Sustainability and agricultural and food chemistry*. Journal of Agricultural and Food Chemistry, 2011. **59**(1): pp. 1–2.

33. Sheldon, R.A., *Green and sustainable manufacture of chemicals from biomass: State of the art*. Green Chemistry, 2014. **16**(3): pp. 950–963.

34. Sashidhar, P., et al., *Biochar for delivery of agri-inputs: Current status and future perspectives*. Science of the Total Environment, 2020. **703**.

35. De Marchi, L., et al., *Engineered nanomaterials: From their properties and applications, to their toxicity towards marine bivalves in a changing environment*. Environmental Research, 2019. **178**.

36. Whitesides, G.M., *Reinventing chemistry*. Angewandte Chemie – International Edition, 2015. **54**(11): pp. 3196–3209.

37. Bexell, M. and K. Jönsson, *Responsibility and the United Nations' sustainable development goals*. Forum for Development Studies, 2017. **44**(1): pp. 13–29.

38. Takhar, S. and K. Liyange. *Understanding the implications of chemical regulations, circular economy and corporate social responsibility for product stewardship.* in *17th International Conference on Manufacturing Research, ICMR 2019, incorporating the 34th National Conference on Manufacturing Research, NCMR 2019*. 2019. IOS Press BV.

4 Life cycle assessment with circularity

4.1 INTRODUCTION

Life cycle-based methodologies are increasingly used as a tool to support the transition towards sustainable production and consumption patterns.[1] Agricultural activities have many effects on the environment (soil quality and quantity, air, contamination of wildlife habitat) and for this reason, there is a need to provide robust indicators capable of providing information on the environmental conditions in the agri-food sector and their trends.[2] The challenge to achieve sustainability in agricultural chemistry (AC) requires metrics and a means by which progress is measured. Researchers and scientists in AC face a bewildering and confusing landscape of data management requirements, recommendations and regulations. This is rendered more difficult due to lack of access to data management training or possessing a clear understanding of practical approaches that can assist in data management in their particular research domain.[3] Life cycle assessments (LCAs), when done rigorously, will lead to understanding the components or stages of products or processes and how they contribute to the ultimate goal of sustainability. (See Figure 4.1) When LCA is coupled with circularity (CIR), a clear set of metrics for agriculture in the twenty-first century emerges.

The life cycle assessment (LCA) is a key tool for performing an environmental sustainability analysis of products and technologies,[4] providing a systematic path for measuring improvements in resource productivity and serving as an effective tool for promoting cleaner production.[5] During an LCA there are several stages that are measured and analyzed. Each stage, or assessment, has a Goal and Scope Definition; Life Cycle Inventory Analysis; Life Cycle Impact Assessment; Interpretation, Reporting and Critical Review.[6] LCA is a method for measuring and monitoring the environmental impacts of products or production processes, and it allows an in-depth evaluation of the ecological dimensions of each stage. When applied to AC, LCAs can provide the environmental impacts of agricultural and food systems, revealing the impact of mitigating interventions and monitoring progress towards sustainable development goals. In pursuit of agricultural sustainability and food security, research and development will benefit from the insights provided by LCA. This is especially true when considering the

DOI: 10.1201/9781003157991-4

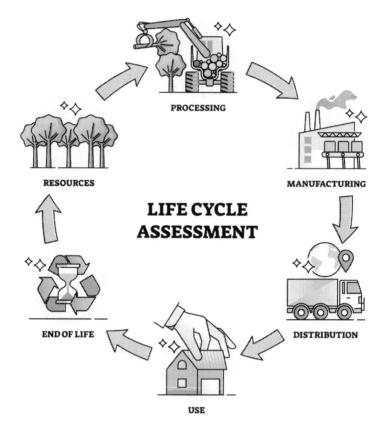

FIGURE 4.1 Life cycle assessment. (See www.shutterstock.com/image-vector/life-cycle-assessment-explanation-all-process-2180139369)4

future effects associated with the large-scale implementation of new programs or of any impact-mitigating changes in processes. For example, promising interventions might involve the promotion of waste circularization strategies, which could also improve the profitability of agriculture.[7]

There are many Life Cycle (LC) methodologies: LCA (or environmental life cycle assessment E-LCA), life cycle costing (LCC), and social life cycle assessment (S-LCA). Each of these result in the measurement of environmental, economic, and social impacts. To be effective they also must be systemic, multidisciplinary, and multicriterial. The LC framework can potentially capture all sustainability dimensions, and can be adapted to evaluate any circularity strategies employed to achieve a "cradle-to-cradle" circular vision.[8] The LCA methodology can be applied to assess numerous environmental dimensions in the value chain of fruits and vegetables,[9] and many studies have shown the significant impact of the agri-food system in terms of energy consumption, greenhouse gas emissions, and other environmental impacts.[10]

4.2 LIFE CYCLE ASSESSMENT

LCA broadly looks at products or processes and analyzes global mass and energy as they relate to all aspects of the product.[2] The product "life cycle" includes all materials and methods that go into making the product. This will include, for example, extracting materials used, products made, product use, transport involved, wastes generated, and end-of-life issues such as reuse or disposal. The close relationship between agricultural activities and the impact on the three dimensions of sustainability (environmental, social, and economic) makes the use of indicators for measuring agricultural activities a *sine qua non* condition to ensure real sustainability based on CE principles.[11] LCA studies can capture these.

LCA began in the 1960s and 1970s by providing energy analysis. It is now a wide-ranging tool used to determine impacts of products or systems over several environmental and resource issues. The approach finds its most beneficial use in research, industry, and global economic policy. Its use continues to expand as it provides insights on impacts as diverse as resource accounting and social wellbeing. Carbon policy for bioenergy has driven many of these changes. Performing an assessment of complex issues over a life cycle basis is beneficial, but the process is difficult, labor-intensive, and expensive. The politically charged environment surrounding biofuels and bioenergy exacerbates all of these. The real value from LCA will come as its use transitions from a reactive perspective to a more pro-active approach. A key example comes from the use of LCA to analyze bioenergy deployment.[12]

4.2.1 COMPONENTS

The LCA process is a systematic, phased approach to analyzing any system or activity. It consists of four components: goal definition and scoping, inventory analysis, impact assessment, and interpretation.[13] These components clearly delineate essential steps that compose viable progress toward sustainability. These stages of LCA (listed below) are also a means to evaluate circularity.

- Goal definition and scoping – The product, process or activity is defined and described. This sets the context in which the assessment is to be made and identifies the boundaries and environmental effects to be reviewed for the assessment.
- Inventory analysis – The materials and quantities involved in the system are identified. This will include the energy, water and materials usage and environmental releases (e.g., air emissions, solid waste disposal, wastewater discharges).
- Impact assessment – The potential human and ecological effects of energy, water, and material usage and the environmental releases identified in the inventory analysis are assessed.

- Interpretation – Lastly, the LCA produces an evaluation of the results of the inventory analysis and impact assessment to select the preferred product, process, or service with a clear understanding of the uncertainty and the assumptions used to generate the results.

By performing an LCA during any stage of sustainability or circularity, increased confidence can result in the value of the process. The results of these analyses are critical to decision-makers to assist them in accomplishing a variety of goals. Tangible results from an LCA include

- Development of a systematic evaluation of the environmental consequences of a given product or process.
- Analysis of the environmental trade-offs associated with one or more specific product/process.
- Quantification of environmental releases to air, water, and land in relation to each life cycle stage and/or major contributing process.
- Assessment of the human and ecological effects of material consumption and environmental releases.
- Comparison of the health and ecological impacts between two or more rival products/processes.

These results will help to ascertain the appropriateness of any movement toward sustainability. This will be important as a gauge as to how an innovation in AC and agriculture helps the achievement of the UN Sustainability goals.

4.2.2 Purpose of LCA

The use of the LCA as a tool for targeting and monitoring any developments in the agri-food industry is valuable. There is the ever-present challenge of reducing the environmental burdens in agri-food production that would benefit greatly from an LCA-based analysis. The LCA analyses could also focus on the processes and chemicals used in food production. The possibility of integrating the LCA with economic and social impact assessments (e.g., under the life cycle sustainability assessment framework) makes LCA an excellent tool for monitoring business or actual farming activities with respect to UN 2030 Sustainable Development Goals.[14] Looking forward, LC methods can serve as important tools to monitor the environmental appropriateness of adopting or adapting low-carbon technologies or to the appropriate use of Smart Agriculture to enhance the agricultural processes.[12]

The benefits of LCA results can extend into multiple areas of environmental remediation which involve AC and agriculture. They may, for instance, indicate that selection of an appropriate phosphorus recovery method should consider both local conditions and other environmental impacts, including global warming, ozone depletion, toxicity, and salinization, in addition to eutrophication and mineral depletion impacts in agricultural systems.[14] An LCA could be used to evaluate the use and function of biochar. (Biochar is a valorized product made from agricultural waste.) It is claimed that the proposed biochar system provides relevant primary energy savings of non-renewable sources and a strong reduction of greenhouse gases emissions without worsening the abiotic resources depletion. An LCA should support this. One more example is found in the use of corn stovers for mulch. These stovers are critical when considering acidification and eutrophication impacts; LCA supports this conclusion. Therefore, removal of corn stovers from the fields must be planned carefully.[14] How this meshes with current practice and promotes both increased agricultural productivity and sustainability will be seen.

LCA methodologies may be used for environmental planning, since the energy sector is the major contributor to greenhouse gas emissions (GHG). As a potential remedy, biomass has been proposed as a replacement feedstock. LCA methodology can be used to carry out a comprehensive, holistic evaluation of biomass-to-energy systems based on indigenous biomass supply chains optimized to reduce production and transportation GHG emissions. Areas assessed within LCA include global warming, acidification, eutrophication potentials, and energy demand. An example study shows that while biomass systems produce lower environmental impacts than coal systems in isolation, a more comprehensive study reveals overall environmental impacts requires a detailed LCA of the reference energy systems which are displaced. In addition, the best outcome may not be the one which achieves the greatest environmental impact reductions.[14] An environmental LCA can be carried out in order to compare the environmental impacts and the energy efficiency of various treatment options of biogas waste production.[15]

Land use is one of the main drivers of biodiversity loss. It is also one of the largest challenges to agricultural sustainability. However, many LCA studies do not yet assess this effect because of the lack of reliable and proven methods. The impacts of land use on biodiversity show the highest values in regions where the greatest natural habitat had been converted in earlier years. As the population continues to increase and food demand also increases, this will occur more frequently. The approach thus far has been retrospective and only able to highlight the impacts in highly disturbed regions. However, better LCA methodologies should be applied to prospective assessments using scenarios of future land use.[16]

4.2.3 LIMITATIONS OF LCA

Sustainability has become the most important path for survival for the entire world. Many processes, including renewable energy, are the backbone of the ongoing initiatives. Sustainability will result from the wise selection of processes that will maximize yields of food and minimize waste emissions. LCA on these areas should focus more attention on the toxic effects of different processes and methods, evaluate the technical ways of adding improved technology to the agrochemical process, and further explore how to markedly reduce environmental damage in order to maximize energy and nutrient recovery from the LCA perspective.[17] LCA in this process becomes a necessary "check-and-balance" for AC and agriculture in the twenty-first century. While important, accessibility to LCA for these purposes is limited.

LCA application to agricultural systems must be operational, accurate, and exhaustive. This is particularly challenging for the newly developing LCA and ISO-compliant processes, with many methods only recently developed, and not containing any dedicated inventory method fully approved. To increase the value of LCA, a variety of tools are needed, ranging from databases (e.g. World Food LCA Database) to complex agro-hydrological models. The selection of the appropriate method and tool for the inventory of material flows in agri-food LCA studies depends on the objectives of the LCA study, data, and resources available (time and skills). Agricultural applications of LCA have limitations that are not unique to the sector. Current LCA databases provide estimates of theoretical material processes by a crop and rely on data and methods that have uncertainties, making them suitable only for background agricultural LCAs.[18]

4.2.4 COST OF LCA

LCA is a technique used to assess the environmental impact of products, processes, or materials. The costs of conducting a full LCA are particularly high. Recently, its importance as a decision-making tool to help evaluate current inventories and innovation of environmentally responsible products has grown. This helps to justify its costs. The amount of information needed to completely assess even the simplest product's environmental impact may require significant time and resources. Myriad quantitative and qualitative effort-reducing strategies have been considered to accelerate the pace and reduce the cost of LCA. Although these streamlining methodologies reduce the time and effort of conducting LCA, they introduce variability and uncertainty into the results, creating a challenge for stakeholders who may need to make decisions based on the information.[19]

4.2.5 LCA AND SUSTAINABLE CIRCULARITY

Included in the movement toward sustainability is the construction of a circular bioeconomy and enhancing the value of material flows. LCA must capture this. Since circular bioeconomy aims to achieve sustainable consumption and production with reduction of greenhouse gas emission, research must show how circular bioeconomy can be achieved through sustainable food waste management by comparing the similarities and differences in concepts of bioeconomy and circular economy. This is enhanced by reviewing the benefits and limitations of the existing policies and evaluating the global situations of food waste. Lastly, the management of household and commercial wastes can promote circular bioeconomy. Future development on food waste management will rely on LCA analyses of the multi-functionality of products, boundary, and allocation in a circular system, and trade-off between food waste and resources.[20]

Unsustainable agricultural management causes environmental impacts. For example, although rice straw has a high potential for bioenergy generation, the whole production cycle and application may cause environmental damage that is not fully understood. Hence, environmental performance studies are required to determine the most effective rice straw utilization options. A comprehensive approach, such as life-cycle assessment (LCA), can give comprehensive information on the possible environmental effects of rice straw utilization for bioenergy. Other impact categories in LCA should be evaluated in the bioenergy production from rice straw research to determine the overall sustainability of the production.[21]

4.3   CIRCULARITY: CIRCULAR ECONOMY AND CIRCULAR CHEMISTRY

The current socioeconomic conditions require emerging solutions to solve challenges such as limited energy supplies, global warming, food security, environmental conservation, and water quality as the world population grows. Circular bioeconomy is the potential solution to address these complex, interconnected issues. This sustainable model reduces the use of natural resources and the growth of waste, cutting carbon (greenhouse-gas) emissions in the atmosphere. Within the food industry, waste generation is unavoidable due to poor food management. Proper waste management necessitates additional ecological issues.[22]

Circularity (CIR) is an approach to achieving sustainability. Within it emerge both circular economy (CE) and circular chemistry (CC). Combining CE and CC will lead to achieving sustainable agricultural chemistry. One of the key issues is the understanding of the link between CIR and sustainability[15] and, this will allow linking it with the chemistry of agriculture. CE deals with economic systems that focus on reusing, recycling, and recovering materials in production/ distribution and consumption processes. As with the generally accepted definition of sustainability, it is tied to the benefit of current and future generations.[23]

The concepts embedded in CIR can and should be incorporated into LCA. The development and implementation of resource efficient and circular economies require critical analyses of the whole supply chain from feedstock to end-use. Initial and current forms of LCA are the logical approach to accomplish this purpose. Most principles of green chemistry (GC) and resource (material and energy) circularity can provide flexible metrics to complete an extensive LCA, incorporating the hybridized indicators including hazardous chemical use, waste generated, resource CIR and energy efficiency. This comes from the bio-based case studies and their petro-derived commercial counterparts. The information derived from merging these methods will provide necessary data for operational optimization. The benefit will be transparency in sustainability reporting and practices to a significant number of manufacturers, policy makers and consumers. [24]

4.3.1   CIRCULAR THINKING AS A COMPLEMENT TO LCA

LCA is a robust and valuable approach that produces a sustainability analysis which measures the environmental impacts associated with the use and reuse of resources. It is insufficient to incorporate the potential contributions of agricultural and biobased chemical processes. CIR approaches the current models of production and consumption and agri-food systems differently. Through new models, it maps new pathways that the field must necessarily adopt for survival. LCA can evaluate these pathways. This approach will help address the need to improve the resource efficiency of agri-food system activities through technical innovations to ensure more sustainable use of renewable resources, the reduction of environmental damages, and the depletion of non-renewable resources. CE is useful in agri-food sectors[16] because it provides a way to solve traditional and systemic problems, by reducing environmental damage or optimizing the use or resources through returning waste and side-products back into the original processes. CE contributes to agri-food systems by aligning them with CIR; it will be necessary to unify ways CE can help to improve specific economic, and environmental aspects of sustainability.

4.3.2 LCA ADAPTING TO AGRICULTURAL CHEMISTRY IN THE TWENTY-FIRST CENTURY

The challenge for agricultural production in the twenty-first century is to produce sufficient high-quality food while minimizing impact. We have seen that LCA can help by identifying the parts of a supply chain with the greatest environmental impact, and to determine which technologies may be most appropriately employed to minimize negative impacts. However LCA cannot be the definitive tool to guide scientific dimensions of agriculture, such as plant breeding, and therefore this approach needs to complement and draw upon other methods.[25] AC will utilize new technologies and new approaches to meet the sustainability demands of the twenty-first century. LCA will be an indispensable tool in this effort.

4.3.2.1 Land use

Understanding the consequences of changes in land use and land cover is among the greatest challenges in sustainability science. It affects food production and therefore it will be a major consideration for agricultural chemistry. Key themes related to land cover change are often left out of sustainability assessment tools. LCA can fill this void. As an example, two approaches show conflicting assessments for the carbon footprint of palm biodiesel: a sustainable endeavor when short-term global warming potential is evaluated yet highly unsustainable when rates of forest loss are measured. It is important to include both historic and future land cover changes into sustainability assessments. This also shows the importance of using a plurality of approaches from different disciplines when evaluating sustainability, and highlights the unique role that landscape ecological approaches can play in sustainability assessments such as LCA.[26,27]

The global concern over the increasing loss of usable land affects agriculture. Since land use is a main driver of global biodiversity loss, its environmental relevance is widely recognized in research on LCA. While biodiversity and its non-uniform land use is documented by regions, LCA applications can analyze this on a global scale. An overall negative land use impact has been found for all analyzed land use types, but results varied considerably. A critical starting point will be the identification of specific land and agricultural activities that result in the changes in biodiversity. Data on land use impacts were very unevenly distributed across the globe and considerable knowledge gaps on cause-effect chains remain. This is a result of both the level of effort and the quality of data. A first rough quantification of land use impact on biodiversity in LCA on a global scale show that as biodiversity is inherently heterogeneous and data availability is limited. Even though there is some uncertainty, the effects are obvious. Therefore, more accurate and regionalized data on regeneration times of ecosystems are needed. [26] A further goal would be for LCA to validate the data results.

4.3.2.2   Crop rotation

Application of LCA to crop rotations offers a potential solution to problems of land use. Methodological changes can evaluate the effects occurring between the crops grown in the same agricultural field in temporal succession. These so-called crop-rotation effects are caused by changes in physical, chemical, and biological properties of the agricultural land over time (presence and availability of different micro and macronutrients, soil structure, soil texture, phytosanitary conditions, and presence of weeds) due to the rotation of crops. These are AC challenges and are a function of CIR. LCA studies with system boundaries containing several vegetation periods are needed. AC can help provide this data to LCA. Many crop-rotation effects between crops are not covered in LCA methodology. New AC methodologies might include improved phytosanitary conditions, stabilization of yields via reduction of harvest failures, improved yields via improved soil texture, soil structure and improved conditions for soil organisms. If new LCA methodologies including sustainability can include soil fertility, a new approach for the modeling of crop-rotation effects is possible. This approach integrates crop-rotation systems into LCA, enabling LCA practice to be able to depict crop rotations more accurately and to avoid the current practice of ignoring the effects between individual crops. Thus, the ability to consider the entire spectrum of crop rotation effects should be integrated into agricultural LCAs.[28]

4.3.2.3   Biodiversity loss

Halting current rates of biodiversity loss can only be achieved if appropriate metrics are used. Indicators and tools are required that monitor the driving forces of biodiversity loss, the changing state of biodiversity, and evaluate the effectiveness of policy responses. The use of indicators and approaches to model biodiversity loss in LCA must be improved. There are serious conceptual shortcomings in the way models are constructed, with scale considerations largely absent. Further, there is a disproportionate focus on indicators that reflects changes in biodiversity, mainly changes in species richness. Functional and structural attributes of biodiversity are largely neglected. The majority of models are restricted to one or a few taxonomic groups and geographic regions. Methods across all drivers can be greatly improved. Future work needs to be done to reflect biodiversity loss in LCA.[29]

4.4 AGRICULTURAL CHEMISTRY RESPONDING TO NEEDS OF THE TWENTY-FIRST CENTURY

The implementation of sustainability and circular economy (CE) models in agri-food production can promote resource efficiency, reduce environmental burdens, and ensure improved and socially responsible systems. (See Figure 4.2) In this context, indicators for the measurement of sustainability play a crucial role. Indicators can measure CE strategies aimed to preserve functions, products, components, materials, or embodied energy.[11] Among the major forces pushing sustainability is the attitude of the general populace. Consumers' concerns towards an environmentally friendly food production will continue to drive the development of agricultural chemistry in the twenty-first century. For example, the dairy industry contributes to the production of important greenhouse gases such as methane. The LCA method quantifies the emissions and the use of resources throughout the entire life cycle of this industry. The key areas where there are important environmental impacts (global warming potential, freshwater eutrophication, terrestrial acidification, and agricultural land occupation) can be associated with milk production. Although this provides additional information relating the most important determinants of

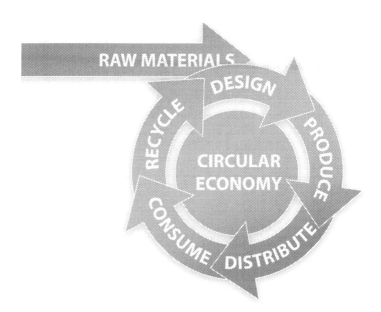

FIGURE 4.2 Agricultural circularity. (See www.shutterstock.com/image-illustration/illus tration-concept-circular-economy-1197087262)

different environmental impacts, further analyses on specific soil characteristics and their impact on environmental efficiency as mentioned in the previous section, are needed.[30]

4.4.1 AGRICULTURAL SUSTAINABILITY THROUGH LCA

Life cycle (LC) methodologies have attracted a great interest in agricultural sustainability assessments, even if they have sometimes been criticized for making unrealistic assumptions and subjective choices. Contemporary Multi-Criteria Decision Analysis and/or participatory methods can be used to balance and integrate aspects of LCA. In the twenty-first century, LCA must address agricultural sustainability. Using the above metrics, LC tools can provide studies with local and global information on how to reduce negative impacts and avoid burden shifts, while other methods can help LC practitioners deal with subjective assumptions, to take into consideration values and to overcome trade-offs among the different dimensions of sustainability. These efforts will lead to the further development of LCA sustainability and CIR.[31]

As was shown in the previous chapter, agricultural sustainability will need to address waste and crop residues. These are potentially significant sources of feedstock for biofuel production in the United States. LCA must capture them. However, there are concerns with maintaining a sufficient quantity to meet environmental functions of these residues while also serving as a feedstock for chemical production. For example, maintaining soil organic carbon (SOC) along with its functional benefits is considered a stronger priority than restricting soil erosion losses to an acceptable level. The avoidance of fossil fuel emissions to the atmosphere by utilizing the cellulose and hemicellulose fractions of crop residue to produce ethanol might reduce the overall greenhouse gas (GHG) emissions because most of this residue carbon would normally be lost during microbial respiration. Therefore, by returning the high-lignin fraction of crop residue to the land after production of ethanol at the biorefinery, soil carbon levels could be maintained along with the functional benefit of increased mineralized nitrogen (N), and more GHG emissions could be offset compared to leaving the crop residues on the land. [32] These are considerations for how to accomplish sustainability in the twenty-first century.

Net carbon management of agro-residues has been an important pathway for reducing the environmental burdens of agricultural production. The combination of life cycle assessment and circular economy modelling is critical to achieve greener and sustainable agricultural production.[33]

4.4.2 AGRICULTURAL CHEMISTRY EXPRESSED IN THE BIOREFINERY

The biorefinery (BR) can utilize crops, waste, and crop residues. The Green biorefinery (GBR) is an additional biorefinery concept that converts fresh biomass into value-added products. Working together, the BR and the GBR can help address sustainability. LCA can be used to evaluate the technical and environmental performance of different GBR configurations and the potential of GBR. LCA results show that the environmental profile of the GBR is highly affected by the choice of conventional product being replaced by bio-product. Furthermore, different BR and GBR configurations shows that higher benefits can be achieved by increasing product yields rather than lowering energy consumption. Green biorefining is shown to be an interesting biorefining concept. Biorefining of green biomass is technically feasible and can bring environmental savings, when compared to conventional production methods. However, the savings will be determined by the processing involved in each conversion stage and on the utilization of the different platform products.[34] AC will need to work in the BR and GBR to optimize products. This process of optimization will take full advantage of the tools in CE, CIR, and LCA.

4.4.3   LCA, AC, AND WATER

As has been mentioned, a major area of concern in global sustainability is the importance of water. The conversion of lignocellulosic biomass to biofuel (as well as other processes in the biorefinery) requires water. Although there has been considerable focus on the assessment of GHG emissions, there is limited work on the assessment of the life cycle water footprint in emerging processes. For example, the results of an LCA study on water consumption during lignocellulosic biomass pyrolysis and hydrothermal liquefaction processes show that differences in water use indicate that the choices of biomass feedstock and conversion pathway water efficiency are crucial factors affecting water use efficiency of biorefinery activities.[35] Given the increasing population and shrinking water supplies, this will become a critical LCA focus.

Water shortage and water contamination necessitate adopting a reuse logistics and a closed-loop supply chain approach, which aligns well with the need for CIR. Applied to water, this is the process of moving wastewater from its typical final destination back into the water supply chain with different levels of treatment for reuse. Hence, the need for incorporation of sustainability concepts through LCA for selecting reclaimed water applications considering reverse logistics and closed-loop systems. Although increasing the degree of treatment for water reuse may increase the implementation and operation costs of the new process, the value of resource recovery is important. This can offset the capital and investment costs associated with the treatment and distribution for water. Improving the reclaimed water quality can reduce the environmental footprint (eutrophication) to almost 50%.[36]

One final area might be the incorporation of soil moisture regenerated by precipitation, or green water, into LCA. Given the global importance of this resource for terrestrial ecosystems and food production, this should be included in water calculations. This is a complementary approach to existing contributions to the water cycle.[37]

4.4.4 LCA, AC, AND ENERGY

LCA can be used to gauge energy (production and usage) contribution to global sustainability. LCA was conducted to evaluate the energy efficiency and environmental impacts of a bioethanol production system that uses sweet sorghum stem on saline-alkali land as feedstock. Typically, agrochemical production consumes 76.58% of the life cycle fossil energy used in agricultural processes. This can be further broken down into sub-categories. The category with the most significant impact on the environment is eutrophication, followed by acidification, freshwater aquatic ecotoxicity, human toxicity, and global warming. Several factors, including waste recycling approach and soil salinity, significantly influence the results. Promoting reasonable management practices to alleviate saline stress and increasing agrochemical utilization efficiency can further improve environmental sustainability.[38] Producing energy from agriculture products and/or waste is insufficient to meet the increasing demand.

4.5 LCA, CIR, AND THE BIOECONOMY

The challenge of developing a sustainable production system includes the reduction of emissions, the efficient use of resources, and the transition to renewable energy. This is the ideal setting for CIR. The bioeconomy proposes a development model aimed at reducing impacts and risks associated with the use of non-renewable resources considering the LC of products. Different analyses (life-cycle assessment (LCA), life-cycle costing (LCC), and externality assessment (ExA)) will continue to be used to assess the impacts of both the bioeconomy and CIR along the production chain. The introduction of bioeconomy based on circularity and guided by LCA is the first step toward the use of renewable resources with a low impact on society. [39] This is the direction of sustainable agriculture in the twenty-first century.

For stakeholders and decision makers within the bioeconomy, it is important that sustainability assessment methodologies be holistic, reliable, and accurate. Life cycle assessment (LCA) methodologies are well-known for their ability to avoid burden shifting by considering the impacts of a product, process, or system throughout the full life cycle. However, when it comes to assessing advanced multifunctional systems within the bioeconomy (i.e., biorefineries), methodological challenges arise. Such issues include the goal, scope, and allocation methods, land use considerations, handling of biogenic carbon and emissions, impacts assessed, simplification of feedstocks and processes, regionality, and future foreground and background systems. In the future challenges remain in capturing social and economic impacts with LCA methodologies, with social assessments lacking data and appropriate quantitative indicators and economic assessments lacking diversity in stakeholder and cost inclusivity.[40]

REFERENCES

1. Tragnone, B.M., M. D'Eusanio, and L. Petti, *The count of what counts in the agri-food Social Life Cycle Assessment.* Journal of Cleaner Production, 2022. **354**: pp. 1–19.

2. Wheaton, E. and S. Kulshreshtha, *Agriculture and climate change: Implications for environmental sustainability indicators.* 9th International Conference on Ecosystems and Sustainable Development, ECOSUD 2013, 2013. **175**: pp. 99–110.

3. Griffin, P.C., et al., Best practice data life cycle approaches for the life sciences [version 2; peer review: 2 approved]. *F1000Research*, 2018. **6**: pp. 1618.

4. Guinée, J.B., et al., *Life cycle assessment: Past, present, and future.* Environmental Science and Technology, 2011. **45**(1): pp. 90–96.

5. Strazza, C., et al., *Resource productivity enhancement as means for promoting cleaner production: Analysis of co-incineration in cement plants through a life cycle approach.* Journal of Cleaner Production, 2011. **19**(14): pp. 1615–1621.

6. Klöpffer, W. and B. Grahl, *Life Cycle Assessment (LCA): A Guide to Best Practice.* Life Cycle Assessment (LCA): A Guide to Best Practice. 2014: Wiley Blackwell. 1–396.

7. Gava, O., et al., *Improving policy evidence base for agricultural sustainability and food security: A content analysis of life cycle assessment research.* Sustainability (Switzerland), 2020. **12**(3): pp. 1–29.

8. Stillitano, T., et al., *Sustainable agri-food processes and circular economy pathways in a life cycle perspective: State of the art of applicative research.* Sustainability (Switzerland), 2021. **13**(5): pp. 1–29.

9. Accorsi, R., L. Versari, and R. Manzini, *Glass vs. plastic: Life cycle assessment of extra-virgin olive oil bottles across global supply chains.* Sustainability (Switzerland), 2015. **7**(3): pp. 2818–2840.

10. Sanyé-Mengual, E., et al., *Eco-efficiency assessment and food security potential of home gardening: A case study in Padua, Italy.* Sustainability (Switzerland), 2018. **10**(7): pp. 1–25.

11. Silvestri, C., et al., *Toward a framework for selecting indicators of measuring sustainability and circular economy in the agri-food sector: a systematic literature review.* International Journal of Life Cycle Assessment, 2022.

12. McManus, M.C. and C.M. Taylor, *The changing nature of life cycle assessment.* Biomass and Bioenergy, 2015. **82**: pp. 13–26.

13. Sieverding, H., et al., *A life cycle analysis (LCA) primer for the agricultural community.* Agronomy Journal, 2020. 112(2): pp. 3788–3807.

14. Gava, O., et al., *A reflection of the use of the life cycle assessment tool for agri-food sustainability.* Sustainability (Switzerland), 2018. **11**(1): pp. 1–16.

15. Geissdoerfer, M., et al., *The Circular Economy – A new sustainability paradigm?* Journal of Cleaner Production, 2017. **143**: pp. 757–768.

16. Barros, M.V., et al., *Mapping of research lines on circular economy practices in agriculture: From waste to energy.* Renewable and Sustainable Energy Reviews, 2020. **131**: pp. 1–12.

17. Ding, A., et al., *Life cycle assessment of sewage sludge treatment and disposal based on nutrient and energy recovery: A review.* Science of the Total Environment, 2021. **769**: pp. 1–19.

18. Payen, S., et al., *Inventory of field water flows for agri-food LCA: critical review and recommendations of modelling options.* International Journal of Life Cycle Assessment, 2018. **23**(6): pp. 1331–1350.

19. Olivetti, E., S. Patanavanich, and R. Kirchain, *Exploring the viability of probabilistic under-specification to streamline life cycle assessment.* Environmental Science and Technology, 2013. **47**(10): pp. 5208–5216.

20. Mak, T.M.W., et al., *Sustainable food waste management towards circular bioeconomy: Policy review, limitations and opportunities.* Bioresource Technology, 2020. **297**: pp. 1–11.
21. Harun, S.N., M.M. Hanafiah, and N.M. Noor, *Rice Straw Utilisation for Bioenergy Production: A Brief Overview.* Energies, 2022. **15**(15).
22. Chakrapani, G., M. Zare, and S. Ramakrishna, *Biomaterials from the value-added food wastes.* Bioresource Technology Reports, 2022. **19**: pp. 1–10.
23. Kirchherr, J., D. Reike, and M. Hekkert, *Conceptualizing the circular economy: An analysis of 114 definitions.* Resources, Conservation and Recycling, 2017. **127**: pp. 221–232.
24. Lokesh, K., et al., *Hybridised sustainability metrics for use in life cycle assessment of bio-based products: Resource efficiency and circularity.* Green Chemistry, 2020. **22**(3): pp. 803–813.
25. McDevitt, J.E. and L. Milà i Canals, *Can life cycle assessment be used to evaluate plant breeding objectives to improve supply chain sustainability? A worked example using porridge oats from the UK.* International Journal of Agricultural Sustainability, 2011. **9**(4): pp. 484–494.
26. De Baan, L., R. Alkemade, and T. Koellner, *Land use impacts on biodiversity in LCA: A global approach.* International Journal of Life Cycle Assessment, 2013. **18**(6): pp. 1216–1230.
27. Eddy, I.M.S. and S.E. Gergel, *Why landscape ecologists should contribute to life cycle sustainability approaches.* Landscape Ecology, 2014. **30**(2): pp. 215–228.
28. Brankatschk, G. and M. Finkbeiner, *Modeling crop rotation in agricultural LCAs – Challenges and potential solutions.* Agricultural Systems, 2015. **138**: pp. 66–76.
29. Curran, M., et al., *Toward meaningful end points of biodiversity in life cycle assessment.* Environmental Science and Technology, 2011. **45**(1): pp. 70–79.
30. Drews, J., I. Czycholl, and J. Krieter, *A life cycle assessment study of dairy farms in northern Germany: The influence of performance parameters on environmental efficiency.* Journal of Environmental Management, 2020. **273**.
31. De Luca, A.I., et al., *Life cycle tools combined with multi-criteria and participatory methods for agricultural sustainability: Insights from a systematic and critical review.* Science of the Total Environment, 2017. **595**: pp. 352–370.
32. Adler, P.R., et al., *Integrating biorefinery and farm biogeochemical cycles offsets fossil energy and mitigates soil carbon losses.* Ecological Applications, 2015. **25**(4): pp. 1142–1156.
33. Zhu, X., et al., *Life-cycle assessment of pyrolysis processes for sustainable production of biochar from agro-residues.* Bioresource Technology, 2022. **360**: pp. 1–13.
34. Corona, A., et al., *Techno-environmental assessment of the green biorefinery concept: Combining process simulation and life cycle assessment at an early design stage.* Science of the Total Environment, 2018. **635**: pp. 100–111.
35. Wong, A., H. Zhang, and A. Kumar, *Life cycle water footprint of hydrogenation-derived renewable diesel production from lignocellulosic biomass.* Water Research, 2016. **102**: pp. 330–345.
36. Rezaei, N., et al., *A multi-criteria sustainability assessment of water reuse applications: a case study in Lakeland, Florida.* Environmental Science: Water Research and Technology, 2019. **5**(1): pp. 102–118.

37. Lathuillière, M.J., C. Bulle, and M.S. Johnson, *Land Use in LCA: Including Regionally Altered Precipitation to Quantify Ecosystem Damage.* Environmental Science and Technology, 2016. **50**(21): pp. 11769–11778.

38. Wang, M., et al., *Environmental sustainability of bioethanol produced from sweet sorghum stem on saline-alkali land.* Bioresource Technology, 2015. **187**: pp. 113–119.

39. Blanc, S., et al., *Use of bio-based plastics in the fruit supply chain: An integrated approach to assess environmental, economic, and social sustainability.* Sustainability (Switzerland), 2019. **11**(9).

40. Vance, C., J. Sweeney, and F. Murphy, *Space, time, and sustainability: The status and future of life cycle assessment frameworks for novel biorefinery systems.* Renewable and Sustainable Energy Reviews, 2022. **159**: pp. 1–15.

5 Use of natural resources affecting sustainability

In addition to climate change, the earth is experiencing environmental pressures from declining natural resources. This is being driven by the growth of the global population and the expansion of large urban areas with increasingly higher expectations regarding the quality of life. Agricultural Chemistry (AC) can play a role in addressing this problem. The efficient use and replacement of the agriculture resources, such as water, soil and energy, is crucial to guarantee the competitiveness of the farming sector, food self-sufficiency, and security.[1] As simple as it may seem, there are many dimensions to the global challenge of sustainability. (See Figure 5.1) The current agricultural production model is still mostly based on linear approaches using fossil resources, where energy, daily use products, and materials are not of renewable origin. The circular agricultural vision presented in this book establishes a sustainable recipe that ties agriculture inextricably to water, soil, air, and waste generated by any production or activity. This circular model builds upon and is inspired by natural processes. In order to integrate circularity (CIR) with AC, smart agriculture (modern agricultural practices based on the new technologies, that combine scientific research and innovation) and the principles of Green Chemistry (GC) will prove indispensable.[2] This will have a positive impact on the environment, and will build long-term resilience, generating business, new technologies, and jobs.[3] It must start with finding value in both renewable natural resources and in waste.

There will be an inevitable transition from a petroleum-based towards a more sustainable biomass-based economy. This will require a synergy among academia, industry, politics, and every society. This will need the support of various scientific fields such as microbiology, molecular biology, chemistry, genetics, and chemical engineering. This will need to involve agriculture to meet the future global challenges. As new methods of growing and protecting plants emerge, they will have to be incorporated into the normal practices. Applying environmentally friendly and efficient biological processes, terrestrial biomass can be converted to high value products e.g., chemicals, building blocks, biomaterials, pharmaceuticals, food, feed, and biofuels.[4] This chapter introduces the natural processes and the new directions in agriculture that support sustainability.

DOI: 10.1201/9781003157991-5

FIGURE 5.1   Renewable resources. (See www.shutterstock.com/image-photo/hand-plant ing-trees-technology-renewable-resources-2057392145

Fundamentally, changes in current methods of using natural resources are needed. These new approaches must involve better understanding and utilization of the critical elements, focusing on conserving and recycling available supplies of the rare ones and finding new substitutes for specific applications, where possible. The changes must involve finding ways to reduce the use of more abundant elements, while much more consideration must be given to the entire cycle of use, repair, upgrading, repurposing, by-products, waste, and disposal, in order to prevent damage to the planetary environment.[5] This will go far beyond traditional chemistry and will involve radical changes. The entire agricultural enterprise will now be best envisioned as a circular process.

5.1   AGRICULTURE IN THE TWENTY-FIRST CENTURY

Today, approximately a billion people are chronically malnourished, even as our agricultural systems are degrading land, water, biodiversity, and climate on a global scale. Feeding everyone while maintaining the Earth's ecosystems are critical goals. Furthermore, agriculture plays a key and direct role in achieving the Sustainable Development Goal (SDG 2) to end hunger, achieve food security and improved nutrition, and promote sustainable agriculture. However, to meet the world's future food requirements, agriculture must also be a component of strategies to address other SDGs on poverty, water, biodiversity, sustainable cities, sustainable energy, and climate change.

A new paradigm of agriculture involving green growth is needed, and this can be achieved through four actions:

- considering agriculture a key contributor to multiple SDGs, not just SDG 2 on zero hunger,
- building cross-sector coalitions,
- transforming our financial systems as they impact agriculture, and
- advancing research and education.

Further, they are critical for agriculture's role in the twenty-first century.

5.1.1 THREE ROLES OF AGRICULTURE

Within the movement of global economy is the movement toward sustainability. Agriculture and AC will have definite roles.

- The first role for agriculture is sustaining ecosystems, thus shifting agriculture from a source of degradation to a driver of restoration and ecosystem health. Sustaining ecosystems calls for more than action at the local farm scale. We need to expand our frame to look at whole landscapes, and how the soil, water, and vegetation resources used by the many different land stakeholders can be managed in a more coordinated way.
- Second, agriculture will play a central role in local development. During this century, agriculture must create a foundation for this development.
- Third, agriculture has a fundamental role to play as a partner in sustainable city and regions. Cities are becoming key catalysts for change in agriculture. This is critical considering that two-thirds of the global population is projected to live in cities by 2050 – though large rural populations remain, especially in Africa.

5.1.2 GLOBAL PARTNERSHIPS

Global policy commitments, and their application at the local level, will drive agriculture green growth. Understanding and implementing sustainable food systems should be ensured at the national level and at the local level. Achieving agricultural green growth will require action by the finance community – public, private and philanthropic – which should establish financial mechanisms to fund agricultural investments that sustain ecosystems, support integrate local investment programs, finance rural-urban partnerships, and fund integrated landscape management.

The challenges facing the attainment of sustainable food systems are a daunting task. This new era of science will shift agriculture from net emissions to net reductions in emissions, from major energy users to energy providers, from a major threat to biodiversity to major habitat for biodiversity. To accomplish these goals will demand new knowledge and education systems that effectively link specialists together to understand and impact complex biological systems. This will require dedicated socioeconomic systems.

5.2   SYSTEMS OF CYCLES AND SPHERES

The concept of the earth's environment being composed of "four spheres" provides a valuable image for our discussion. The four overlapping subsystems of the earth's environment set the initial context, as they contain all the world's land masses, water sources, living organisms, and gases. These four systems are in a dynamic equilibrium. Importantly, the chemical elements comprising these spheres are in continuous, sustainable movement. Unlike energy, elements are not lost and replaced as they pass through ecosystems maintaining homeostasis. They are recycled repeatedly. All chemical elements that are needed by living things are recycled in ecosystems, including carbon, nitrogen, hydrogen, oxygen, phosphorus, and sulfur. Water is also recycled. These processes are effective, efficient, and are intended for sustainability.

As critical as the soil is to sustainability, it is only part of the total environmental system. Earth's social, environmental, and economic fabric is being threatened by the forces described above, including the unsustainable use of earth's resources. Natural systems work to correct this imbalance, as nature itself is "programmed" to achieve its sustainability. The role of science and technology is to support sustainability. This effort must include environmental and green chemistry, industrial ecology, and green (sustainable) science and technology. The anthrosphere (discussed later) is the portion of the environment constructed and operated by humans.[6] Human activity results in Global Environmental Change, causing alterations in the structure and functioning of earth systems. This results from the detrimental effects of human activities in the biophysical and socioeconomic spheres.[7] Exploring our interconnectedness with the earth, we need to learn from nature how to work toward sustainability. Land-use policies, ecological restoration, forest management, local living, and sustainability thinking all involve land and agriculture.[8] Therefore, they also involve AC.

The biogeochemical cycle is composed of a vast closed loop through which chemical elements and water move through ecosystems. Chemicals cycle through both biotic and abiotic components of this ecosystem. Through this process of cycling, elements or water may be held for various lengths of time by different components of a biogeochemical cycle. Components (or compartments) that hold elements or water for a relatively short period of time are called exchange pools. For example, the atmosphere is an exchange pool for water. It holds water for several days at the longest. This is a very short time compared with the thousands of years the deep ocean can hold water. Even though this substance is held in a compartment, it is usually in dynamic equilibrium with the surrounding environment. The ocean is an example of a reservoir for water. Reservoirs are components of a geochemical cycle that hold elements or water for a relatively long period of time. The effect of human activity (in these natural/environmental settings and exchange pools) has been recorded and measured. This activity defines a proposed epoch known as the Anthropocene.[9]

5.2.1 IMPORTANCE OF SOIL FOR AGRICULTURE AND GLOBAL ISSUES

Demands on soil resources from a densely populated and rapidly industrializing world of the twenty-first century are huge. In addition to food supply, modern societies have insatiable demands for energy, water, wood products, and land area for urbanization, infra-structure, and disposal of urban and industrial wastes. There is also a need to alleviate rural poverty and raise the standard of living of masses dependent on subsistence farming. In addition, there are several environmental issues which need to be addressed such as climate change, eutrophication, and contamination of natural waters, land degradation and desertification, and loss of biodiversity. To a great extent solutions to these issues lie in sustainable management of world's soil resources (Figure 5.2), through adoption of agronomic techniques which are at the cutting edge of science. This is a central concern for agriculture and AC in the twenty-first century.

5.2.2 ADVANCING FOOD SECURITY

The future population dynamics reveal the fact that most of the projected population increase of about 3.5 billion will occur in developing countries of Asia (mostly South

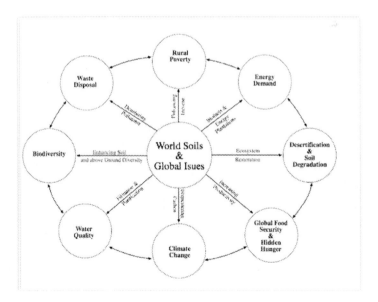

FIGURE 5.2 World soils and global issues of the twenty-first century.[10]

Asia) and Africa (mostly sub-Saharan Africa). These are also the regions where soil resources are limited in extent (per capita), fragile to natural and anthropogenic perturbations, and prone to degradation by the projected climate change and the increase in demographic pressure. Thus, any future increase in agronomic/food production will have to occur through vertical increase in production per unit area, time and input (e.g., nutrients, water, energy) of the resources already committed to agriculture.[11]

The management and methods of farming will need to adjust to this new reality. Soil management techniques must be chosen to ensure:

- liberal use of crop residues, animal dung and other biosolids,
- minimal disturbance of soil surface to provide a continuous cover of a plant canopy,
- judicious use of sub-soil fertigation techniques to maintain adequate level of nutrient and water supply required for optimal growth,
- an adequate level of microbial activity in the rhizosphere for organic matter turnover and elemental cycling, and
- use of complex cropping/farming systems which strengthen nutrient cycling and enhance use efficiency of input.

The use of soil for production of food will necessarily be conducted in parallel with, and sometimes in competition with, other necessary societal needs.(See Figure 5.2) [10] This underscores the necessity for crop residues to be considered as a source of energy and other chemical products. Crop residues must not be considered a waste, because they could have multiple but competing uses, including conservation of soil and water, cycling of nutrients, enhancement of the use efficiency of fertilizers and irrigation water, and above all, as a food of soil organisms which are essential to making soil a living entity. The issue of using crop residues for energy as cellulosic ethanol production must not be determined on the basis of short-term economic gains.

5.3   THE EARTH AS A SYSTEM

Understanding the natural world and carefully controlling the effects that human activities have on the environment are essential when moving toward sustainability. The stronger, more persistent winds that are expected to accompany a warming climate will affect agriculture through such phenomena as increased periods of drought and their prolonged duration. The atmosphere is the key to both life and death as it circumnavigates the globe. This illustrates one of the many natural resources challenging sustainability. Air as the key to life carries the spores of fungi and the pollen of various trees and grasses to the reproductive benefit of the species. It also transports dust and water. In addition to the carbon dioxide affecting the world's oceans, toxins from such areas with polluted air are transported by the wind toward other populated areas. Forest edges in fragmented landscapes function as significant traps for airborne nutrients and pollutants. These forest edges effectively concentrate these chemical species below the canopy, where they can have cascading effects on soil-nutrient cycling, microbial activity, seedling dominance, and other ecological processes.[12]

Assume that there is a large, but finite, amount of natural resources.(See Figure 5.1) Sustainable development must draw from this reliable supply of natural resources. Traditional usage and management have broadly failed to avoid environmental degradation. Unless new processes emerge, challenges to resource systems are projected to increase over time and will challenge sustainability. This situation will be exacerbated by a significant expansion in engineering deployments and progressive loss of natural resource capital. To an extent we can prolong its viability via a heavy reliance on engineering-based approaches.[13] However, this is merely delaying the inevitable loss of the natural resources and is an example of "borrowing from the future." This is not unlike trying to keep a sink filled by constantly running water, yet leaving the drain open at the same time.

5.3.1 IMPORTANCE OF SOIL

Now the soil, the "earth's thin skin," is the primary sphere for the activity of agriculture and it can become a means to help correct the problem. It serves as the active interface between the biosphere, hydrosphere, atmosphere, and lithosphere. It is a dynamic and hierarchically organized system of various organic and inorganic constituents and organisms, the spatial structure of which defines a large, complex, and heterogeneous interface. Biogeochemical processes at soil interfaces are fundamental for the overall soil development, and they are the primary driving force for key ecosystem functions such as plant productivity and water quality. Ultimately, through the soil, these processes control the fate and transport of contaminants and nutrients throughout the other spheres. This provides our model for circularity (CIR).

The importance of sustainability lies in how AC manages and works with the soil. A mechanistic understanding of the complexity of these biogeochemical interfaces in soils and of the interplay and interdependencies of the physical, chemical, and biological processes acting at and within these dynamic interfaces in soil provides a necessary key to unlock the problem of agricultural sustainability. The major challenges are

- to identify the factors controlling the chemical activities at biogeochemical interfaces,
- to link the processes operative at the individual molecular and/or organism scale to the phenomena active at the aggregate scale in a mechanistic way, and
- to explain the behavior of organic chemicals in soil within a general mechanistic framework.

5.3.2 Transport processes

We have seen that materials and energies freely move among the spheres. However, the term circular bioeconomy, also known as biobased circular economy, is an integrated concept of circular economy and bioeconomy. In other words, it denotes the circular movement of biomass beginning as biological resources into a systemic approach for economic development.[14] This will help understand how the movement toward sustainability occurs and the role of AC. It is critical to see that, as the spheres interact to effect earth's systems and processes, they are constantly using/reusing the material components, and so must our agricultural systems. The main spheres of the earth system are interconnected by flows (also known as pathways or fluxes) of energy and materials. This flow of material is analogous to communication and flow of information.

5.3.2.1 Energy and material flows

The earth is a vast, complex system powered by two sources of energy: an internal source (the decay of radioactive elements in the geosphere, which generates geothermal heat) and an external source (the solar radiation received from the sun); most of the energy in the earth system comes from the sun. This constant flow of energy is necessary to maintain agriculture. The reliability of these flows is critical to the mission of agriculture. Energy both drives and flows through environmental systems, even though energy pathways may be highly complex and difficult to identify.

Analogously, materials flow throughout the earth's system. These materials are comprised of elements and molecules. Utilizing energy, this matter is transformed as it flows through the spheres. Keeping environmental equilibrium, nature utilizes energy and material flows to replenish and maintain the environment. Since nature is neither creating nor destroying matter, it accomplishes all its tasks by a circular process, using processes that have been optimized over millions of years. The Biorefinery (BF) and GC attempt to mimic these processes so as to move the Anthropocene toward sustainability.

Water is a critical material flow. It is important to consider the concept of a water movement in relation to irrigated food production. A close analysis of water cycles affords opportunities to reduce its environmental impact. An important component of a water cycle is the environmental impact of the water use. Using geographical information system (GIS) data on the spatial distribution of irrigation usage when added to information on water resource availability highlights where irrigated food production is likely to be having an environmental impact. This will lead to ways to assess the options for managing water better in irrigated agriculture, both on-farm and from a catchment (water regulation) perspective. Measures to reduce the water footprint on-farm include better management (scheduling) and the adoption of new technologies to improve irrigation application uniformity and water efficiency. Water regulation measures could include limiting pollution in water where environmental damage is known to be occurring, using water trading to reallocate water to higher value agricultural uses and changing the timing of irrigation to encourage winter reservoir storage.[15]

5.3.2.2   Biogeochemical cycles

Essential to the earth is the activity in the biosphere, and especially the activity of humans. The earth system contains several cycles in which key elements and materials are continually transformed and transported through the environment. These cycles occur in closed systems, and the key materials that cycle through the biogeochemical cycles are carbon, oxygen, hydrogen, nitrogen, phosphorous and sulfur – all of which are essential for life. Nature accomplishes most of its work with few elements. The biogeochemical cycles operate at the global scale and facilitate the transportation of the main components of the earth system continually among the geosphere, atmosphere, hydrosphere and biosphere.

Since the biogeochemical cycles involve elements that are essential for life, living organisms play a vital part in those cycles. Typically then, the biogeochemical cycles involve an inorganic component (the abiotic part of the cycle, including sedimentary and atmospheric phases) and an organic component (comprising plants and animals, both living and dead). The compounds are transported through cycles, being often modified to facilitate its movement through the environment.

5.4   SUSTAINABILITY MODEL

The sum of the processes and flows ensures environmental sustainability. The balance of nature is a result of ecological systems that are usually in a stable equilibrium or homeostasis. The balance is sometimes depicted as easily disturbed and delicate, while other times it is inversely portrayed as powerful enough to correct any imbalances by itself because "it is supposed to be balanced".[16] The view that natural systems in balance seems to be at odds with constant disturbances leading to chaotic and dynamic changes being the norm in nature.[17] It is still valuable to view the ways in which nature and the world have evolved and how they maintain and recover from disasters.

The harmony among the four spheres suggests a path to sustainability. A necessary condition for sustainability is the law of conservation of mass [18]: If a system gains in mass, its environment must lose in mass by the same amount. This is absolutely true, but CIR requires that the mass continues to be functional in the system. Animals live in complex environments in which they are constantly confronted with short- and long-term changes due to a wide range of factors, such as environmental temperature, photoperiod, geographical location, nutrition, and socio-sexual signals. Homeostasis, the state of relative physiological stability in an organism, is a prerequisite for survival. As nature and the environment conform to this conservation law, it uses the flow of energies and materials through the spheres to maintain a balance, demonstrating a model of sustainability. Agriculture and AC will build upon this sustainability model for agricultural sustainability. Nature will also be used to evaluate its current state and future promise of sustainability.

Awareness of the finite nature of many resources in our world, including the issue of increasing water scarcity, has grown tremendously in the past few decades. It has become obvious that the linear route of production, in which scarce resources are consumed and their value-added products are discarded as waste, is a root cause of several impending global crises such as climate change, diminished biodiversity, as well as food, water and energy shortages.[19] Nature, through the efforts of a circular agricultural chemistry, can be assisted to maintain environmental sustainability in the twenty-first century.

While sustainability is an essential concept to ensure the future of humanity and the integrity of the resources and ecosystems on which we depend, identifying a comprehensive yet realistic way to assess and enhance sustainability may be one of the most difficult challenges of our time.[20] There is not a uniform pathway to sustainability, and therefore the approaches are generally not comprehensive and are subject to unintended consequences. There is the need for a comprehensive definition of sustainability (that integrates environmental, economic, and social aspects) with a unified system-of-systems approach that is causal, modular, tiered, and scalable, as well as new educational and organizational structures to improve systems-level interdisciplinary integration.

5.4.1 MATERIAL CIRCULATION

The development and maintenance of soil depends on multiple self-reinforcing feedback loops. Biologically, soil microorganisms provide the nutrients for plants to grow, and plants in turn provide the carbon, in the form of organic material, that selects for and alters the communities of soil organisms. One influences the other, and both determine the soil's development and health. There are additional paths (e.g., air or water) whereby the materials are moved and made available. Even though protection of soil, its fertility, and all the supportive pathways can be justified economically, our human dependence upon the soil escapes most people. This all stems from the view of agriculture through the lens of the traditional, linear economic system.[12]

5.4.2 RESOURCES RECOVERY

Resource renewability alone is not a measure of sustainability, as nature demonstrates. The popular opinion that oil, gas and coal are harmful, whereas renewables are clean, removes attention from the true sustainability problem: material circulation. How can we configure our agriculture so that our processes are circular and that we accomplish the goals of a sustainable system? Evidence strongly suggests that sustainable transformations of agricultural systems will help meet World and Earth needs.[9] In order for this to occur, the processes of agriculture must assume a leadership-by-example role in the path to sustainability.

5.5 AGRICULTURE, HUMAN ACTIVITY, AND GLOBAL SUSTAINABILITY

Sustainable is what can be sustained or maintained. The only way to attain sustainable balance between civilization and nature is through technological development that incorporates both humans and nature as part of the same solution. Sustainability implies and includes development and growth in civilization. The latter involves public and private choices that allow a free market/socially responsible society. The bottlenecks of circular development are food, water, energy (including transportation) and waste disposal. The solution to these challenges will also address second-order problems such as global warming.[21]

Convincing evidence has emerged that humanity has entered the Anthropocene, where human activities have precipitated effects on a planetary scale in the ecosystem, depleted available resources, and raised risks of environmental shocks and large-scale tipping points.[22] Agriculture is a primary driver of global change and is the single largest contributor to the rising environmental risks of the Anthropocene.[23] Therefore, agriculture can and should play a leading role in the drive to sustainability.

5.5.1 ANTHROPOCENE

The Anthropocene is a proposed geological epoch dating from the commencement of significant human impact on earth's geology and ecosystems, including anthropogenic climate change and impacts on biodiversity.[24] This period is marked by effects resulting only from human activities. The current problems we face as a global community are also problems we can resolve.

Human activity is leaving a pervasive and persistent signature on earth. The forces responsible for many of the anthropogenic changes are a product of the three linked challenges facing sustainability: accelerated technological development, rapid growth of the human population, and increased consumption of resources. These have combined to result in increased use of metals and minerals, fossil fuels, and agricultural fertilizers and increased transformation of land and nearshore marine ecosystems for human use. Carbon emissions from fertilizer and pesticide production in chemistry contribute significantly to the emissions from the agricultural sector, therefore agricultural policies should properly limit the use of those chemical productions to reduce the emission.[6] Consequently, to minimize these challenges and improve sustainability, agricultural chemistry and AC must play a leading role in their solution.

Agriculture and food production lie at the heart of these anthropogenic effects. Food production is the world's single largest driver of global environmental change and, at the same time, is most affected by these changes.[25] To maintain planetary health, human activities must limit the use of earth's resources within finite boundaries and limit further environmental degradation. At present, food systems account for a substantial use of natural resources and contribute considerably to climate change, degradation of land, water use, and other impacts, which in turn threaten human health through food insecurity.[26] The transformation of agricultural practices to draw society toward sustainability is imperative. Studies recommend need for adaptation strategies to overcome the vulnerabilities of climate change through evaluation and modification of agricultural practices and AC.[27]

5.5.2 Human sustainability

Agriculture is deeply connected to the drive for human sustainability. Human development is often considered as the increase in education or improvement in human capital. It is considered important to economic growth and plays a crucial role in technological progress of countries.[7] As the wellbeing or development of people improve, they not only seek improvement in education and material wealth but also a better living environment.[8] Land usage encompasses all processes and activities related to the benefits gained from the land and the unintended social and ecological outcomes of societal activities. Land systems are thus essential to the functioning of both social and ecological systems. Among other ecosystem services, land systems support food production together with material and energy resources.[28] This, in turn, is tied to human sustainability.

The Ecological Footprint (EF) represents a measure of how human pressure on the environment is measured by the collective effects of human activities on the environment from all the production and consumption of goods and services expended to satisfy human needs. The cumulative effects provide the recorded results of the Anthropocene. The global practice of agriculture must include human development and meet sustainability criteria that enables food and all other agricultural ecosystem services (i.e., climate stabilization, flood control, support of mental health, nutrition, etc.) to be generated within a stable and resilient earth system, which in turn can be defined from earth science applying the planetary boundary framework. The means by which this can be achieved is through sustainable intensification of agriculture in the Anthropocene.

Human sustainability will require maintaining a healthy balance with nature, including a high-quality food supply and general good health, which are fundamental needs that have to be satisfied if individuals within society are to attain a high standard of living. In the twenty-first century the efforts to guarantee a high quality of food are being done in tandem with heightened production to meet a growing world population. To truly accomplish this will require massive production values, which at present is harmonized with societal sustainability issues.[29]

Sustainable intensification of agriculture (SIA) will be developed in the next chapter, but an introduction is given here. Since agriculture plays a central role in determining and regulating earth's resilience, the sustainability methodology is needed for agriculture. Adopting SIA as the strategy to meet twin objectives for people and the planet is an imperative. Agriculture is the key to attaining the UN Sustainable Development Goals of eradicating hunger and securing food for the projected world population of 9–10 billion by 2050. This may require an increase in global food production of between 60 and 110 % in a world of rising global environmental risks.[30] Additionally, agriculture is also the direct livelihood of 2.5 billion smallholder farmers. SIA, in conjunction with circularity and other twenty-first century developments, will increase the likelihood of success.

SIA will require a shift in paradigms.[9] There is a strong scientific justification for a shift from our current practice of agriculture of focusing on productivity first and sustainability as a question of reducing environmental impacts, to the new paradigm where sustainability constitutes the core strategy for agricultural development. The remaining material in this book will define what constitutes sustainable agriculture. Without this shift and how agriculture and AC are positioned to lead, it is unlikely the twin objectives of feeding humanity and living within ecological boundaries can be attained.[31]

5.6 PRESERVATION OF NECESSARY RESOURCES THROUGH CIRCULAR CHEMISTRY

Sustainable agriculture is the primary strategy to ensure productivity to meet rising food needs and enable the earth to develop. The emerging structures provide a picture of its form. The challenge is to both produce more food while managing the entire food supply chain much more efficiently. Reducing waste, which has reached unacceptable proportions (estimated at 30%), is necessary for promoting better distribution, access, and nutrition. This requires a planetary food revolution which will largely be driven by the 2.5 billion smallholders that control 500 million small farms and which provide most of the food supply to most of the world's population. What is the form of this revolution? It is an intensification of agriculture with an emphasis on sustainability. (See Figure 5.3) SIA aims at meeting food production goals through biodiversity conservation which will provide ecological functions in agriculture. It will require well-informed regional and targeted solutions[32] drawing upon the strengths of both land-sparing and land-sharing approaches underpinned by strategic land-use planning and allocation[33] across local, regional, and regional scales.[34]

Soil is the most diverse and important ecosystem on the planet. The dynamic nature of land forms make them difficult to quantify. Creating an understanding of the near-constant change in plant and soil will require blending traditional soil, plant, and animal scientists with scientists from other disciplines (e.g., engineers, computer scientists, geographers, economists, human nutritionists, sociologists, and psychologists). This can bring new perspectives about what makes a pasture good.[35] The soil seems "invisible" and thus indestructible, it's a seamless system due to the complexity of which seems all but unknown and beyond comprehension, even to soil scientists. Because of the ever-changing complexities of soil, it is necessary to accept that we will never fully understand it; only then will we have the requisite patience to protect the organisms that perform the functions through which soil is kept healthy. It is also important to understand the underlying chemistries. Soil health cannot be sustained over time through applications of

FIGURE 5.3   Sustainable intensive agriculture. (See www.shutterstock.com/image-photo/
huge-yellow-tractor-working-on-cereal-2008020998)

inorganic fertilizer, which not only disrupts the biophysical network of the soil's infrastructure but also "addicts" soil to petrochemicals in order to grow the desired plants. This occurs, for example, when massive monocultures of corn are grown in the same fields for many sequential years. Furthermore, much of the fertilizer is lost as it leaches downward through the soil into the groundwater. It's therefore much wiser to work in harmony with the soil and the organisms and environmental processes that govern its infrastructure because they are responsible for the flows that provide nutrients to the plants and thus support entire food web.

5.6.1 INTRODUCTION TO SIA

Food security has determined the history of human civilizations. The increasing global population during the next four decades will result in increasing food and feed demands, which will further increase the pressure on the use of land, water and nutrients. Social policies and sustainable technologies are required that can better balance the economy, environment, and society. The increasing demand for food, feed and fuel will require transitions in land and water management, improving crop productivity and resource-use efficiencies. In order to accomplish this we will need sustainability and ecosystem services at regional and global scale in a cost-effective way.[36]

The new structure of agricultural production is one of more circular and sustainable systems to address the simultaneous challenges of resource depletion, environmental degradation, and the growing global demand for food under the threat of climate change. Specifically, circular production enabled by novel digital, mechanical, and biological technologies will allow closing loops of nutrient and energy flows within the farm, through the optimization of land-use choices and crop management.[37]

SIA will require adoption of practices during the entire value chain of the global food system. These changes will meet rising needs for nutritious and healthy food through practices that build resilience and enhance natural capital within the global environment. SIA requires a refocusing of food production that encapsulates the twofold aims of increasing yields and the ecosystem services provided by agriculture.[38]

SIA will use strategic land-use planning to maintain and improve the interacting resources and flows involving water, nutrients, energy, carbon, and biodiversity across the entire spectrum of natural, semi-natural, and agricultural land uses. The goal will be to be able to manage the whole of agriculture across scales from local to regional to national levels.

5.6.2   COMPONENTS OF SIA

From a production perspective, SIA will entail a three-step approach:

- To improve agriculture to be as resource efficient as possible by combining locally relevant crop and animal genetic improvement and practices that minimize inputs and close nutrient, carbon, and water cycles,
- To encourage the adoption of practices that build large-scale resilience by sustaining ecosystem functions and services, such as water flows and biodiversity, and
- To incorporate new concepts and ideas by connecting thinking, planning, and practice across scales to fully align the soil to environmental health and global concerns in the four spheres. This will include improved and more equitable access to knowledge and resources.

5.7   AGRICULTURAL CHEMISTRY CONTRIBUTIONS TO PROMOTING SUSTAINABILITY

The importance of AC, guided by both Green Chemistry (GC) and Circular Chemistry (CC) to the accomplishment of SIA is huge. AC will provide the fundamental science to drive this effort. Awareness of the finite nature of many resources as well as the environmental intolerance towards some products of the chemical industry has grown tremendously in the past few decades. The processes and products of GC will address both directly and indirectly these environmental intolerances. A circular economy will aim to keep products, components, and materials at their highest utility and value at all times. Chemistry is crucial for achieving these goals. Chemists understand their role in designing and developing indispensable materials and technologies, but also recognize the potentially detrimental effects that this may have on their practice. Chemists are increasingly aware of the necessity to design with sustainability in mind.[19] As a result, they are invaluable and essential in delivering a sustainable solution in twenty-first-century agriculture.

We close with three example areas where agriculture and AC will improve the use of natural resources toward sustainability.

5.7.1 SUSTAINABLE BIOMASS

Our increasing dependence on a small number of agricultural crops, such as corn, is leading to reductions in agricultural biodiversity. The belowground implications of simplifying agricultural plant communities are not fully understood; however, agroecosystem sustainability will be compromised if severe changes in biodiversity reduce soil carbon (C) and nitrogen (N) concentrations, alter microbial communities, or degrade soil ecosystem functions in natural communities. Crop rotations, especially those that include cover crops, sustain soil quality and productivity by enhancing soil C, N, and microbial biomass, making them a cornerstone for sustainable agroecosystems.[39]

Through work in AC, scientific and practical evidence clearly indicates that agriculture can shift from "foe," in terms of being the single large contributing sector to global environmental risks, to "friend," thereby contributing to global sustainability, and, in so doing, build natural capital and resilience. This is done while increasing productivity and improving livelihoods of the agricultural community.[28] The sources of sustainable practices range across all areas of agricultural development, in soil tillage systems, water resource management, crop and nutrient management, integrated landscape management, pest management, and management of ecosystem services. The next challenge will be required during a scaling up. The tools and methodologies to accomplish these goals are found in AC.

5.7.2 BIOCHAR

Biochar application to promote soil sustainability has attracted extensive attention worldwide due to its carbon (C) sequestration and fertility-enhancing properties. However, the lack of biochar accumulation in highly disturbed agroecosystems challenges the perceived long-term stability of biochar in soil.[40] Biochar is a charcoal-like material consisting largely of recalcitrant carbon. Its use to assist agriculture in meeting productivity, sustainability, and economic goals is important. Raw biochar is unpredictable in its performance, as it accumulates compounds that have positive, negative, or neutral influences on plants and microbes, depending the chemistry involved. Closing the circular economy by using biochar to resuse waste, manage nutrients shows great promise.

5.7.3 PLANT HEALTH PROTECTION

AC and sustainable agriculture must address plant health protection as a necessary requirement for agricultural sustainability. This must include nutrients and protection from pests and insects. Global increase in agricultural production from a gradually decreasing and degrading land resource has placed immense pressure on the agroecosystems. Soil microbial populations are engaged in a web of interactions affecting plant fitness as well as soil quality. They are engaged in core activities ensuring the productivity as well as stability encompassing agricultural systems and natural ecosystems. Agricultural sustainability can be improved through optimal use and management of soil fertility along with physical environmental properties, which altogether depends upon soil biological processes and biodiversity. Soil fertility in addition to other properties, e.g., texture, aeration, available moisture, etc., known to support agricultural production has been found to depend on the biomass, metabolites, and activities of microorganisms. Hence, an understanding of microbial diversity in agricultural scenario is not only important but also useful to identify measures which may indicate soil quality and plant productivity. Soil microbial community mediate the biogeochemical cycling of carbon, nutrients, and trace elements by catalyzing redox reactions which moderate atmospheric composition, water chemistry, and the bioavailability of elements in soil. Positive plant-microbe interactions in the rhizosphere are the core determinants of plant health and soil fertility. This complex system can stimulate germination and growth, help plants to develop disease resistance, promote stress resistance, and influence plant fitness.[41]

REFERENCES

1. Martinho, V.J.P.D. and R.P.F. Guiné, *Integrated-smart agriculture: Contexts and assumptions for a broader concept.* Agronomy, 2021. **11**(8): pp. 1–21.
2. Ragaveena, S., A. Shirly Edward, and U. Surendran, *Smart controlled environment agriculture methods: a holistic review.* Reviews in Environmental Science and Biotechnology, 2021. **20**(4): pp. 887–913.
3. Maschmeyer, T., R. Luque, and M. Selva, *Upgrading of marine (fish and crustaceans) biowaste for high added-value molecules and bio(nano)-materials.* Chemical Society Reviews, 2020. **49**(13): pp. 4527–4563.
4. Krüger, A., et al., *Towards a sustainable biobased industry – Highlighting the impact of extremophiles.* New Biotechnology, 2018. **40**: pp. 144–153.
5. Matlin, S.A., et al., *The Periodic Table of the Chemical Elements and Sustainable Development.* European Journal of Inorganic Chemistry, 2019. **2019**(39–40): pp. 4170–4173.
6. Zhang, Q., Y. Tian, and X. Ma, *Identifying the impetus of Chinese household consumption of carbon emissions in structural path analysis.* Beijing Daxue Xuebao (Ziran Kexue Ban)/Acta Scientiarum Naturalium Universitatis Pekinensis, 2019. **55**(2): pp. 377–386.
7. Teixeira, A.A.C. and A.S.S. Queirós, *Economic growth, human capital and structural change: A dynamic panel data analysis.* Research Policy, 2016. **45**(8): pp. 1636–1648.

8. Opoku, E.E.O., K.E. Dogah, and O.A. Aluko, *The contribution of human development towards environmental sustainability*. Energy Economics, 2022. **106**: pp. 1–15.

9. Rockström, J., et al., *Sustainable intensification of agriculture for human prosperity and global sustainability*. Ambio, 2017. **46**(1): pp. 4–17.

10. Lal, R., *Soils and sustainable agriculture. A review*. Agronomy for Sustainable Development, 2008. **28**(1): pp. 57–64.

11. Roberts, J.M., et al., *Vertical farming systems bring new considerations for pest and disease management*. Annals of Applied Biology, 2020. **176**(3): pp. 226–232.

12. Maser, C., *Earth in our care: Ecology, economy, and sustainability*. Earth in Our Care: Ecology, Economy, and Sustainability. 2009: Rutgers University Press. 1–276.

13. Vörösmarty, C.J., et al., *A green-gray path to global water security and sustainable infrastructure*. Global Environmental Change, 2021. **70**: pp. 1–13.

14. Leong, H.Y., et al., *Waste biorefinery towards a sustainable circular bioeconomy: a solution to global issues*. Biotechnology for Biofuels, 2021. **14**(1): pp. 1–15.

15. Hess, T., et al., *Managing the water footprint of irrigated food production in England and Wales*, in *Issues in Environmental Science and Technology*. 2011, Royal Society of Chemistry. pp. 78–92.

16. Werth, A. and D. Allchin, *Teleology's long shadow*. Evolution: Education and Outreach, 2020. **13**(1): pp. 1–11.

17. Simberloff, D., *The "Balance of Nature"—Evolution of a Panchreston*. PLoS Biology, 2014. **12**(10): pp. 1–4.

18. Lewis, G.N., *The fundamental laws of matter and energy [1]*. Science, 1909. **30**(759): pp. 84–86.

19. Keijer, T., V. Bakker, and J.C. Slootweg, *Circular chemistry to enable a circular economy*. Nature Chemistry, 2019. **11**(3): pp. 190–195.

20. Little, J.C., E.T. Hester, and C.C. Carey, *Assessing and Enhancing Environmental Sustainability: A Conceptual Review*. Environmental Science and Technology, 2016. **50**(13): pp. 6830–6845.

21. Bisk, T. and P. Bołtuć, *Sustainability as growth*, in *Technology, Society and Sustainability: Selected Concepts, Issues and Cases*. 2017, Springer International Publishing. pp. 239–250.

22. Waters, C.N., et al., *The Anthropocene is functionally and stratigraphically distinct from the Holocene*. Science, 2016. **351**(6269).

23. Foley, J.A., et al., *Solutions for a cultivated planet*. Nature, 2011. **478**(7369): pp. 337–342.

24. Andermann, T., et al., *The past and future human impact on mammalian diversity*. Science Advances, 2020. **6**(36): pp. 1–17.

25. Struik, P.C., et al., *Deconstructing and unpacking scientific controversies in intensification and sustainability: Why the tensions in concepts and values?* Current Opinion in Environmental Sustainability, 2014. **8**: pp. 80–88.

26. Fresán, U. and J. Sabaté, *Vegetarian diets: planetary health and its alignment with human health*. Advances in Nutrition, 2019. **10**: pp. S380–S388.

27. Gul, F., D. Jan, and M. Ashfaq, *Assessing the socio-economic impact of climate change on wheat production in Khyber Pakhtunkhwa, Pakistan*. Environmental Science and Pollution Research, 2019. **26**(7): pp. 6576–6585.

28. Kastner, T., et al., *Global agricultural trade and land system sustainability: Implications for ecosystem carbon storage, biodiversity, and human nutrition*. One Earth, 2021. **4**(10): pp. 1425–1443.

29. Hofmeister, S. and K. Kümmerer, *Sustainability, substance-flow management, and time, Part II: Temporal impact assessment (TIA) for substance-flow management.* Journal of Environmental Management, 2009. **90**(3): pp. 1377–1384.

30. Pardey, P.G., et al., *A bounds analysis of world food futures: global agriculture through to 2050.* Australian Journal of Agricultural and Resource Economics, 2014. **58**(4): pp. 571–589.

31. Steffen, W., et al., *The trajectory of the anthropocene: The great acceleration.* Anthropocene Review, 2015. **2**(1): pp. 81–98.

32. Hanjra, M.A., et al., *Global food security: Facts, issues, interventions and public policy implications,* in *Global Food Security: Emerging Issues and Economic Implications.* 2013, Nova Science Publishers, Inc. pp. 1–35.

33. Law, E.A., et al., *Better land-use allocation outperforms land sparing and land sharing approaches to conservation in Central Kalimantan, Indonesia.* Biological Conservation, 2015. **186**: pp. 276–286.

34. Fischer, J., et al., *Mind the sustainability gap.* Trends in Ecology and Evolution, 2007. **22**(12): pp. 621–624.

35. Kallenbach, R.L., *Describing the dynamic: Measuring and assessing the value of plants in the pasture.* Crop Science, 2015. **55**(6): pp. 2531–2539.

36. Spiertz, H., *Avenues to meet food security. The role of agronomy on solving complexity in food production and resource use.* European Journal of Agronomy, 2012. **43**: pp. 1–8.

37. Basso, B., et al., *Enabling circularity in grain production systems with novel technologies and policy.* Agricultural Systems, 2021. **193**: pp. 1–8.

38. Fróna, D., J. Szenderák, and M. Harangi-Rákos, *The challenge of feeding the world.* Sustainability (Switzerland), 2019. **11**(20): pp. 1–18.

39. McDaniel, M.D., L.K. Tiemann, and A.S. Grandy, *Does agricultural crop diversity enhance soil microbial biomass and organic matter dynamics? A meta-analysis.* Ecological Applications, 2014. **24**(3): pp. 560–570.

40. Yi, Q., et al., *Temporal physicochemical changes and transformation of biochar in a rice paddy: Insights from a 9-year field experiment.* Science of the Total Environment, 2020. **721**: pp. 1–10.

41. Ratnakar, A. and Shikha, *Role of microbial genomics in plant 10 health protection and soil health maintenance,* in *Microbial Genomics in Sustainable Agroecosystems: Volume 2.* 2019, Springer Singapore. pp. 163–179.

6 Programs and processes that define agricultural chemistry in the twenty-first century

6.1 CHARACTERISTICS OF TWENTY-FIRST CENTURY AGRICULTURAL CHEMISTRY

Human security, which entails both the qualities of material freedom and the ability to live with dignity, is the main challenge for the twenty-first century and beyond. In the global society there exist many threats to human security in areas such as health, food, economics, the environment, and sustainable development. However, the human individual must still remain at the center of attention. To meet the goal of human security, there must be a re-evaluation of many of the existing systems, especially as they relate to agriculture and agricultural chemistry (AC). The price for making these changes is high, but the cost for not doing so is overwhelming.

Balancing our need for food, clothing, housing, and amenities against the associated disturbances to the natural environment involves a range of pragmatic (short- and long-term) choices and ethical dilemmas about the environment and our responsibility for it.[1] Whether or not we can devise more effective ways for living life in accord with the environment is not merely theoretical, it is a moral imperative. We do not have the luxury of a foolproof and complete plan of how to accomplish this, yet we must continue to act wisely and with due diligence.

Agriculture and AC are central to addressing these challenges, but surprisingly their role and contributions to human security have hitherto not been explicitly set out. In the twenty-first century new philosophies and processes are becoming solidified in the field of food production. Human health will be impacted by the new directions being taken, as it lays bare the interconnections between human and environmental health. (See Figure 6.1)

Six words can capture the essence of twenty-first-century AC. The words (Smart, Intense, Green, Circular, Renewable, and Sustainable) express the ways this field has adapted to meet the goals of sustainability. From the words emerge the necessary bold measures that will affect required changes necessary to achieve sustainability.

6.1.1 SMART

When the term smart is applied to agriculture it refers to the use of technologies to make the entire system work more smoothly. The ultimate goal is increasing

DOI: 10.1201/9781003157991-6

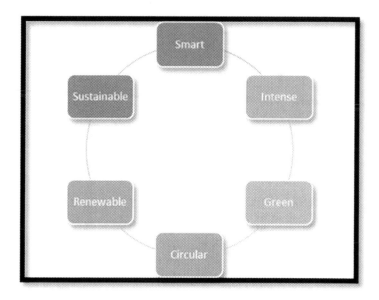

FIGURE 6.1   Essence of agricultural chemistry

the quality and quantity of the crops while optimizing the human labor used. It involves using the internet to gather information, sensors to monitor production, optimizing location systems, robotics and artificial intelligence.[2]

6.1.2   INTENSE

Intensive agriculture is necessary to meet the increasing demands of the increasing population. It is characterized by a low fallow ratio, higher use of inputs such as capital and labor, and higher crop yields per unit land area. This approach to agriculture, greater use of both crop plants and animals, demands higher levels of input and output per unit of agricultural land area. It tends to negatively affect environmental sustainability.[3]

6.1.3   GREEN

Green AC is not just the color; it is a principled approach to practicing AC. New systems in agriculture integrate sustainable and environmentally friendly concepts of Green Chemistry (GC) with intelligent and integrated farming processes, optimizing the agricultural production. Benefits include the production of new classes of renewable biofuels and feedstock, as well as new ways to use biomass to generate bio-based chemicals, focusing on GC concepts towards sustainable agriculture and processing models.[4]

6.1.4   CIRCULAR

AC must now fully embrace material circularity, ensuring wise use of resources. The circular approach emphasizes least amounts of material necessary are consumed, that the design of products maximize opportunities for recycling and reuse, and that all by-products, waste products, and post-use products are recaptured and reused. This approach is tied to life cycle analysis techniques to account for both material cycling efficiencies and environmental impacts.[5]

6.1.5   RENEWABLE

In the twenty-first century there is the increasing need for an immediate and dramatic increase in agricultural production. For long-term sustainability, agriculture must have the capacity for renewal.

Increased production often relies on non-renewable fuel to produce and transport. Additionally, most seed in industrial agriculture is non-renewable due to legal and genetic mechanisms that make it problematic for farmers to save and replant what they have grown on their farms. Is a renewable agriculture with a high level of productivity possible?[6]

6.1.6   SUSTAINABLE

Sustainable agriculture and AC are usually described as using a resource so that the resource is not depleted or permanently damaged. This should include sustainable agricultural systems that do not deplete the soil, water, and biotic resources; and of sustainable food supplies.[7] Agriculture and AC in the twenty-first century might add to these new ideas of green or organic agriculture, or industrial agriculture with its use of synthetic pest control and fertilizers. In agriculture and food production, sustainability will be expressed through minimal use of limiting resources; recovering tillable land, recycling fresh water, and helping produce alternative fuel and energy.

Agriculture and AC will conform itself to the above six defining words during the twenty-first century. The extent to which it is successful will be the measure of its contribution to the larger question of human survival.

6.2   CENTRALITY OF CHEMISTRY

Chemistry has matured over time through several stages of development, and it has many useful applications in AC. The basic techniques that have been developed are now applicable to new classes of opportunities and services to society.[8] The simple categories of the traditional fields of chemistry ("atoms, molecules, and reactions") no longer seems to capture its uses, its obligations to society, or the complexity of the challenges it faces. The field of agriculture has many problems that chemical sciences can address. Chemistry in its roles in agriculture and AC may now have the most important role in its potential to impact society.[8]

New opportunities are, however, much broader in scope and greater in complexity than the simpler, previous problems. There will be enhanced, new requirements for new compounds and methods. This will necessarily involve all aspects of AC and will also entail great novelty. There are a wide range of problems in AC that will involve environmental, sustainable, and human survival in which global societies may have substantial interests, and in which it may be necessary to develop new approaches to reaching familiar goals. Chemistry applications are essential in this area.

Agricultural chemists need to be aware of and increasingly guided by the principles of human security while applying their knowledge and skill to developing solutions to the oncoming global challenges. The changes that AC will make are inevitable. The areas of chemical innovation will be important for involvement in human security and, in particular, for the first time explicitly frames the role, contributions and opportunities for chemistry in this area, focusing mainly on four dimensions (health, food, economic, and environmental security). The new applications in the field of chemistry are indispensable and their future roles need to be strengthened and re-imagined.[9]

6.2.1 EXPLAINING ACTIVITIES, PROCESSES, AND TRANSPORT IN SPHERES

Chemistry tends to study systems at, or moving toward, thermodynamic equilibrium. They have seemed complicated enough. But the most interesting systems in the world around us are fluid so that their characteristics and most interesting features only emerge when there is a flux of energy through them.[10] Studying dissipative systems such as these has, of course, been a subject of physical science for decades, but, unlike equilibrium systems, understanding these systems is still at early stages. These systems are not at rest, but are in motion. The environment and nature mimic this; chemistry is needed to explain this behavior.

Understanding enzyme activity and catalysts in AC will always be challenging. Enzymes are normally an example of fluid chemistry. The immobilization of enzymes is an excellent alternative to overcome the drawbacks of using these biocatalysts in solution. This process plays a significant role in cost-effective recovery, increased catalyst productivity and in simplifying process operations. In the context of sustainability, the process of using enzymes for material production constitutes a clean route for the development of more sustainable biocatalysts capable of applications in various areas.[10]

Chemistry can help to attain a better understanding of soil organic matter (SOM) biogeochemistry. SOM is critical to meeting the environmental challenges that currently accompany the changing climate. Humic substances obtained from alkaline extractions of soils have played a predominant role in developing our knowledge of SOM chemistry and behavior during the twentieth century. Recently, there have been new important developments that, according to some researchers, may be used to improve our appreciation of the value of humic substances in soils as they are presently defined. Continued progress in understanding SOM and carbon (C) biogeochemistry is essential to achieve long-term ecological, environmental, and agricultural sustainability through the development of management strategies that conserve soil C and reduce the reliance on chemical inputs for meeting crop nutrient needs.[11] This is work that is in the realm of AC.

Chemistry is critical in minimizing soil damage due to intense farming. Dissolved organic carbon (DOC) through leaching into the soils is another mechanism of net C loss. It plays an important role in impacting the environment and is exacerbated by soil and crop management practices. However, since little is known about the impacts of landscape positions and nitrogen (N) fertilizer rates on DOC leaching, more work is needed. Data show that there is a moderate positive relationship between the total average DOC contents and the total average biomass yields. Overall, the DOC contents from leachate in the land are significantly influenced by landscape positions and N rates. Switching crops could retain soil and environment sustainability to some extent. These findings will assist in understanding the mechanism of changes in DOC contents with various parameters in the natural environment and crop management systems.[12]

In the need for sustainability, the question of crop quality is an important concern. The main objectives for environmental sustainability are a decrease in chemical inputs, a reduction in the level of pollutants, and an improvement in the soil's biological activity. For example, among inorganic pollutants emitted by vehicle traffic and some industrial processes in urban areas, antimony (Sb) is observed on a global scale. While this metalloid is known to be potentially toxic, it can transfer from the soil or the atmosphere to plants and accumulate in their edible parts. Urban agriculture is developing worldwide, and could therefore increasingly expose populations to Sb, as well as other metals. Gaining knowledge of Sb uptake and bioaccumulation by crops, to reveal investigative fields on which to focus is important. While there is still no legal maximal value for Sb in plants and soils, light has to be shed on its accumulation and the factors affecting it. A relative absence of data exists about the role of soil flora and fauna in the transfer, speciation and compartmentation of Sb in vegetables. Moreover, little information exists on Sb ecotoxicity for terrestrial ecosystems.[13]

In general, understanding the environmental chemistry of metals by chemists will further help predict their impact on health, agriculture, and sustainability. For example, nickel (Ni)-contamination impairs soil ecosystem, threatening human health.[14] The impact of redox potential (EH), pH, iron (Fe), manganese (Mn), chloride (Cl^-), aliphatic and aromatic dissolved organic carbon (DOC), and sulfate (SO_4^{2-}) on the release of dissolved arsenic (As), cadmium (Cd), cobalt (Co), and vanadium (V) have been studied using an automated biogeochemical microcosm apparatus. Release of As and V appeared to be related to changes of EH/pH, co-precipitation with Fe oxides, and the release of dissolved aromatic carbon compounds. Concentrations of soluble Cd (4.8–11.2 µg L^{-1}), Mn, SO_4^{2-}, and Cl^- increased under oxidizing conditions. Release of Co (166.6–258.2 µg

L-1) was related to the chemistry of Fe, Mn and DOC. Phospholipid fatty acids analysis indicated the potential for the microbial community to be involved in biogeochemical processes such as the formation of sulfides, oxidation and reduction of compounds, and the bio-methylation of elements such as As.[15]

Phosphorus (P) is an important biogeochemical element and the environmental fate of P is receiving increasing attention.[16] Introduction of chemicals in the environment may lead to undesirable consequences. Over-fertilization has resulted in high phosphorus levels in agricultural soils, but the accumulated reserves can be depleted due to under-fertilization. It is possible to map the spatial pattern of phosphorus levels and its change in the last few decades to document the effect of fertilization and underlying socio-economic conditions on P concentrations, to identify the role of soil properties in changing soil soluble P and to quantify the total amount of soluble phosphorus level change in agricultural areas. The results confirmed that phosphorus levels in agricultural areas depend mainly on agricultural use, while soil physical characteristics play a smaller role. The total loss (caused by harvesting, fixation and erosion) is ~1.5 million tons of soluble phosphorus, which is twice as much as the reported phosphorus balances indicated. Results show that approximately 50% of agricultural areas are possibly characterized by a very low supply of phosphorus (according to the latest data), posing a risk of nutrient depletion in these areas.[17]

When P leeches into water, the buildup of it can upset the mineral balance. The addition of iron is a convenient way for removing phosphorus from wastewater, but this often limits P recovery. The current poor understanding of iron and phosphorus chemistry in wastewater systems is preventing processes being developed to recover phosphorus from iron-phosphorus rich wastes like municipal wastewater sludge. Methods are needed for manipulating iron-phosphorus chemistry in wastewater treatment processes to allow phosphorus to be recovered.[18]

Inorganic nitrogen contaminants (INC) (NH^{4+}, NO_3^-, NO_2^-, NH_3, NO, NO_2, and N_2O) also pose a growing risk to the environment, and their remediation methods are highly sought after. Recently developed in situ NO_3 sensors provide new opportunities to measure changes in stream concentration at high temporal frequencies that historically have not been feasible.[19] In addition, AC is studying application of carbon materials (CM), such as biochar and activated carbon, to remediate INC from agricultural fields and wastewater treatment plants over the past few years. Understanding the role of surface chemistry of CM in adsorption of various INC is highly critical to increase adsorption efficiency as well as to assess the long term impact of using these highly recalcitrant CM for remediation of INC. Critical reviews of adsorption studies related to INC have revealed that carbon surface chemistry (surface functional groups, pH, EH, elemental composition, and mineral content) has significant influence on adsorption of INC. Compared to basic functional groups, oxygen containing surface functional groups have been found to be more influential for adsorption of INC. However, basic sites on carbon materials still play an important role in chemisorption of anionic

INC. Apart from surface functional groups, pH and EH of CM and elemental and mineral composition of its surface are important properties capable of altering INC interactions with CM.[20]

Chemistry is valuable in understanding both the cause and prevention of chemical runoff. Sorption and degradation are the primary processes controlling the efficacy and runoff contamination risk of agrochemicals. AC is examining the use of biochar soil to achieve a reduction in agricultural damage (minimum impact on efficacy) and environmental (minimum runoff contamination) benefits. The results are encouraging and suggest the utility of biochar for remediation of sites where concentrations of highly stable and mobile agrochemicals exceed the water-quality benchmarks.[21]

6.2.2 HUMAN HEALTH AND AGRICULTURAL CHEMISTRY

Agriculture and health are linked in many ways. While agriculture is essential for good health: it can be linked with poor health, including malnutrition, malaria, foodborne illnesses, human immunodeficiency virus/acquired immunodeficiency syndrome (HIV/AIDS), livestock-related diseases, chronic diseases, and occupational ill-health. Health also affects agriculture: people's health status influences the demand for agricultural outputs, and in agricultural communities, poor health reduces work performance, reducing income and productivity and perpetuating a downward spiral into ill-health. Health and agricultural sectors remain poorly coordinated. Health and agricultural researchers likewise need to work more closely together to achieve common goals.[22]

6.2.3 CLIMATE INSTABILITY

The largest contributor to the predicted anthropogenic climate change arises from the burning of fossil fuels that generates carbon dioxide, a greenhouse gas. Increases in CO_2 concentration will not only influence climate but also the acidity of the oceans. While acid–base equilibria and their changes are at the heart of the latter issue, in the atmosphere, CO_2 is not very chemically active.

In addition to CO_2, there are many other emissions of chemically active species that directly or indirectly affect earth's climate. They include CH_4, halocarbons, N_2O, nonmethane hydrocarbons (NMHC), and nitrogen oxides. Together, these non-CO_2 emissions contribute almost as much as human-produced CO_2 to today's climate changes, Unlike the greenhouse gases, aerosols (a suspension of liquid or solid matter in the air) and clouds are expected to exert a global negative forcing and they are currently estimated to be offsetting positive forcing by the greenhouse gases by as much as 50% of the forcing by CO_2. Finally, most emissions are removed from the atmosphere by the oxidants in the atmosphere such as OH radicals, nitrate radicals, and ozone; these determine the crucial "cleansing" capacity of the atmosphere. Evidently, chemically active agents are a large part of the influence of human activities on climate.[23]

6.2.4 ENERGY USAGE

Agriculture requires energy as an important input to production. Agriculture uses energy directly as fuel or electricity to operate machinery and equipment, to heat or cool buildings, and for lighting on the farm, and indirectly in the fertilizers and chemicals produced off the farm. U.S. farm production has become increasingly mechanized and requires timely energy supplies at stages of the production cycle to achieve optimum yields. Energy's share of agricultural production expenses varies widely by activity, production practice, and locality.

6.2.5 ANALYTICAL TECHNIQUES THAT OPEN NEW AREAS OF SCIENCE

Analytical chemistry is a much more important area than it may seem. One of the most important steps in opening new areas of science is developing new analytical techniques that make possible relevant measurements.[17] Chemistry is still in a period of very active development of new analytical techniques, and probing molecular behaviors and motions at the subcellular level, and (at the other end of scales of sizes) developing the measurement infrastructure for managing atmospheres, all represent enormously interesting and challenging problems.

Thiabendazole (TBZ), has been extensively employed as a pesticide and/or a fungicide in agriculture, while its residues would be a threat to public health and safety. Simple, rapid and sensitive probes for detection of TBZ in real food samples is significantly desirable.[24] The referenced research demonstrated a reliable and rapid method. This type of analytical work is necessary for AC to ensure the security of the food produced and to maintain trust from consumers.

Surface-enhanced Raman spectroscopy (SERS) is an emerging technique for the detection of pesticide residues on food surfaces, permitting quantitative measurement of pesticide residues without pretreating the sample. However, previous studies have mainly involved the single Raman spectrum of samples, but have given little information on pesticide residue distribution. A study shows that this method can achieve rapid and quantitative detection and obtain basic information about the distribution of pesticide residues during pesticide application, which has the potential to be applied to the studies of the diffusion and absorption processes of pesticides in agricultural products.[25]

Chemical tools can assist in understanding the effects of climate change, which will also benefit agriculture. An analytical method has been developed for the pressurized liquid extraction (PLE) of a wide range of semi-volatile organic compounds (SVOCs) from atmospheric particulate matter. Approximately 130 SVOCs from eight compound classes were selected as molecular markers of (1) agricultural activity (30 current and historic-use pesticides), (2) industrial activity (18 PCBs), (3) consumer products and building materials (16 PBDEs, 11 OPEs), and (4) motor vehicle exhaust (22 PAHs, 16 alkanes, 9 hopanes, 8 steranes). Currently, there is no analytical method validated for the extraction of all eight compound classes in a single automated technique.[26]

6.2.6   MICROBE CHEMISTRY

Soil microbes have critical influence on the productivity and sustainability of agricultural ecosystems, yet the magnitude and direction to which management practices affect the soil microbial community remain unclear. Bacterial and fungal communities in three cropping systems clustered and separated, suggesting that management practices as such played minor roles in shaping the soil microbial community compared to plant type (i.e., woody vs. herbaceous plants). The data indicates that while moderately affecting the overall structure of the soil microbial community, management practices, particularly fertilization and the source of N (synthetic vs. organic), were important regulating the presence and abundance of habitat-specific microbes.[27]

6.3   EXISTING AGRICULTURAL STRUCTURES

Traditional farming is often regarded as a method of farming, which is still being used by half of the world's farming population. (See Figure 6.2) It involves the application of indigenous knowledge, traditional tools, natural resources, organic fertilizers, and cultural beliefs of the farmers. The production of a variety of household crops and livestock are possible through this farming. Common traditional farming practices include agroforestry, intercropping, crop rotation, cover cropping, traditional organic composting, integrated crop-animal farming, shifting cultivation, and slash-and-burn farming.

FIGURE 6.2   Current agricultural practices. (See www.shutterstock.com/image-photo/har vesting-hay-bales-field-under-blue-1508433593)

Industrial agriculture is a form of modern farming that refers to the industrialized production of crops. The methods of industrial agriculture include innovation in agricultural machinery and farming methods, genetic technology, techniques for achieving economies of scale in production, the creation of new markets for consumption, the application of patent protection to genetic information, and global trade. These methods are widespread in developed nations and increasingly prevalent worldwide. Most of the fruits and vegetables available in supermarkets are produced using these methods of industrial agriculture.

6.3.1 SMALL SCALE FARMING METHODS

6.3.1.1 Characteristics of small-scale farming[28,29]

These farms are only 1–10 acres, sometimes even less considering backyard farms in cities. The small-scale farming is usually a more sustainable way of farming land, compared to large scale factory farms. These farms can often adopt multiple methods used in sustainable farming, to save money and provide better long term stability.

Most family-run farms are on a very tight budget, which forces them to optimize in every possible way, for example by abstaining from expensive technology and equipment. Because of this, small-scale farms usually include a lot of manual labor. Tractors and other machinery is rarely used and most of the work around crops is done by hand. This way of farming can be characterized as a very efficient way of producing crops often surpassing production per land unit when compared to regular factory farming. Small farms often use crop rotation systems, making them less vulnerable to diseases.

They also often use organic fertilizer and avoid oil-based soil improvement.

6.3.1.2 Main differences between small-scale and conventional farming

While there are a lot of differences comparing these two farming methods the biggest ones have to be operation size and budget. Most small-scale farms run on a small property with a very tight budget and infrequent use of machinery. Factory farming turns this completely around.

There are also some less obvious differences, for example, small-scale farms tend to produce for local communities, therefore eliminating transportation and distributions costs. Bigger factory farms produce one crop for many different states, exporting large amounts to a lot of different places.

Also, the way farms are operated is vastly different. Factory farms run on a strict business plan, and are only run to improve profit. Small-scale farms, on the other hand, are often run by a family, providing a more personal connection to the farm.

In general, small farms improve soil quality, while in larger factory farming, very little thought is given to soil health. Small farms often try to not only maintain the quality of the soil but improve it over time, so that future generations will profit even more from this farming method. Small farms tend to grow a more diverse crop selection. Bigger farms often grow only a single crop on a very large scale.

6.3.1.3   Challenges and problems

A small-scale farmer tends to work long days of hard manual labor. The work is high risk, as health and safety are key factors in production. Weather is one of the most important, yet most unreliable factors in farming. A bad storm can wipe out an entire harvest. Preparing for these kinds of situations can be very stressful. Since most small farms operate on a very tight budget, the farmers survive with very little to no luxury.[30]

6.3.1.4   Greenhouse farming

Similar to small scale farms is greenhouse vegetable production. The selection of cultivated vegetables in greenhouses is mainly based on farmers' own experience rather than site-dependent soil conditions. For sustainable development of greenhouse vegetable production systems, there are two key aspects. First, it is imperative to reduce environmental pollution and subsequent health risks through integrated nutrient management. Second, a conversion of cooperative and small family business models can stabilize vegetable yields and increase farmers' benefits.[31]

Greenhouse vegetable production (GVP) has become a growing source of public vegetable consumption and farmers' income. However, various pollutants can be accumulated in GVP soils due to the high cropping index, large agricultural input, and closed environment. Ecological toxicity caused by excessive pollutants' accumulation can then lead to serious health risks. To guarantee a sustainable GVP development, several strategies are needed. Implementation of various strategies not only requires the concerted efforts among different stakeholders, but also the whole lifecycle assessment throughout the GVP processes as well as effective enforcement of policies, laws, and regulations.[31]

6.3.2   INDUSTRIAL INTENSE FARMING

In response to the global need for a sustainable quantity of food, industrial agriculture provides cheap and plentiful food. The current large scale agriculture also provides convenience for the consumer and it contributes to the global economy on many levels, from growers to harvesters to processors to sellers.

6.3.2.1   Description

Large scale farming grew from small scale proportions due to primarily three factors. The identification of nitrogen, potassium, and phosphorus as critical factors in plant growth, which led to the manufacture of synthetic fertilizers. Chemicals developed for use in World War II gave rise to synthetic pesticides, and this allowed healthy plants during large scale farming. Developments in shipping networks and technology have made long-distance distribution of agricultural produce feasible.

The ability to grow and distribute massive amounts of crops has resulted in the number of workers on U.S. farms shrinking to 1.5 percent in 2002. This means that the larger farms service more consumers. During this time, the number of small and medium scale farming operations have also decreased due to the increased production and farmland costs Many farmers were reluctant to mobilize because of the affect this would have on their family business. The separation between the management styles of farmers can be described by two approaches; farming as a lifestyle versus farming solely for profit.

6.3.2.2   Problems

The challenges and issues of industrial agriculture for global and local society, for the industrial agriculture sector and for the individual industrial agriculture farm include the costs and benefits of both current practices and proposed changes to those practices. This is a continuation of thousands of years of the invention and use of technologies in feeding ever growing populations.

The goals of industrial agriculture are lower cost products and greater productivity. Industrial methods have side effects both good and bad. Industrial agriculture is composed of numerous separate elements, each of which can be modified, and in fact is modified in response to market conditions, government regulation, and scientific advances. So the question then becomes for each specific element that goes into an industrial agriculture method or technique or process: What bad side effects are bad enough that the financial gain and good side effects are outweighed? Different interest groups not only reach different conclusions on this, but also recommend differing solutions, which then become factors in changing both market conditions and government regulations.

6.3.2.3   Solutions

U.S. agriculture is both a major source of global food and a key contributor to multiple interconnected situations. Climate change, biodiversity loss, and severe impacts on soil and water quality are among the challenges caused by U.S. industrial agriculture. Regenerative methods of farming are necessary to confront all these challenges simultaneously, in addition to addressing the increasing challenges to farm labor conditions. Transforming U.S. agriculture to a regenerative system will require a focus on creating traction for the values, beliefs, worldviews, and paradigms that effectively support such transformation while decreasing the friction that works against them.[32]

6.3.3   ENVIRONMENT

Industrial agriculture uses huge amounts of water, energy,[33] and industrial chemicals. This increases pollution in the arable land, usable water and atmosphere. Herbicides, insecticides, fertilizers, and animal waste products are accumulating in ground and surface waters. Many of the negative effects of industrial agriculture are remote from fields and farms. But other adverse effects are showing up within agricultural production systems -- for example, the rapidly developing resistance among pests is rendering our arsenal of herbicides and insecticides increasingly ineffective.[11]

The industrialization and modern agricultural practices have polluted the environment with toxic heavy metals such as Cr(VI), Cu^{2+}, Cd^{2+}, Pb^{2+}, and Zn^{2+}. Among the hazardous heavy metal(loid)s contamination in agricultural soil, water, and air, hexavalent chromium [Cr(VI)] is the most virulent carcinogen. Toxic chromium [(Cr(VI)] in food chain has created an alarming situation for human life and ecosystems. Various methods have been employed to reduce the concentration of Cr(VI) contamination with nano and bioremediation being the recent advancement to achieve recovery at low cost and higher efficiency. Bioremediation is a process using biological sources such as plant extracts, microorganisms, and algae to reduce the heavy metals while the nano-remediation uses nanoparticles to adsorb heavy metals.[34] The application of the natural organic material, as a land management technique, seems to be a cost-effective method consistent to related protocols for the protection of the soil quality.[35]

6.3.4 Pesticides

The term pesticide covers a wide range of compounds including insecticides, fungicides, herbicides, rodenticides, molluscicides, nematicides, plant growth regulators and others. Among these, organochlorine (OC) insecticides, used successfully in controlling a number of diseases, such as malaria and typhus, were banned or restricted after the 1960s in most of the technologically advanced countries. The introduction of other synthetic insecticides – organophosphate (OP) insecticides in the 1960s, carbamates in 1970s and pyrethroids in 1980s and the introduction of herbicides and fungicides in the 1970s–1980s contributed greatly to pest control and agricultural output. Ideally a pesticide must be lethal to the targeted pests, but not to non-target species, including man.[36]

Pesticides are important inputs for enhancing crop productivity and preventing major biological disasters. However, more than 90 percent of pesticides transfer into the environment and reside in agricultural products during the process of application because of the nature of conventional pesticide formulation, such as the use of a harmful solvent, poor dispersion, dust drift, etc. In recent years, using nanotechnology to create novel formulations has shown great potential in improving the efficacy and safety of pesticides. The development of nano-based pesticide formulation aims at precise release of necessary active ingredients in responding to environmental triggers and biological demands through controlled release mechanisms.[37]

Quaternary Ammonium Herbicides (QUATs) are nonselective contact herbicides, widely used at weed emergence to protect a wide range of crops. The benefits achieved by the use of these herbicides are indisputable with regard to plant protection. However, several environmental dangers have emerged from the overuse of these compounds. Therefore, there has been a great concern in the presence of these compounds in soils, water, and food. Once in the soil, the mobility of these agrochemicals plays an important role in their fate and transport in the environment.[38]

Detoxification (detox) plays a major role in pesticide action and resistance. The mechanisms involved are sometimes part of the discovery and development process in seeking new biochemical targets and metabolic pathways. Genetically modified and chemical-modified crops are a marked exception and often involve herbicide detox by design to achieve the required crop tolerance. There is a role for detox by design or chance and target-site-based selectivity in insecticide, herbicide, and fungicide action and human health and environmental effects.[39] This is an area of continued Research and Development for agriculture and AC in the twenty-first century.

Herbicides are pesticides used to eradicate unwanted plants in both crop and non-crop environments. The chemistries of these products are toxic to weeds due to inhibition of key enzymes or disruption of essential biochemical processes required for weedy plants to survive. Crops can survive systemic herbicidal applications through various forms of detoxification. Disruptions in biochemical pathways in plants due to the application of herbicides or other pesticides have the potential to alter the nutrient quality, taste, and overall plant health associated with edible crops.[40]

6.4   NEW FACE OF AGRICULTURAL CHEMISTRY

Agriculture is facing an enormous challenge: It must ensure that enough high-quality food is available to meet the needs of a continually growing population. Current and future agronomic production of food, feed, fuel, and fiber requires innovative solutions for existing and future challenges, such as climate change, resistance to pests, increased regulatory demands, renewable raw materials or requirements resulting from food chain partnerships.(See Figure 6.3) Modern agricultural chemistry must support farmers to manage these tasks.[41] Control technologies for most of the environmental issues are chemical in nature, and this underlines the central importance of chemistry in causing, monitoring, correcting, and eliminating these problems.[42]

Chemistry (including knowledge derived from analytical, physical, and synthetic aspects) has been fundamental to food science, helping to ensure the safety of food, build understanding of the composition of foodstuffs and elucidate how food constituents contribute to human nutrition, as well as playing central roles in enabling large expansions in food production and preservation during the twentieth century.[43] It is essential for chemistry to identify new solutions to ensuring food security in a greatly changed twenty-first-century context,[44] by addressing the four pillars considered intrinsic to food security:

- availability of food,
- stability of the food supply,
- access to adequate food and
- utilization of food.[45]

FIGURE 6.3   Modern agricultural practices. (See www.shutterstock.com/image-photo/ smart-robotic-farmers-agriculture-futuristic-robot-1155509476)

6.4.1   Genetically Modified Crops

Genetic modification (GM) in plants was first recorded 10,000 years ago in Southwest Asia where humans first bred plants through artificial selection and selective breeding. Since then, advancements in agricultural science and technology have brought about the current GM crop revolution. GM crops are promising to mitigate current and future problems in commercial agriculture, with proven case studies in Indian cotton and Australian canola. However, controversial studies along with current problems linked to insect resistance and potential health risks have jeopardized its standing with the public and policymakers, even leading to full and partial bans in certain countries.[46]

Can GM crops be a part of the drive toward sustainability? Sustainable development serves as the foundation for a range of international and national policymaking. Traditional breeding methods have been used to modify plant genomes and production. Genetic engineering is the practice of assisting agricultural systems in adapting to rapidly changing global growth by hastening the breeding of new varieties. On the other hand, the new developments of genetic engineering have enabled more precise control over the genomic alterations made in recent decades. Genetic changes from one species can now be introduced into a completely unrelated species, increasing agricultural output or making certain elements easier to manufacture. Harvest plants and soil microorganisms are just a few of the more well-known genetically modified species. GMOs increase yields, reduce costs, and reduce agriculture's terrestrial and ecological footprint. Modern technology benefits innovators, farmers, and consumers alike.[47]

Biofortification is the process of adding essential micronutrients and other health-promoting compounds to crops or foods to improve their nutritional value. This is imperative as the diets of over two-thirds of the world's population lack one or more essential mineral elements and the three staple crops, rice, maize, and wheat, which provide nearly half of the calories consumed by humans, are deficient in micronutrients. A large body of information exists on augmentation through breeding approaches, both conventional and molecular, or through agronomic management practices. Other options include dietary diversification, mineral supplementation, food fortification, or increase in the concentrations and/or bioavailability of mineral elements in the produce. More options in biofortification strategies need to be included in agronomic and breeding approaches to develop effective biofortification strategies for the staple crops.[48]

Closely tied to sustainability are the environmental impacts associated with using crop biotechnology (specifically GM crops) in global agriculture. Specifically, are there environmental impacts associated with changes in pesticide use and greenhouse gas emissions arising from the use of GM crops since their first widespread commercial use 22 years ago? The adoption of GM insect resistant and herbicide tolerant technology has reduced pesticide spraying by 775.4 million kg (8.3%) and, as a result, decreased the environmental impact associated with herbicide and insecticide use on these crops. The technology has also facilitated important cuts in fuel use and tillage changes, resulting in a significant reduction in the release of greenhouse gas emissions from the GM cropping area. In 2018, this was equivalent to removing 15.27 million cars from the roads.[49]

6.4.2 Artificial Intelligence

Agriculture plays a significant role in the economic sector. The automation in intensive agriculture is a concern and is an emerging subject across the world. The growing population is increasing the demand of food and employment. The traditional methods used by the farmers are not sufficient to fulfill these requirements. The implementation of new methods continues to move agriculture toward meeting both the food requirements and also provided employment opportunities to billions of people. Artificial Intelligence (AI) serves a vital role in AC. This technology is being used to protect crop yields from various factors like the climate changes, population growth, employment issues, and the food security problems. AI in agriculture can be used in such functions as irrigation, weeding, spraying with the help of sensors and other means employing robots and drones. These technologies save the excess use of water, pesticides, herbicides, maintains the fertility of the soil, also help in the efficient use of labor, while increasing both productivity and quality.[43]

One area that AI is assisting AC and agricultural systems to address is the unprecedented challenges in the age of changing climate. As mentioned earlier, there has been a movement toward utilizing emerging technologies, such as nanotechnology. Nanotechnology with plant systems can be involved with agriculture processes to develop environmental remediation strategies. AI can be used to guide the development of plant-safe synthesized nanoparticle (NPs) that are eco-friendly, less time consuming, less expensive, and provide long-term product safety. These designer NPs provide tools that could contain nutrients, fungicides, fertilizers, herbicides, or nucleic acids that target specific plant tissues and deliver their payload to the targeting location of the plant to achieve the intended results for environmental monitoring and pollution resistance.[44]

While global agriculture is poised to benefit from the rapid advance and diffusion of AI technologies, there are risks. AI in agriculture could improve crop management and agricultural productivity through plant phenotyping, rapid diagnosis of plant disease, efficient application of agrochemicals and assistance for growers with location-relevant agronomic advice. However, the ramifications of machine learning (ML) models, expert systems and autonomous machines for farms, farmers and food security are poorly understood and under-appreciated. Risk-mitigation measures are needed, including inviting rural anthropologists and applied ecologists into the technology design process, applying frameworks for responsible and human-centered innovation, setting data cooperatives for improved data transparency and ownership rights, and initial deployment of agricultural AI in test settings.[50]

6.4.3 CHEMICAL PRODUCTS FROM AGRICULTURE

Biorefineries (BR) are a new concept in chemical manufacturing in which naturally occurring, sustainable biomass resources such as forestry and agricultural waste are converted to diverse fuel and chemical product streams. It is analogous to the processing of non-renewable fossil fuels by petrochemical refineries. Biomass derived from waste agricultural and forestry materials or non-food crops offers the most easily implemented and economical solutions for transportation fuels, and the only nonpetroleum route to organic molecules for the manufacture of bulk, fine, and specialty chemicals necessary to secure the future needs of society. The successful implementation of biorefineries can address concerns over dwindling oil reserves, carbon dioxide emissions from fossil fuel sources and associated climate change. This chemistry needed to run biorefineries will draw heavily upon catalytic processes to produce platform chemicals and biofuels.[51]

In contrast to fossil-derived crude oil, which has low oxygen content, the high oxygen and water content of biomass feedstocks presents challenges for their utilization and requires innovations in catalyst and process design for the selective conversion of these hydrophilic, bulky feedstocks into fuels or high-value chemicals. The production of biofuels as an efficient source of renewable energy has received considerable attention due to increasing energy demands and regulatory incentives to reduce greenhouse gas emissions. Second-generation biofuel feedstocks, including agricultural crop residues generated on-farm during annual harvests, are abundant, inexpensive, and sustainable. Crop residues contain recalcitrant polysaccharides, including cellulose, hemicelluloses, pectins, and lignin and lignin-carbohydrate complexes. In addition, their cell walls can vary in linkage structure and monosaccharide composition between plant sources. There are methods to characterize agricultural residues and the microbial communities that digest them provides promising streams of research to maximize value and energy extraction from crop waste streams.[52]

Lignocellulosic feedstocks provide valuable raw material for BRs. They are readily available and are bio-renewable. Using lignocellulosic materials, especially from agricultural and forestry sectors could help reduce the over-dependence on petrochemical resources while providing a sustainable waste management alternative. There are many novel industrial applications of lignocellulosic biomass, which includes biorefining for biofuel and biochemical production, biomedical, cosmeceuticals and pharmaceuticals, bioplastics, multifunctional carbon materials and other eco-friendly specialty products. The potential industrial utility of cellulose and lignin-based specialty materials such as cellulose fiber, bacterial cellulose, epoxides, polyolefins, phenolic resins, and bioplastics are being explored. The cutting-edge industrial utilization of lignocellulosic biomass suggests its major role in establishing a circular bioeconomy that consists of innovative design and advanced production methods to facilitate industrial recovery and reuse of waste materials beyond biofuel and biochemical production.[53]

6.4.4 GREEN CHEMISTRY

The concept of green chemistry, as primarily conceived by Paul Anastas and John Warner, is commonly presented through the Twelve Principles of Green Chemistry. [54] Indeed, the object of green chemistry is the reduction of pollution and risks by chemicals by avoiding their generation or their introduction into the biosphere. The distinction between pollutant chemicals and dangerous chemicals, along with the consideration of the exhaustion of fossil resources and the acknowledgment of the harmful effects of the chemicals employed in a great variety of activities, leads to the recognition of general objectives for green chemistry. Progress in the green chemical research areas and their application through successive approaches will certainly provide safer agricultural chemicals and much more satisfactory processes for the agricultural industry.[55]

The sustainability of agriculture is the core area that requires green chemistry strategies in the agrochemical field for implanting the judicious use of pesticides and fertilizers. The principles of green chemistry are especially relevant to the manufacturing of agrochemicals due to their direct impact on human and environmental health. However, current agricultural practices are still based on intensive production methods using unsustainable technologies.

Green chemistry does seek farm profitability, community prosperity, and improving soil quality by reducing the dependence use of non-renewable resources like synthetic fertilizers and pesticides, minimizing the adverse effects on water quality, wildlife, and safety. There are various alternatives to existent chemical farming such as biological agriculture, organic farming, natural farming, bio-dynamic agriculture, and ecological agriculture. Bio-pesticides are organic in nature, so these can be employed in farming for controlling pests, insects, and weeds and for plant physiology and productivity. These biopesticides are bio-degradable in the environment. Therefore, for sustainable developments, it is desirable to shift agricultural farming into green chemistry manufacturing processes, use crop protection and production, and develop green agrochemicals.

6.4.5 SUSTAINABLE INTENSE AGRICULTURE

Agriculture is critical to the rise of civilization. Agriculture encompasses a wide range of skills and techniques, including ways to expand the lands suitable for crops, digging canals and various forms of irrigation. In today's world, due to concerns about lack of resources, agriculture must become sustainable agriculture with intensive farming, so that we can meet our needs in the future. The history of agriculture shows that modern agronomy, plant breeding, pesticides, fertilizers and technological advances significantly increase crop yields (intensive) are great steps have been taken in the last century, but they are at odds with sustainability.[56]

In the twenty-first century, after a long period of intensive farming on a wide scale, high yields of crops have been obtained. However, there has also been a negative impact on the environment threatening the very future of agriculture. The methods of intensive agriculture have exerted great pressure on the natural resources that underpin agriculture. Among the negative effects intensive agriculture has produced include environment degradation, long-term and very high soil salinization, excessive irrigated areas, pollution of surface and groundwater, increasing plant resistance to pesticides, and loss of biodiversity in these areas. Intensive agriculture has also affected the environment considerably by massive deforestation that have been made to obtain new arable land but also the emission into the atmosphere of greenhouse gases and other polluting substances.[51] In order to meet the sustainability goals, intense agriculture must be transformed into Sustainable Intense Agriculture (SIA).

To understand the need for sustainable agriculture, it is critical to understand the importance of soil. Soil a complex, living, fragile medium that must be protected and nurtured to ensure its long-term productivity and stability. Because the maintenance of soil structure and fertility is of paramount importance for plant growth and because this in turn is dependent on the activities of soil organisms, soil microbiology has a significant role to play in SIA. Consideration of this role will focus on those aspects of soil microbiology that contribute directly and indirectly to plant growth. It is very important that soil micro-organisms are included in the maintenance of soil structure, their role in nutrient recycling, and their beneficial and detrimental interactions with plants. This must occur if intensive agriculture is to become sustainable.[57]

6.4.6 CIRCULARITY

Circularity in agriculture focuses on the production of agricultural commodities using a minimal amount of external inputs, closing nutrient loops and reducing negative discharges to the environment (in the form of wastes and emissions). AC can contribute to the development of a circular economy, by incorporating the 12 principles for a "circular chemistry" In doing so, a framework analogous to that of green chemistry is established, which can be adapted to facilitate the transition to a circular economy. This approach aims to make chemical processes truly circular by expanding the scope of sustainability from process optimization to the entire lifecycle of chemical products.[58]

Planet globalization, population growth and its consequent need to produce large amounts of food, or individual economic benefits and the prioritization of this over environment health, are factors that have contributed to the development, in some cases, of a linear-producing modern agricultural system. Modern agriculture currently produces tons of waste, creating controversial consequences, instead of having them reintroduced into the production chain with a novel purpose. The re-utilization of waste compounds not only represents numerous potential applications, such as food and feed additives, functional foods, nutraceuticals, cosmeceuticals, and so forth, but also represents a favorable measure for the environment, and results in the formation of value-added products. Compound recovery from agricultural wastes generated in the agro-food industries, and their potential applications create a circular and sustainable bioeconomy.[59]

6.5 AGRICULTURAL CHEMISTRY IN THE TWENTY-FIRST CENTURY

As this chapter has shown, the challenges facing the world sustainability goals do not end with food production, but many do involve food. A third or more of all the food produced in the world goes to waste. In addition to the lost nutrition, food waste decomposing in landfills releases CH_4. The carbon footprint of food produced but not eaten is equivalent to 3.3 billion tons of CO_2 per year – the world's third-largest aggregate source of greenhouse gases.[60] Chemistry has made many contributions to helping limit post-harvest food losses, including through control of chemical environments for storage and ripening, inhibition of decay processes, and packaging materials to preserve quality and extend shelf life. But much more can be done, and greater efforts made to ensure that innovations to reduce food wastage or to capture value in food not consumed (e.g., by energy-generating incineration or biogas production) do not cause other harm to human security, including to health or the environment.

Chemistry now has a major opportunity to contribute by taking a comprehensive approach, amalgamating quantity and quality aspects to enhance security in food, health, and the environment together.

REFERENCES

1. Matlin, S.A., et al., *Re-imagining Priorities for Chemistry: A Central Science for "Freedom from Fear and Want"*. Angewandte Chemie – International Edition, 2021. **60**(49): pp. 25610–25623.

2. Agrimonti, C., M. Lauro, and G. Visioli, *Smart agriculture for food quality: facing climate change in the 21st century*. Critical Reviews in Food Science and Nutrition, 2021. **61**(6): pp. 971–981.

3. Gaffney, J., et al., *Science-based intensive agriculture: Sustainability, food security, and the role of technology*. Global Food Security, 2019. **23**: pp. 236–244.

4. Perlatti, B., M.R. Forim, and V.G. Zuin, *Green chemistry, sustainable agriculture and processing systems: a Brazilian overview*. Chemical and Biological Technologies in Agriculture, 2014. **1**(1).

5. Matlin, S.A., et al., *Material circularity and the role of the chemical sciences as a key enabler of a sustainable post-trash age*. Sustainable Chemistry and Pharmacy, 2020. **17**.

6. Saleem, M., *Possibility of utilizing agriculture biomass as a renewable and sustainable future energy source*. Heliyon, 2022. **8**(2).

7. Seiber, J.N., *Sustainability and agricultural and food chemistry*. Journal of Agricultural and Food Chemistry, 2011. **59**(1): pp. 1–2.

8. Whitesides, G.M., *Reinventing chemistry*. Angewandte Chemie – International Edition, 2015. **54**(11): pp. 3196–3209.

9. Hopf, H., et al., *The chemical sciences and the quest for sustainability*. Nachrichten aus der Chemie, 2021. **69**(9): pp. 18–22.

10. Torres, J.A., et al., *Novel eco-friendly biocatalyst: soybean peroxidase immobilized onto activated carbon obtained from agricultural waste*. RSC Advances, 2017. **7**(27): pp. 16460–16466.

11. Ohno, T., N.J. Hess, and N.P. Qafoku, *Current understanding of the use of alkaline extractions of soils to investigate soil organic matter and environmental processes*. Journal of Environmental Quality, 2019. **48**(6): pp. 1561–1564.

12. Lai, L., et al., *Evaluating the impacts of landscape positions and nitrogen fertilizer rates on dissolved organic carbon on switchgrass land seeded on marginally yielding cropland*. Journal of Environmental Management, 2016. **171**: pp. 113–120.

13. Pierart, A., et al., *Antimony bioavailability: Knowledge and research perspectives for sustainable agricultures*. Journal of Hazardous Materials, 2015. **289**: pp. 219–234.

14. Xia, X., et al., *Toxic responses of microorganisms to nickel exposure in farmland soil in the presence of earthworm (Eisenia fetida)*. Chemosphere, 2018. **192**: pp. 43–50.

15. Shaheen, S.M., et al., *Redox effects on release kinetics of arsenic, cadmium, cobalt, and vanadium in Wax Lake Deltaic freshwater marsh soils*. Chemosphere, 2016. **150**: pp. 740–748.

16. Wang, L. and T. Liang, *Effects of exogenous rare earth elements on phosphorus adsorption and desorption in different types of soils*. Chemosphere, 2014. **103**: pp. 148–155.

17. Kassai, P. and G. Tóth, *Agricultural soil phosphorus in Hungary: High resolution mapping and assessment of socioeconomic and pedological factors of spatiotemporal variability*. Sustainability (Switzerland), 2020. **12**(13): pp. 1–12.

18. Wilfert, P., et al., *The relevance of phosphorus and iron chemistry to the recovery of phosphorus from wastewater: a review.* Environmental Science and Technology, 2015. **49**(16): pp. 9400–9414.

19. Rode, M., et al., *Continuous in-stream assimilatory nitrate uptake from high-frequency sensor measurements.* Environmental Science and Technology, 2016. **50**(11): pp. 5685–5694.

20. Sumaraj and L.P. Padhye, *Influence of surface chemistry of carbon materials on their interactions with inorganic nitrogen contaminants in soil and water.* Chemosphere, 2017. **184**: pp. 532–547.

21. Uchimiya, M., L.H. Wartelle, and V.M. Boddu, *Sorption of triazine and organophosphorus pesticides on soil and biochar.* Journal of Agricultural and Food Chemistry, 2012. **60**(12): pp. 2989–2997.

22. Hawkes, C. and M. Ruel, *The links between agriculture and health: An intersectoral opportunity to improve the health and livelihoods of the poor.* Bulletin of the World Health Organization, 2006. **84**(12): pp. 985–991.

23. Ravishankara, A.R., Y. Rudich, and J.A. Pyle, *Role of chemistry in Earth's climate.* Chemical Reviews, 2015. **115**(10): pp. 3679–3681.

24. Peng, X.X., et al., *Highly sensitive and rapid detection of thiabendazole residues in oranges based on a luminescent Tb3+-functionalized MOF.* Food Chemistry, 2021. **343**: pp. 1–17.

25. Chen, J., D. Dong, and S. Ye, *Detection of pesticide residue distribution on fruit surfaces using surface-enhanced Raman spectroscopy imaging.* RSC Advances, 2018. **8**(9): pp. 4726–4730.

26. Clark, A.E., et al., *Pressurized liquid extraction technique for the analysis of pesticides, PCBs, PBDEs, OPEs, PAHs, alkanes, hopanes, and steranes in atmospheric particulate matter.* Chemosphere, 2015. **137**: pp. 33–44.

27. Chen, H., et al., *Eighteen-year farming management moderately shapes the soil microbial community structure but promotes habitat-specific taxa.* Frontiers in Microbiology, 2018. **9**(AUG): pp. 1–14.

28. McDougall, R., P. Kristiansen, and R. Rader, *Small-scale urban agriculture results in high yields but requires judicious management of inputs to achieve sustainability.* Proceedings of the National Academy of Sciences of the United States of America, 2019. **116**(1): pp. 129–134.

29. Lorat, D. *Farm it Yourself.* [cited 2022 22 May]; Small scale farming].

30. Jouzi, Z., et al., *Organic Farming and Small-Scale Farmers: Main Opportunities and Challenges.* Ecological Economics, 2017. **132**: pp. 144–154.

31. Hu, W., et al., *Soil environmental quality in greenhouse vegetable production systems in eastern China: Current status and management strategies.* Chemosphere, 2017. **170**: pp. 183–195.

32. Day, C. and S. Cramer, *Transforming to a regenerative U.S. agriculture: the role of policy, process, and education.* Sustainability Science, 2022. **17**(2): pp. 585–601.

33. Karamian, F., A.A. Mirakzadeh, and A. Azari, *The water-energy-food nexus in farming: Managerial insights for a more efficient consumption of agricultural inputs.* Sustainable Production and Consumption, 2021. **27**: pp. 1357–1371.

34. Azeez, N.A., et al., *Nano-remediation of toxic heavy metal contamination: Hexavalent chromium [Cr(VI)].* Chemosphere, 2021. **266**: pp. 1–11.

35. Raptis, S., et al., *Chromium uptake by lettuce as affected by the application of organic matter and Cr(VI)-irrigation water: Implications to the land use and water management.* Chemosphere, 2018. **210**: pp. 597–606.

36. Aktar, W., D. Sengupta, and A. Chowdhury, *Impact of pesticides use in agriculture: Their benefits and hazards.* Interdisciplinary Toxicology, 2009. **2**(1): pp. 1–12.

37. Zhao, X., et al., *Development Strategies and Prospects of Nano-based Smart Pesticide Formulation.* Journal of Agricultural and Food Chemistry, 2018. **66**(26): pp. 6504–6512.

38. Pateiro-Moure, M., M. Arias-Estévez, and J. Simal-Gándara, *Critical review on the environmental fate of quaternary ammonium herbicides in soils devoted to vineyards.* Environmental Science and Technology, 2013. **47**(10): pp. 4984–4998.

39. Casida, J.E., *Pesticide detox by design.* Journal of Agricultural and Food Chemistry, 2018. **66**(36): pp. 9379–9383.

40. Cutulle, M.A., et al., *Several pesticides influence the nutritional content of sweet corn.* Journal of Agricultural and Food Chemistry, 2018. **66**(12): pp. 3086–3092.

41. Jeschke, P., *Progress of modern agricultural chemistry and future prospects.* Pest Management Science, 2016. **72**(3): pp. 433–455.

42. Mahaffy, P.G., et al., *Integrating the molecular basis of sustainability into general chemistry through systems thinking.* Journal of Chemical Education, 2019. **96**(12): pp. 2730–2741.

43. Talaviya, T., et al., *Implementation of artificial intelligence in agriculture for optimisation of irrigation and application of pesticides and herbicides.* Artificial Intelligence in Agriculture, 2020. **4**: pp. 58–73.

44. Agrawal, S., et al., *Plant development and crop protection using phytonanotechnology: A new window for sustainable agriculture.* Chemosphere, 2022. **299**: pp. 1–23.

45. Sachs, J., et al., *Monitoring the world's agriculture.* Nature, 2010. **466**(7306): pp. 558–560.

46. Raman, R., *The impact of Genetically Modified (GM) crops in modern agriculture: A review.* GM Crops and Food, 2017. **8**(4): pp. 195–208.

47. Sharma, P., et al., *Genetic modifications associated with sustainability aspects for sustainable developments.* Bioengineered, 2022. **13**(4): pp. 9508–9520.

48. Prasanna, R., et al., *Biofortification with microorganisms: Present status and future challenges,* in *Biofortification of Food Crops.* 2016, Springer India. pp. 249–262.

49. Brookes, G. and P. Barfoot, *Environmental impacts of genetically modified (GM) crop use 1996–2018: impacts on pesticide use and carbon emissions.* GM Crops and Food, 2020. **11**(4): pp. 215–241.

50. Tzachor, A., et al., *Responsible artificial intelligence in agriculture requires systemic understanding of risks and externalities.* Nature Machine Intelligence, 2022. **4**(2): pp. 104–109.

51. Dorneanu, M., *Intensive farming versus-agriculture environmentally sustainable.* Quality – Access to Success, 2017. **18**: pp. 195–197.

52. Tingley, J.P., et al., *Combined whole cell wall analysis and streamlined in silico carbohydrate-active enzyme discovery to improve biocatalytic conversion of agricultural crop residues.* Biotechnology for Biofuels, 2021. **14**(1): pp. 1–19.

53. Okolie, J.A., et al., *Chemistry and Specialty Industrial Applications of Lignocellulosic Biomass.* Waste and Biomass Valorization, 2020.

54. Anastas, P.T. and J.C. Warner, *Green Chemistry: Theory and practice.* 1998, Oxford: Oxford University Press.

55. Mestres, R., *Green chemistry: views and strategies.* Environmental Science and Pollution Research, 2005. **12**(3): pp. 128–132.

56. Sharifi, P., *Sustainable agriculture: And introduction to extensive and intensive agriculture*. Journal of Engineering and Applied Sciences, 2017. **12**(10): pp. 2747–2751.

57. Pankhurst, C.E. and J.M. Lynch, *The role of soil microbiology in sustainable intensive agriculture*, in *Advances in Plant Pathology*. 1995. pp. 229–247.

58. Keijer, T., V. Bakker, and J.C. Slootweg, *Circular chemistry to enable a circular economy*. Nature Chemistry, 2019. **11**(3): pp. 190–195.

59. Jimenez-Lopez, C., et al., *Agriculture waste valorisation as a source of antioxidant phenolic compounds within a circular and sustainable bioeconomy*. Food and Function, 2020. **11**(6): pp. 4853–4877.

60. Adelodun, B., et al., *Assessment of environmental and economic aspects of household food waste using a new Environmental-Economic Footprint (EN-EC) index: A case study of Daegu, South Korea*. Science of the Total Environment, 2021. **776**: pp. 1–10.

7 Unsustainable agricultural waste streams

7.1 MAGNITUDE OF AGRICULTURAL WASTE

Agricultural and food wastes constitute a significant global problem.(See Figure 7.1) There are adverse effects that agricultural waste has on the environment, economy, and society.[2] The agricultural enterprise must deliver adequate quantities of food, which meet national needs and provide health benefits. These must be delivered in an environmentally sustainable manner. With the global population projected to reach 9.8 billion by 2050, the goal of a sustainable food supply is a daunting challenge. Climate change, urbanization, and soil degradation will continue to hinder this goal by affecting the availability of agricultural land. It is further a fact that about one third of foods primarily produced from land has been wasted. This is equivalent to an annual generation of about 1.6 billion tons of uneaten food. This could be classified as food loss and waste (FLW). Given its extremely huge quantity, FLW is not negligible in determining a path to achieve agricultural sustainability.[3] This global agricultural picture stresses the paradox that a huge quantity of food is lost or wasted while there are over 800 million people worldwide that are affected by hunger and malnutrition.

The current linear agricultural system results from the planetary globalization of commerce, population growth with its food requirements, and rising individual economic benefits over the prioritization of environment health. The modern agriculture system results in the generation of tons of waste. The challenge will be to reintroduce this waste into the production chain in order to lessen FLW.[4] We begin by presenting the problem, from which point we can envision the approach of sustainable agriculture and agricultural chemistry (AC) to address it.

FLW results from a reduction of food, which is destined for human consumption, along the food chain. FLW take place from the time when a food product is suitable for harvesting or it is just harvested up to the consumption stage or when a product is lost from the supply chain.[5] Surprisingly, the amount of food produced at the farm level is much higher than the quantity that is required for human consumption, but food losses from the many sources are very high and aggravate the current state of food availability.

DOI: 10.1201/9781003157991-7

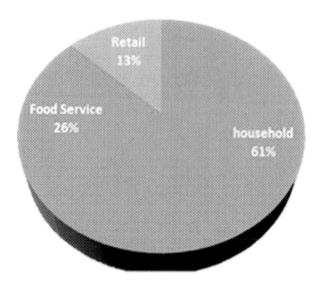

FIGURE 7.1   Sources of food waste.[1]

7.1.1 TYPES OF FLW

The residues of raw agricultural products make up the majority of the agricultural wastes.[6] Such residues could include manure and animal carcasses (animal waste); corn stalks, sugar cane bagasse, drops and culls from fruits and vegetables, pruning (crop waste); pesticides, insecticides, and herbicides (hazardous and toxic agricultural waste); and food processing waste obtained during growing and processing in liquid, slurry, or solid forms. The main food groups contributing to nutrient and food waste or loss are cereals and pulses, fruits and vegetables, meat and animal products, roots, tubers, and oil-bearing crops.[7] Roots, tubers, and oil-bearing crops (26%) and fruits and vegetables (22%) are the first two groups, based on food loss.[1] In Table 7.1 these areas are broken down by percentages.

The approximate food waste is 307 g per capita per day for high-income countries, corresponding to twice that of the upper-middle income countries. Food waste is obtained primarily at later stages of the supply chain in industrialized countries. However, because of lacking financial and technical properties during harvesting, storage, and cooling, most of the waste in developing countries is obtained at early stages in the food supply chain.[1]

7.1.2 CAUSES OF FOOD LOSSES AND WASTE

Three broad areas encompass where most FLW occurs. These FLW are influenced by agricultural production, domestic infrastructure, distribution channels, marketing chains, and consumer food-related practices. (see Figure 7.1)[8] In the developing world, FLW can result from the lack of suitable or efficient facilities along the

TABLE 7.1
Global Annual Food Waste[1]

Food group	Percentage
Cereals	30%
Root crops/fruits/vegetables	40–50%
Oilseeds	20%

TABLE 7.2
Sources of Food Loss and Waste (FLW)[12]

Category	Details
Food production	• Excessive planting • Damage by weather, pests and disease • Market conditions • Inedible food • Un-harvested crops (low demand) • Cosmetic imperfections • Food safety scares and improper refrigeration
Processing	• Losses while trimming off edible portions • Overproduction • Product damage and • Technical problems at manufacturing facilities • Loss in Transportation and Distribution Networks
Consumer	• Overstocked product displays • Expectation of cosmetic perfection of fruits, vegetables and other foods • Oversized packages • Availability of prepared food until closing • Expired "sell by" dates • Damaged goods • Outdated seasonal items • Over purchasing of unpopular foods • Under staffing

food supply chain (harvest, storage, transport, processing) and ineffective, or even inexistent, coordination along the supply chain.[9] In developed countries, FLW are caused downstream of the food supply chain, that is, at the intersection among retail, food service/catering, and consumers, but also within the households. In Table 7.2 these areas are broken down in more detail.

In developing countries, the production and postharvest handling cause a greater percentage of the total food losses.[10] The causes of FLW in poor, low-income countries are largely related to technical and managerial deficiencies in harvesting, cooling, and/or storage.[5] In industrialized regions (e.g., Europe, Industrialized

Asia, North America), losses in agricultural production are mainly due to postharvest grading (especially of fruit and vegetables) related to quality standards set by the retailers.[11] Food may also be lost because poor farmers sometimes harvest crops too early due to desperate need for cash or food deficiency.

When food is deemed unsafe (does not comply with minimum food safety standards) it is discarded and is regarded as waste, since it is not fit for human consumption. A variety of factors can determine food safety, including unhygienic handling and storage conditions, inadequate temperature control, naturally occurring toxins, contaminated water as well as improper use of pesticides and veterinary drugs.[5] This type of classification is more prevalent in developed countries, but it can be significant in developing countries as well.

At the retail stage, large amounts of food are discarded because of quality standards (especially private ones) that highlight appearance (i.e., cosmetic specifications). Some agricultural products are excluded at the farm level by supermarkets, which apply rigorous standards regarding size, weight, appearance, and shape.[13] Food retailers put forward their commitment to reducing food wastage while at the same time they set standards resulting in the large-scale wastage of still edible agri-food products based on rigorous cosmetic specifications.[14] Therefore large portions of harvested food are never brought to market, and so therefore fall into the "food loss" category. [13]

7.1.3 ECONOMIC CONSEQUENCES OF WASTE

Food waste and loss have economic consequences as well. The global cost of food waste has been estimated to approach 1000 billion dollars annually. This number can swell to 2600 billion dollars when environmental costs are considered. Regarding the total loss and waste, ~1.3 billion tons annually, that corresponds to about one-third of the food produced for human consumption. Using a different perspective, the food waste results from using nearly 30% of the agricultural land area in the world.[15] Food waste has environmental and economic impacts right along with its social impacts. Food waste has devastating effects on climate change.[16]

As Gustavsson highlighted, avoidable FLW have negative impacts on farmers' and consumers' income. In fact, FLW represent an unprofitable investment that decreases farmers' incomes, due to lower marketable yield not followed by a reduction of production inputs (e.g., fertilizers, plant protection products, fuel, workforce, etc.) in terms of quantity and costs.[17]

Vegetables primarily contribute to the economic cost of food waste and loss with 23%. Meat, fruits, and cereals contribute to the total cost with the percentages of 21%, 19%, and 18%, respectively.[15] Annually, more than 55 million metric tons of avoidable food wastes are produced in the U.S., and this corresponds to almost 29% of the annual production. The cost of these wasted foods equals 198 billion dollars.[18] At least 18.6 billion dollars are lost in the U.K. with 8.3 billion metric tons of annual household food waste. Food waste reduction can help provide sufficient food for the increasing population globally; it is a critical issue in terms of economy, environment, and society.[19]

Economically, the disposal of FLW is viewed as cheaper than the reuse of food in chains of industrialized countries. However, the proportions of food waste are growing in parallel to the increase of production and consumption. Careless consumer attitude is exacerbated by the abundance of food in developed countries. The amount of available food per capita in both the United States and Europe has increased steadily during the last decades. Even setting aside the recent effects from Covid-19, many restaurants serve buffets at fixed prices (i.e., all-you-can-eat buffets). This may encourage clients to fill their plates with more food than they would actually eat, and this attitude and the restaurant size are key factors in determining the amount of food wasted in restaurants.[20]

7.1.4 Social impacts/nutrition

Wasting food brings about the loss of life-supporting nutrition, in addition to the decline of resources such as water, land, and energy used in food production, processing, and distribution.[21]

Food waste has social impacts associated with nutrient loss and world hunger. The wasted foods could theoretically fill nutritional gaps for millions of people. For example, the annual amount of food loss and waste can provide a diet of 2100 kcal per day for 2 billion people. According to FAO, this potential of food loss and waste is crucial since 690 million people were estimated to be suffering from hunger in 2019. This number has increased dramatically during the COVID-19 pandemic, and it is expected to increase further. Food waste of the U.S. food supply in 2012 at the retail and consumer levels consisted of 33 g of protein, 5.9 g of dietary fiber, 1.7 μg of vitamin D, 286 mg of calcium, 880 mg of potassium, and 1217 kcal per capita per day.[2] It has been reported that the wasted calcium, choline, riboflavin, zinc, and vitamin B12 especially arise from the loss of meat, dairy, and eggs. The daily food waste per person corresponds to the food of 795–840 kcal. And, carotenoids have the highest value with 31%, and vitamin D has the lowest value with 25% within all wasted nutrients.[22] Because of these wasted nutritious foods, food waste reduction can provide more available nutrients for human consumption.[23]

If the necessary precautions for food waste reduction are not taken, and food consumption and production are not planned, the world in 2050, with a population of more than nine billion, will need 60% more food, which is equivalent to at least 2 billion tons of additional annual food production. The predictions show that the food gap in 2050 can be reduced by ~20% via decreasing the global food waste by half.[24]

7.1.5 Climate change

Food waste has devastating effects on climate change.[16] The estimated carbon footprint contribution of food waste to greenhouse gas (GHG) emissions equals $\sim$3.3 billion tons of CO_2 accumulation in the atmosphere per year.[25] A provocative example of the food consumption and waste generation trends of family members in China was reported.[26] The investigation used the household's carbon, water, and ecological footprint quantification. The annual wastage of food at home (16 kg) caused 40 kg CO_2 eq of carbon, 18 m^3 of water, and 173 gm^2 of ecological footprints. Food waste can also create many other environmental problems, since it is disposed without any appropriate pretreatment by landfill or incineration in the dumping sites. As is the case with the incineration of food waste, toxic byproducts contaminate groundwater and cause corrosive gases, such as methane and hydrogen sulfide, to be generated by food waste landfilling.[27] The landfilling of food waste reduces its energy content.

7.2 TREATMENTS

There are generally five different methods for food waste treatment. Food waste, as a special category of municipal solid waste, has the feature of high moisture, salinity, organic and oil content, which requires different methods of treatment from common municipal solid waste.[28] Common treatment methods for handling food waste include anaerobic digestion, landfill, incineration, composting and heat/moisture reaction.[29] To evaluate their effectiveness and their environmental performance a study was conducted.[30] See Table 7.3 for a detailed comparison of the waste treatment methods.

Saving food is the ultimate way to reduce environment impact and the distance for transportation also has huge impacts. The result of comparison shows that landfill contributes most to climate change, which is about 10 times larger than other treatment technologies. For acidification and eutrophication, incineration shows the worst result. Composting has the largest impact in the build-up of carcinogens as its result. On the whole, the waste treatment technologies that are

TABLE 7.3
Comparison of Agricultural Waste Treatment Technologies[30]

Technology	Benefits	Liabilities
anaerobic digestion	• produces methane as the main product • produces fertilizers as the byproduct	• high cost, • strict startup condition • long time for fermentation
composting	• simple operation management • high economic efficiency	• can cause P buildup • increase carcinogens
landfill	• traditional method	• requires large area of land • great amount of GHG emissions
incineration	• produce thermal energy • greatly reduce the volume of waste	• water affects the process • causes acid rain and eutrophication
Heat/moisture	• effectively eliminate bacteria and odor	• costly

recommended in proper sequence are anaerobic digestion, heat/moisture reaction, composting and incineration and landfill.

7.3 VALORIZATION OF WASTE

Food waste can be viewed as a valuable biomass that can be utilized for profitable products instead of an uncontrollable discard. Since food waste is renewable and inexpensive, it can be beneficial for obtaining energy, biofuels, enzymes, antioxidant extracts, novel biodegradable materials, and other commercial products.[31] The value-added products, which include fine chemicals, nutraceuticals, antioxidants, bioactives, biopolymers, biopeptides, antibiotics, industrial enzymes, bio-nanocomposites, single-cell proteins, polysaccharides, activated carbon adsorbent, chitosan, corrosion inhibitors, organic acids, pigments, sugars, wax esters, and xanthan gum, can be recovered by using food wastes as a substrate.[32] In later chapters we will go into more detail, but the technologies in the twenty-first century agriculture and Agricultural Chemistry (AC) are available to accomplish this.

Food wastes have enormous amounts of valuable chemicals. Food waste is actually a significant source of complex carbohydrates, proteins, lipids, and phytochemicals, because of its high contents of polysaccharides, dietary fibers, oils, vitamins, phenolics, carotenoids, and other pigments. The recovery of these will provide the potential health benefits of wasted foods with their high contents of biologically active compounds.[33]

Several available valorization strategies, which are more sustainable and profitable to manage food waste, arise as alternative options to isolate the value-added products mentioned above. There are also special chemicals that can be refined from food waste, ranging from solvents to antioxidant materials, which are essential for nutraceutical and biomaterial applications.[34] The combined methods, which include biochemical, chemical, and physical steps, can be applied to separate the potentially marketable compounds found in food wastes and byproducts to selectively extract and modify the preferred components and change them to higher-value food products and additives. These technologies are practiced in the biorefineries.

7.3.1 FOOD WASTE MANAGEMENT GAPS

Existing FLW management technologies have two major gaps. First, many technologies such as gasification, pyrolysis, etc., are still in the laboratory stage or economically infeasible. Given the urgency of food security and environment sustainability[35], these technologies will need to be developed further. Second, some widely used technologies, including landfill and incineration, also provide negative impacts on both food security and environmental sustainability. A more focused and rigorous assessment on recent progress in FLW management strategies with commercial potentials and on the relationship between strategies and circular economy should be conducted.

7.4   FOOD WASTE MANAGEMENT IN THE TWENTY-FIRST CENTURY

The need to both avoid waste production and find new renewable resources has led to new and promising research based on the possibility of revalorizing the waste biomass to produce sustainable chemicals and/or materials. (See Figure 7.2) This may play a major role in replacing non-renewable systems traditionally obtained from non-renewable sources. Most of the low-value biomass is lignocellulosic, referring to its main constituent biopolymers: cellulose, hemicellulose, and lignin. Using different extraction methods optimized over the years for the extraction of these components from natural fibers and sources, it is possible to use these techniques for the extraction of lignocellulosic components from agricultural and forest wastes.[36]

A real example is an urgent need globally to find alternative sustainable steps to treat municipal solid wastes (MSW) originated from mismanagement of urban wastes with increasing disposal cost. Vermicomposting offers excellent potential to promote safe, hygienic, and sustainable management of biodegradable MSW. It has been demonstrated that, through vermicomposting, MSW such as city garbage, household and kitchen wastes, vegetable wastes, paper wastes, human faeces and others could be sustainably transformed into organic fertiliser or vermicompost that provides great benefits to agricultural soil and plants. Apart from setting the optimum operational conditions for the vermicomposting process, other

FIGURE 7.2   Valorizing waste. (See www.shutterstock.com/image-photo/motivation-inspiring-sayings-about-life-this-1978281446)

approaches such as pre-composting, inoculating micro-organisms into MSW and redesigning the conventional vermireactor could be introduced to further enhance the vermicomposting of MSW.[37]

7.4.1 REUSING AND RECYCLING

The first option to manage food waste is to prevent waste generation. Reusing and recycling are considered as secondary options in food waste management.[38] Other methods, such as reduce–reuse–recycle, extended producer responsibility, and sustainable management to reduce the wasted food, have been developed to manage the food wastage.[39] Food losses can also be prevented by local investments, educations, improving packaging and market facilities in low-income countries. For high-income countries, enhancing communication in the supply chain, improving purchase/consumption planning, and awareness of the best-before-dates can be options for food loss prevention. To manage or reduce food waste, the countries apply several policies for individuals, organizations, and businesses, depending on their consumer behaviors, income levels, and development levels.[19]

7.4.2 VALORIZING WASTE

While the best options are prevention or reduction of the food waste, according to the food waste management methods, the valorization of food waste can be considered one of the best methods to reduce or manage food waste that is produced when prevention or reduction is not possible. Valorization refers to the diversion of former food waste to food and feed products. It also includes converting food waste to extracted food and feed ingredients, considering food waste quality, robustness, and composition.

On a first level, large amounts of organic wastes, which pose a severe threat to the environment, can be thermally pyrolyzed to produce biochar. Biochar has many potential uses owing to its unique physicochemical properties. Objectively and undeniably, there are still negative or ineffective cases of biochar amendment on crop yield and heavy metal immobilization, which is worthy of further attention. The medium-long term field monitoring of biochar-specific agricultural functions, as well as the exploration of wider sources for biochar feedstocks, are still needed. [40]

7.4.2.1 Biofuel

Food waste can be converted into biofuel. Biofuel, which can be in solid, liquid, and gaseous forms, is defined as the energy originated from biomass and refined products of biomass, consisting of bioethanol, biodiesel, biokerosene, natural gas, etc. From the beginning of human civilization, biofuel has been widely used in daily human activities like cooking, lighting, and heating.[24] Biofuels are just some of the beneficial products from biorefineries.

Biowastes are continuously generated worldwide in very large quantities per annum, primarily from lignocellulosic and food wastes. Waste disposal is a costly environmental problem, but wastes can serve as renewable resources for the sustainable production of fuels and chemicals. Hot compressed water (HCW), a benign method of biowaste pre-treatment, hydrolyzes waste into bioprocess feedstock for onward bioconversion into fuels and chemicals. A major degradation product (DP) component, 5-hydroxymethyl furfural (5-HMF), was upgraded to 2,5-dimethyl furan (DMF), a platform chemical and 'drop-in' fuel using a novel metallic catalyst made on waste bacteria sourced from the biohydrogen or alternative microbial processes. The impact of this dual benefit on process economics is important within a biorefinery concept.[41]

7.4.2.2 Valuable biomaterials

This is an area where there is great opportunity for exploration and expansion. Biopolymers, bioplastics, biofertilizers, enzymes, organic acids, single-cell protein (microbial biomass) can be obtained from agricultural food wastes/byproducts with the application of different treatments like fermentation and composting. These valuable biomaterials are used in the cosmetics, pharmaceutical, chemical, food, and beverage industries.[25] Enzymes are significant biocatalysts for different products and processes, and they have a significant role in the industry, since they exhibit specificity against substrate and product, moderate reaction conditions, formation of byproducts in a minimum amount, and high yield. The raw material costs are responsible for up to 30% of the total production cost of enzymes.[26] Therefore, the use of food wastes and byproducts is a good option for reducing the raw material costs for the production of enzymes.

Another valuable product obtained from waste materials of foods with microbial fermentation is single-cell protein.[27] The demand to formulate innovative and alternative proteinaceous food sources is increasing, because of concerns about population growth and the increasing number of hungry and chronically malnourished people. The most crucial step to respond to this demand is single-cell protein production.[28] Single-cell protein, the extracted protein from microbial biomass like bacteria, yeast, algae, and fungi, can be used as a supplement protein source instead of conventional high-cost protein sources in the staple human diet to alleviate problems related to protein scarcity. Besides the nutritional benefits of using single-cell proteins in human or animal diet, another advantage is reducing the costs of final products during the formulation of food and fodder stocks, rich in protein, by using bioconversion products from wastes of agriculture and industry.[28]

Biopolymers, are important products obtained from food wastes and byproducts. They include a wide variety of products. These biopolymers are used in critical applications in different industries like medicine, cosmetics, pharmaceutical and food industries, water treatment, production and development of biosensors, industrial plastics, and clothing fabrics, because of their biodegradability, biofunctionality, biostability, and biocompatibility.[42]

Food waste can be utilized for bioplastic production including polyhydroxyal-kanoates (PHA) and polyhydroxybutyrate (PHB) as organic polymers. These can completely degrade into carbon dioxide and water within months after they are buried. Therefore, bioplastic production from food waste contributes to reducing both plastic waste and food waste.[43]

Agricultural by-products are often hidden sources of healthy plant ingredients. The investigation of the nutritional values of these by-products is essential towards sustainable agriculture and improved food systems. As an example, in the vine industry, grape leaves are a bulky side product which is strategically removed and treated as waste in the process of wine production. 'Pinot noir' leaves present a high antioxidant capacity, putting grapevine leaves at the top of the list of foods with the highest antioxidative activity. 'Pinot noir' leaves have a high content and diversity of biologically active phytochemical compounds which make it of exceptional interest for pharmaceutical and food industries.[44]

7.4.2.3   Bioactive compounds

Recent work on the valorization of biomass by both conventional methods (solid–liquid extraction after maceration) and novel green methods (extraction with sonication, supercritical fluids, microwaves, and pulsed electric fields) recently attracted attention. The produced extracts have high phytochemical content, which provides antioxidant, anti-inflammatory, and antibacterial properties leading to potential health benefits.[45] Health-promoting bioactive compounds in fruit and vegetable wastes can play an essential role in anticancer, antimutagenic, antiviral, antioxidant, antitumor activities and that can help to reduce the risks of cardiometabolic diseases. Also, recent studies focusing on phytochemicals conclude that the high antioxidant properties of polyphenols from plant sources, including skins, seeds, pulp, or pomace, make polyphenols one of the primary phytochemicals important for health applications.[46]

Prebiotics are nondigestible food components that impact gut microbiota composition and activity, which in turn could result in possible health benefits. Therefore, the production of prebiotics from agro-industrial by-products is another area where waste may have added value. In this regard, polysaccharides are usually found in these food wastes and they have potential use as prebiotics. These compounds act as substrates for the human gut microbiota, and they have the potential to modulate its composition through many mechanisms. Additionally, the use of agricultural by-products is advantageous because waste is both cheap and an abundantly available material. If this is successful, then the utilization of agro-industrial byproducts can compete in the market with the commercial ones or act as a source for new food ingredients.[47]

7.4.2.4   Food waste as additives in food products

Synthetic additives in foods have always been challenged regarding their nutritional value and/or their safety, without reaching a consensus. This situation leads the food industry to search for natural food additives produced using novel, natural, and economical protein sources, dietary fiber, flavoring agents, colorants, antioxidants, and antimicrobials.[48] Since the environmental problems can be reduced, and food additives or supplements with high economic and nutritional value can be utilized, byproducts from food waste (FLW) can serve as a source for natural food additives. With the contribution of food waste components as natural additives into food products to provide nutrients and bioactive compounds, the sensory properties and consumer acceptance will be essential criteria for food products in order for them to have widespread use.[49]

Although synthetic bioactive compounds are approved in many countries for food applications, they are becoming less and less welcome by consumers. Natural bioactive compounds can replace these synthetic compounds. These natural compounds can be used as food additives to maintain the food quality, food safety and appeal, and as food supplements or nutraceuticals to correct nutritional deficiencies, maintain a suitable intake of nutrients, or to support physiological functions, respectively. Food wastes, particularly fruit and vegetables byproducts, are a good source of bioactive compounds. These can be isolated and reintroduced into the food chain as natural food additives or in food matrices.[50]

7.5 FOOD AND NUTRITION SECURITY THROUGH WASTE CIRCULARITY

The non-sustainability of food systems is a significant driver of malnutrition and food insecurity.

To ensure food and nutrition security, all components of the agri-food systems must be made sustainable, efficient, and resilient.(See Figure 7.3) Viewed in this context, FLW are recognized as barriers to moving forward toward more sustainable agrifood systems.[51]

Food waste can be valorized to nutritional, functional, and nutraceutical raw materials that can be used in various applications, as we saw in the previous section. This makes it a potential solution for economic, social, and environmental problems. Since FLW can be used directly as food components or proteins, lipids, vitamins, fibers, starch, minerals, and antioxidants, this is a fertile field to pursue. Other biomolecules within the food waste can be physically or chemically extracted and used as nutritional and functional components.

Since the effective utilization of renewable resources and the valorization of renewable carbon wastes can replace fossil resources to produce chemicals, materials, polymers, fuels, and energy, FLW can also serve as the source of material for other commodity chemicals. The industrial development in this area will be sustainable through specialty product extraction, conversion by green chemical or biotechnological processes, integrated biorefining, industrial bioengineering, cascade processing, and on-site processing of seasonal waste streams. This will require the effective repurposing of agricultural and forestry residues, aquatic biomass, and different waste streams.[52]

FIGURE 7.3 FLW in the circular economy. (See www.shutterstock.com/image-vector/circular-economy-icon-lightbulb-solar-panels-1932366944)

7.5.1 BIOREFINERY FOR AGRICULTURAL FOOD WASTE

Biomass residues or wastes generated in the agricultural sector represent a source of potentially sustainable feedstock for biorefineries. The strategy toward such a bio-based economy will only succeed if enough biomass and adequate qualities can be provided not only to produce adequate feedstock for the biorefinery but also to fulfil the food security and health requirements of the growing population. The majority of the biomass generated comes from agriculture sector, but it needs to be selected judiciously. It is therefore important to have a reliable estimate of the biomass, biowaste, and agro-residue generated to define policies for their valorization as well as identifying technologies which could be used for such purpose.[53]

In a society where the environmental conscience is gaining attention, it is necessary to evaluate the potential valorization options for agricultural biomass to create a change in the perception of the waste agricultural biomass from waste to resource. The biorefinery approach has been proposed as the roadway to increase profit in the agricultural sector and, at the same time, ensure environmental sustainability. The biorefinery approach integrates biomass conversion processes to produce fuels, power, and chemicals from biomass. Anaerobic digestion and composting of agricultural waste serve as the raw material source for the biorefinery. This biorefinery approach is compared to other biorefinery configurations, such as bioethanol production. Many of these processes can be conceived as a single operation unit for agricultural waste valorization. Valuable compound extraction, anaerobic digestion, and composting of agricultural waste, whether they are not, partially, or fully integrated will be critical for the success of the biorefinery.[54]

The economic and social value of food waste will be changed by utilizing this waste more in chemical synthesis to create a closed-loop economy by working it into a renewable supply chain. For example, a recent provisional agreement on the renewable energy policy, which is also targeting the obligation of the development of waste-derived biofuels, is being developed by the EU. The agreement defining a biorefinery based on food wastes is planned to have a crucial role in contributing to a more sustainable and greener society in the future.[34]

Consequently, food waste management involving/including the concept of biorefinery has positive environmental effects, because of less greenhouse gas emissions, environmental burden reduction of their disposal, and being more independent about the use of fossil-based sources for fuel generation.[55]

7.5.2 Circular approaches

Application of a circular economy in the food system can reduce the amount of FLW. This is done by the reuse of food, utilization of by-products and food waste, and nutrient recycling.[56] The five technologies discussed earlier (anaerobic digestion, heat/moisture reaction, composting, incineration and landfill) can now be further categorized as linear or circular to improve our understanding about circular economy in food system. Incineration, which has been regarded as a waste to energy option by a previous review[57], is considered as a part of linear model, because FLW alone is ill-suited for incineration due to high moisture contents.

7.5.2.1 Digestate

The use of digestate in agriculture is an efficient way to recycle materials and to decrease the use of mineral fertilizers. The agronomic characteristics of the digestates can promote plant growth and soil properties after digestate fertilization. Harmful effects can arise due to digestate quality, e.g. pH, organic matter and heavy metal content. Food waste and organic fraction of municipal solid waste digestates have high agronomic value due to the availability of nutrients and low heavy metal load.[58]

7.5.2.2 Composting

The use of compost in urban agriculture offers an opportunity to increase nutrient recycling in urban ecosystems. Compost application often results in phosphorus (P) being applied far in excess of crop nutrient demand, creating the potential for P loss through leachate and runoff. Management goals such as maximizing crop yields or maximizing the mass of nutrients recycled from compost may inadvertently result in P loss, creating a potential ecosystem disservice.

Because of the low N:P (Nitrogen:Phosphorus) ratio of compost relative to crop nutrient uptake, compost application based on crop N demand can result in overapplication of P. Crop yield did not differ among treatments receiving compost inputs, and the mass of P recovered in crops relative to P inputs decreased for treatments with higher compost application rates. Treatments receiving compost targeted to crop N demand had P leachate rates approximately twice as high as other treatments. These results highlight tradeoffs inherent in recycling nutrients, but they also show that targeted compost application rates have the capacity to maintain crop yields while minimizing nutrient loss. If composting is to be scaled up in order to maximize potential social, economic, and environmental benefits, it is especially important to carefully manage nutrients to avoid ecosystem disservices from nutrient pollution.[59]

Combining agricultural wastes or a finished compost (wheat straw, horse manure and bedding, sheep manure, and a wheat straw-SHW finished compost) as compost feedstocks with cattle slaughterhouse wastes (SHW) on a field-scale can further promote circularity. Agricultural by-products and composts are suitable feedstocks for use with SHW to generate a stable final product while meeting regulatory parameters.[60]

7.5.2.3 Anaerobic digestion

Anaerobic digestion (AD) of organic wastes is a promising alternative to landfilling for reducing Greenhouse Gas Emission (GHG). Biogas produced from AD represents a useful source of green energy, and its by-product (digestate) has useful applications. The sustainability of anaerobic digestion plants partly depends on the management of their digestion residues. Changes in soil chemical and biochemical characteristics depend on the source of digestate, the type of fraction and the concentration used. The mainly affected soil parameters were: Soil Organic Matter (SOM), Microbial Biomass Carbon (MBC), Fluorescein Diacetate Hydrolysis (FDA), Water Soluble Phenol (WSP) and Catalase (CAT) that can be used to assess the digestate agronomical feasibility. These results show that the agronomic quality of a digestate is strictly dependent on percentage and type of feedstocks that will be used to power the digester.[61]

7.5.2.4 Land applications

Agricultural wastes from plant and animal operations are often land applied to recycle and manage residues. Compositional variability among wastes is vast. Some waste components can potentially represent a threat to the environment and humans depending on their nature, application loads, and soil type. Biochar, the product obtained by biomass heating under oxygen-limited conditions, has the potential to minimize risks associated with waste characteristics while promoting soil health. However, variation in the residue wastes (feedstocks) used to produce biochar carries over to the resultant biochar. In order to minimize nutrient deficiencies and environmental liabilities of biochars feedstocks, or mixtures of feedstocks, should be matched to the needs of specific crops, and by considering the P retentive capacity of the soil where the biochar is applied.[62]

The use of biochar represents an emerging technology that has a potential role in carbon sequestration, reducing greenhouse gas emissions, waste management, renewable energy, soil improvement, crop productivity enhancement, and environmental remediation. A balanced overview of the advantages and disadvantages of the pyrolysis process of biochar production, end-product quality and the benefits versus drawbacks of biochar on: (a) soil geochemistry and albedo, (b) microflora and fauna, (c) agrochemicals, (d) greenhouse gas efflux, (e) nutrients, (f) crop yield, and (g) contaminants (organic and inorganic) is needed.

Finally, there are reports of agricultural production wastes that can be re-used to improve the quality of the environment. One example is the use of abundant, unmodified agricultural wastes and by-products (AWBs) from grape, wheat, barley and flax production, to reduce the concentration of Cd, a highly toxic and mobile heavy metal, in contaminated water. Some AWBs can be directly (i.e. without pre-treatment or modification) used in bulk to remediate effluents contaminated with heavy metals, without requiring further cost or energy input, making them potentially suitable for low-cost treatment of persistent (e.g. via mine drainage) or acute (e.g. spillages) discharges in rural and other areas.[63]

7.6   CONCLUDING REMARKS

The shift toward sustainable food consumption and production can be achieved by addressing both the product supply and the consumptive demand elements through the adoption of smart, adequate, and efficient food consumption and production patterns. Therefore, the reduction of FLW is essential to enhance the sustainability of the current agri-food systems and achieve food and nutrition security for all.

To achieve the sustainability of food systems (from environmental, economic, and social points of view) as well as sustainable food and nutrition security, a global commitment to the reduction of FLW is necessary. Implementation of a systemic, integrated, and holistic food chain approach that involves circularity should consider the actual interactions between FLW and the dimensions of food security, as well as the connections among FLW and the SDGs related to food and nutrition security is needed.

REFERENCES

1. Rao, P. and V. Rathod, *Valorization of food and agricultural waste: a step towards greener future.* Chemical Record, 2019. **19**(9): pp. 1858–1871.
2. Capanoglu, E., E. Nemli, and F. Tomas-Barberan, *Novel approaches in the valorization of agricultural wastes and their applications.* Journal of Agricultural and Food Chemistry, 2021.
3. Ma, Y., Y. Shen, and Y. Liu, *Food waste to biofertilizer: a potential game changer of global circular agricultural economy.* Journal of Agricultural and Food Chemistry, 2020. **68**(18): pp. 5021–5023.
4. Jimenez-Lopez, C., et al., *Agriculture waste valorisation as a source of antioxidant phenolic compounds within a circular and sustainable bioeconomy.* Food and Function, 2020. **11**(6): pp. 4853–4877.

5. Palmisano, G.O., et al., *Food losses and waste in the context of sustainable food and nutrition security*, in *Food Security and Nutrition*. 2020, Elsevier. pp. 235–255.

6. Tilman, D., et al., *Agricultural sustainability and intensive production practices.* Nature, 2002. **418**(6898): pp. 671–677.

7. Chen, C., A. Chaudhary, and A. Mathys, *Nutritional and environmental losses embedded in global food waste.* Resources, Conservation and Recycling, 2020. **160**: pp. 1–12.

8. Adamashvili, N., F. Chiara, and M. Fiore, *Food loss and waste, a global responsibility?!* Economia Agro-Alimentare, 2019. **21**(3): pp. 825–846.

9. Aschemann-Witzel, J., A. Giménez, and G. Ares, *Household food waste in an emerging country and the reasons why: Consumer´s own accounts and how it differs for target groups.* Resources, Conservation and Recycling, 2019. **145**: pp. 332–338.

10. Aragie, E., J. Balié, and C. MoralesOpazo, *Does reducing food losses and wastes in sub-Saharan Africa make economic sense?* Waste Management and Research, 2018. **36**(6): pp. 483–494.

11. Johnson, L.K., et al., *Estimating on-farm food loss at the field level: A methodology and applied case study on a North Carolina farm.* Resources, Conservation and Recycling, 2018. **137**: pp. 243–250.

12. Yadav, V.S., et al., *A systematic literature review of the agro-food supply chain: Challenges, network design, and performance measurement perspectives.* Sustainable Production and Consumption, 2022. **29**: pp. 685–704.

13. Krzywoszynska, A., *Spotlight on... waste: uncovering the global food scandal.* Geography, 2011. **96**(2): pp. 101–104.

14. Devin, B. and C. Richards, *Food waste, power, and corporate social responsibility in the Australian food supply chain.* Journal of Business Ethics, 2018. **150**(1): pp. 199–210.

15. *Food Wastage Footprint: Impacts on Natural Resources.* 2013, FAO (Food and Agriculture Organization of the United Nations).

16. Melikoglu, M., C.S.K. Lin, and C. Webb, *Analysing global food waste problem: Pinpointing the facts and estimating the energy content.* Central European Journal of Engineering, 2013. **3**(2): pp. 157–164.

17. Gustavsson, J. and J. Stage, *Retail waste of horticultural products in Sweden.* Resources, Conservation and Recycling, 2011. **55**(5): pp. 554–556.

18. Venkatramanan, V. and S. Shah, *Climate smart agriculture technologies for environmental management: The intersection of sustainability, resilience, wellbeing and development*, in *Sustainable Green Technologies for Environmental Management*. 2019: Springer Singapore. pp. 29–51.

19. Chen, C.R. and R.J.C. Chen, *Using two government food waste recognition programs to understand current reducing food loss and waste activities in the U.S.* Sustainability (Switzerland), 2018. **10**(8).

20. Principato, L., C.A. Pratesi, and L. Secondi, *Towards zero waste: an exploratory study on restaurant managers.* International Journal of Hospitality Management, 2018. **74**: pp. 130–137.

21. Priefer, C., J. Jörissen, and K.R. Bräutigam, *Food waste prevention in Europe – A cause-driven approach to identify the most relevant leverage points for action.* Resources, Conservation and Recycling, 2016. **109**: pp. 155–165.

22. O'Regan, A. and K. Wolfe, *Assessing the relationship between individual diet quality and food waste.* Nutrition Today, 2021. **56**(2): pp. 85–88.

23. Conrad, Z. and N.T. Blackstone, *Identifying the links between consumer food waste, nutrition, and environmental sustainability: A narrative review.* Nutrition Reviews, 2021. **79**(3): pp. 301–314.

24. *Creating a sustainable food future world resources institute.* Renewable Resources Journal, 2020. **34**(3): pp. 7–17.

25. Paritosh, K., et al., *Food waste to energy: an overview of sustainable approaches for food waste management and nutrient recycling.* BioMed Research International, 2017. **2017**: pp. 1–19.

26. Song, G., et al., *Food consumption and waste and the embedded carbon, water and ecological footprints of households in China.* Science of the Total Environment, 2015. **529**: pp. 191–197.

27. Yukesh Kannah, R., et al., *Food waste valorization: Biofuels and value added product recovery.* Bioresource Technology Reports, 2020. **11**: pp. 1–14.

28. Zhang, Z., et al., *Environmental impacts of hazardous waste, and management strategies to reconcile circular economy and eco-sustainability.* Science of the Total Environment, 2022. **807**: pp. 1–17.

29. Bernstad, A. and J. La Cour Jansen, *Review of comparative LCAs of food waste management systems – Current status and potential improvements.* Waste Management, 2012. **32**(12): pp. 2439–2455.

30. Gao, A., et al. *Comparison between the technologies for food waste treatment.* In *8th International Conference on Applied Energy, ICAE 2016.* 2017. Elsevier Ltd.

31. Plazzotta, S. and L. Manzocco, *Food waste valorization*, in *Saving Food: Production, Supply Chain, Food Waste and Food Consumption.* 2019, Elsevier. pp. 279–313.

32. Kavitha, S., et al., *Introduction: Sources and characterization of food waste and food industry wastes*, in *Food Waste to Valuable Resources: Applications and Management.* 2020, Elsevier. pp. 1–13.

33. Duenas, M. and I. Garciá-Estévez, *Agricultural and food waste: Analysis, characterization and extraction of bioactive compounds and their possible utilization.* Foods, 2020. **9**(6).

34. Xiong, X., et al., *Value-added chemicals from food supply chain wastes: State-of-the-art review and future prospects.* Chemical Engineering Journal, 2019. **375**: pp. 1–24.

35. Birkin, F., J. Margerison, and L. Monkhouse, *Chinese environmental accountability: Ancient beliefs, science and sustainability.* Resources, Environment and Sustainability, 2021. **3**: pp. 1–8.

36. Fortunati, E., A. Mazzaglia, and G.M. Balestra, *Sustainable control strategies for plant protection and food packaging sectors by natural substances and novel nanotechnological approaches.* Journal of the Science of Food and Agriculture, 2019. **99**(3): pp. 986–1000.

37. Sim, E.Y.S. and T.Y. Wu, *The potential reuse of biodegradable municipal solid wastes (MSW) as feedstocks in vermicomposting.* Journal of the Science of Food and Agriculture, 2010. **90**(13): pp. 2153–2162.

38. Giordano, C., et al., *The role of food waste hierarchy in addressing policy and research: A comparative analysis.* Journal of Cleaner Production, 2020. **252**.

39. Teigiserova, D.A., L. Hamelin, and M. Thomsen, *Towards transparent valorization of food surplus, waste and loss: Clarifying definitions, food waste hierarchy, and role in the circular economy.* Science of the Total Environment, 2020. **706**: pp. 1–13.

40. Guo, X.X., H.T. Liu, and J. Zhang, *The role of biochar in organic waste composting and soil improvement: A review.* Waste Management, 2020. **102**: pp. 884–899.

41. Orozco, R.L., et al., *Chapter 3: Biohydrogen Production from Agricultural and Food Wastes and Potential for Catalytic Side Stream Valorisation from Waste Hydrolysates*, in *RSC Green Chemistry*, L.E. Macaskie, D.J. Sapsford, and W.M. Mayes, Editors. 2020, Royal Society of Chemistry. pp. 57–86.

42. Ranganathan, S., et al., *Utilization of food waste streams for the production of biopolymers*. Heliyon, 2020. **6**(9): pp. 1–13.

43. Ramadhan, M.O. and M.N. Handayani. *The potential of food waste as bioplastic material to promote environmental sustainability: A review*. in *1st International Conference on Science and Technology for Sustainable Industry, ICSTSI 2020*. 2020. IOP Publishing Ltd.

44. Maia, M., et al., *Vitis vinifera 'Pinot noir' leaves as a source of bioactive nutraceutical compounds*. Food and Function, 2019. **10**(7): pp. 3822–3827.

45. McClements, D.J. and B. Öztürk, *Utilization of nanotechnology to improve the application and bioavailability of phytochemicals derived from waste streams*. Journal of Agricultural and Food Chemistry, 2021. 70(23): pp. 6884–6900.

46. Kumar, R., et al., *Particle size reduction techniques of pharmaceutical compounds for the enhancement of their dissolution rate and bioavailability*. Journal of Pharmaceutical Innovation, 2021. 17(2): pp. 333–352.

47. Vazquez-Olivo, G., E.P. Gutiérrez-Grijalva, and J.B. Heredia, *Prebiotic compounds from agro-industrial by-products*. Journal of Food Biochemistry, 2019. **43**(6): e12711.

48. Apicella, A., et al., *Valorization of olive industry waste products for development of new eco-sustainable, multilayer antioxidant packaging for food preservation*. Chemical Engineering Transactions, 2019. **75**: pp. 85–90.

49. Torres-León, C., et al., *Food waste and byproducts: an opportunity to minimize malnutrition and hunger in developing countries*. Frontiers in Sustainable Food Systems, 2018. **2**: pp. 1–17.

50. Vilas-Boas, A.A., M. Pintado, and A.L.S. Oliveira, *Natural bioactive compounds from food waste: Toxicity and safety concerns*. Foods, 2021. **10**(7): pp. 1–26.

51. Foley, J.A., et al., *Solutions for a cultivated planet*. Nature, 2011. **478**(7369): pp. 337–342.

52. Lin, C.S.K., et al., *Current and future trends in food waste valorization for the production of chemicals, materials and fuels: A global perspective*. Biofuels, Bioproducts and Biorefining, 2014. **8**(5): pp. 686–715.

53. Cardoen, D., et al., *Agriculture biomass in India: Part 1. Estimation and characterization*. Resources, Conservation and Recycling, 2015. **102**: pp. 39–48.

54. Fermoso, F.G., et al., *Valuable compound extraction, anaerobic digestion, and composting: a leading biorefinery approach for agricultural wastes*. Journal of Agricultural and Food Chemistry, 2018. **66**(32): pp. 8451–8468.

55. O'Connor, J., et al., *A review on the valorisation of food waste as a nutrient source and soil amendment*. Environmental Pollution, 2021. **272**: pp. 1–17.

56. Jurgilevich, A., et al., *Transition towards circular economy in the food system*. Sustainability (Switzerland), 2016. **8**(1): pp. 1–14.

57. Kumar, A. and S.R. Samadder, *A review on technological options of waste to energy for effective management of municipal solid waste*. Waste Management, 2017. **69**: pp. 407–422.

58. Tampio, E., T. Salo, and J. Rintala, *Agronomic characteristics of five different urban waste digestates*. Journal of Environmental Management, 2016. **169**: pp. 293–302.

59. Shrestha, P., G.E. Small, and A. Kay, *Quantifying nutrient recovery efficiency and loss from compost-based urban agriculture.* PLoS ONE, 2020. **15**(4): pp. 1–15.
60. Price, G.W., J. Zeng, and P. Arnold, *Influence of agricultural wastes and a finished compost on the decomposition of slaughterhouse waste composts.* Journal of Environmental Management, 2013. **130**: pp. 248–254.
61. Muscolo, A., et al., *Anaerobic co-digestion of recalcitrant agricultural wastes: Characterizing of biochemical parameters of digestate and its impacts on soil ecosystem.* Science of the Total Environment, 2017. **586**: pp. 746–752.
62. Freitas, A.M., V.D. Nair, and W.G. Harris, *Biochar as influenced by feedstock variability: implications and opportunities for phosphorus management.* Frontiers in Sustainable Food Systems, 2020. **4**: pp. 1–11.
63. Melia, P., et al., *Agricultural wastes from wheat, barley, flax and grape for the efficient removal of Cd from contaminated water.* RSC Advances, 2018. **8**(70): pp. 40378–40386.

8 Agricultural chemistry in the food, energy, and water nexus

The food, energy, and water (FEW) nexus lays out three major areas that must be addressed simultaneously in order to achieve sustainability. (See Figure 8.1) The food system is a major contributor to climate change, depletion of freshwater, and pollution of aquatic and terrestrial ecosystems from nitrogen and phosphorus fertilization. Between 2010 and 2050, the environmental effects of the food system could increase by 50–90%. These changes will be driven by increases in population and improved standards of living and income levels. Without changes and mitigation measures, environmental stresses may reach levels that exceed the planetary boundaries for a safe operating space for humanity.[1] In order to address the situation, enhanced management approaches that utilize a synergistic combination of measures are needed.

The systems approach required to integrate the diversity of ecosystem challenges, environmental factors, magnitude of scale, and stressor-related responses necessary to address the recovery and resilience in agricultural landscapes needs to focus on the FEW nexus. It will identify future challenges in areas of the ecosystem provided by different types of agroecosystems that need to be assessed in concert.[2]

8.1 THE NEXUS OF FOOD, ENERGY, AND WATER (FEW)

The Earth's population is expected to exceed 9 billion by 2050, posing significant challenges in meeting human needs while minimally affecting the environment. To support this increased population, secure and safe sources of food, energy, and water will be needed. The nexus of food, energy, and water addresses all three in an integrated manner. This becomes the most critical issue facing society. There is no more land to exploit, and the dwindling supply of fresh water in some areas of the world limits the use of land for food. All solutions must also deal with the effects of global climate change. This is the present situation. There is no "magic bullet". Meeting current and future populations needs will require security in food, energy, and water supplies. A nexus approach is presented that attempts to improve food, energy, and water security. It will involve integrating the management of

DOI: 10.1201/9781003157991-8

FIGURE 8.1   Food, energy, and water nexus. (See www.shutterstock.com/image-photo/hands-holding-globe-glass-green-forest-1310886638)

the limited resources, while transitioning to a "greener" economy, which provides adequate food, energy, and water for the expanding human population.[1]

Over the next 30 years, the agriculture and food delivery enterprise will face unprecedented pressures from population, climate change, and limited natural resources, particularly water. The ability to engage in Sustainable Intense Agriculture is the subject of the next chapter, but will require addressing the FEW nexus. Here we lay out the practical reality of working with increasingly limited resources. The Green Revolution made major contributions to improved food security and agricultural productivity, which resulted in major reductions in worldwide hunger. However, it did this without consideration of the ultimate limitations of our natural resources. In the future, along with sustainable productivity, there are increasing concerns about access to adequate nutritious food and global access to natural resources. The next evolution will result in a new focus, which will consider food availability, diet quality, and efficient use of resources. It will begin with and always consider the FEW nexus.

As the global population grows and increasingly lives in cities, and does so with lifestyles based on greater material consumption, more attention must be given to the integrated system that supplies our energy, water, and food (FEW nexus). There is also mounting concern about effects on the nexus of climate change and damage to the natural environment that provides essential ecosystem services. Nexus analysis applies existing techniques, such as computational modelling and Life Cycle Assessment, but new frameworks and tools are needed, including those that will integrate societal and technical dimensions. Agricultural chemistry (AC) will provide the chemical tools and understanding to help integrate solutions, while including scientific knowledge to frame them. Studies show the importance of an integrated systems view of the complex interrelationships of the nexus when planning effective remedies. Assessments conclude that transformative social and political change is needed to create new structures, markets, and governance to deal with the nexus if we are to meet agreed upon sustainable development goals.[3]

8.1.1 WHAT IS THE FEW NEXUS

The integrated system that supplies energy, water, and food for use within human society is the FEW nexus. This nexus comes from the Latin *nectere*, to bind or connect. In the current world, the three areas are bound together, in obvious and even subtle ways. But with the growth of the world's population and its concentration in urban areas, which distance people from the sources of their primary energy, water, and food, the supply of these three commodities has become ever more complex and interrelated. Importantly, they must all be satisfied, if sustainability is to be achieved.

The FEW nexus has emerged as a useful concept to focus and address the complex and interrelated nature of our global resource systems, on which we depend to achieve different social, economic, and environmental goals. In practical terms, it presents a conceptual approach to better understand and systematically analyze the interactions between the natural framework around human activities, and to work toward a more coordinated management and use of necessary natural resources across sectors. This can help us to identify and manage trade-offs and to build synergies through our responses, allowing for more integrated and cost-effective planning, decision-making, implementation, monitoring, and evaluation.[4]

8.1.2 INTERACTION AMONG NEXUS

The magnitude of the interaction among the components of the FEW nexus is shown in Figure 8.2. The nexus is analogous to the interactions among the four spheres of the earth's ecosystem.

Food, energy, and water systems are inexorably linked. The challenges of simultaneously harmonizing solutions in the nexus are complicated by Earth's changing climate and increasing competition for finite tillable land resources. The complex relationships within the FEW nexus implies that any solution for one parameter of the nexus must equally consider and account for the other parameters. [5] Population growth and climate change further complicate these interactions. Reduction of greenhouse gas emissions is rendered as imperative, but it will require aggressive mitigation actions and adaptation; these will affect the nexus. There are many approaches to greenhouse gas reductions, and we will only be successful if we use all of them. There is no single solution to the problems obstructing sustainability. Agriculture, AC, and food production are central to the solutions involving the FEW nexus.

Addressing the nexus challenge requires considering the interactions between multiple dimensions and scales, including the natural environment, social

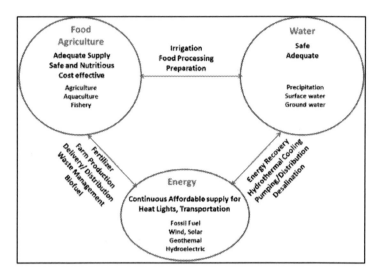

FIGURE 8.2   FEW interactions.[5]

systems, institutions, local communities, nations, business, and technology. Previous discussions of the four spheres in the global environment encourages a multidimensional consideration of the FEW nexus and how it frames the proposed solutions. (See Figure 8.2) This framing includes what dimensions and scales are considered, what interactions are addressed, and what consequences are considered. The International Institute for Sustainable Development emphasizes managing the nexus by the need to consider both natural systems and built systems. [3] It is important to advocate using existing and well-researched concepts to construct integrated solutions to nexus challenges.[6] The benefits of considering the following simultaneously will all be necessary if this is to succeed: [2]

- the interactions between various ecosystem services;
- the ecological footprints of water, carbon, and other materials;
- and planetary boundaries.

A simple example of the latter is the effect of climate change, brought about primarily by the burning of fossil fuel, with an additional contribution from land-use change in developing new agriculture. As carbon dioxide concentrations in the atmosphere rise and weather patterns change owing to the greenhouse effect, the interaction of energy, water, and food systems is being demonstrated on a global scale.[7]

Since each resource is finite, interactions among food, energy, and water are complex and inseparable when we consider the food and agricultural dimension. This is captured in Figure 8.2. Water is directly or indirectly used in the generation of energy and the production of food. Agriculture used to produce food and fiber is the largest consumer of water. However, large volumes of water are required for the generation of electricity, extraction of fossil fuels, mining, refining, residue disposal and processing of fossil fuels, and growing feedstocks for biofuels in addition to the water used for agricultural food and fiber production. Energy is required to transport water and to treat it to be suitable for drinking water and clean wastewater for reuse or releases to the environment. Agricultural food and fiber production utilizes about 70% of the fresh water used for human food production. The need to produce food and energy stresses the limited supply of fresh water on the planet. Food systems include the entire gamut of agricultural production and harvesting as well as animal production, aquaculture, and processing, storage, and distribution of food products through the wholesale and retail systems. It also includes seed, land preparation, labor, energy, soil, animals, equipment, and final food preparation of an adequate, nutritious, and safe food supply.[8] If we now consider the effects of global warming, which is exacerbated by the excess use of carbon fuels for energy, the FEW nexus is critical. In this perspective water is the most critical resource for production of food and energy.

Even given this dependence on water, food output and energy consumption have remained constant in recent years. The food industry uses less energy, but its diverse activities include a wide range of energy usage, depending upon the product. For example, wet corn milling and sugar manufacturing are more energy-intensive than the dairy industry. The U.S. Energy Information Administration (EIA) projections show a declining contribution from both plant-based industries to the economy but an increasing share of energy consumption.[8] The obvious concern is that with increasing population, this will increase.

8.1.3 DIMENSIONS OF THE FEW NEXUS

Managing the interactions in the food-energy-water nexus is essential to continued sustainable development worldwide. The increasing demand for all three is driven by a rising global population, rapid urbanization, changing diets, and economic growth. Agriculture is the largest consumer of the world's freshwater resources, and over 25% of global energy use is expended on food production and distribution. Water is a finite resource; beyond food industry, it will need to sustain more people and usages, and therefore, ensuring everyone has access to a reliable supply is crucial to human survival. Water resources will become more stretched, directly affecting the energy and food sectors, which are critically dependent upon water.

The challenge for the the FEW nexus is how to ensure that any solution for one dimension of the nexus must equally consider the other aspects.[9] Reduction of greenhouse gas emissions is imperative, and it requires aggressive mitigation actions and adaptation.[10] There are many approaches to greenhouse gas reductions, and they will only be successful if all of them are used. There is no single solution to the problem. The GHG problem affects all areas of sustainability, but its solutions must also be in accord with the sectors of the FEW nexus.

There will be competition among resource demands. Heightened demands for water (upstream and downstream) result in trade-offs of water use between agriculture (food production) and electricity generation. Critical needs for water include intensive (hydroelectric power and powerplant cooling) and urban and environmental needs. Climate change represents additional complications and demands on the nexus. Climate change influences water availability and potential agricultural production and some components of energy production and demand.[4]

As water resources become more stretched, the dependence upon water will affect the energy and food sectors. The fact that all three parts of the FEW nexus directly relate to several of the sustainable development goals makes this more important. This means that decision makers in all three domains now increasingly must focus on water resource management, ecosystem protection, and water supply and sanitation as part of their policy and practice.[4]

8.1.4 FOOD

The elements of FEW nexus that directly relate to food are key to the role that agriculture plays in sustainability. From this, food is seen as essential.[11]

- food availability, affected by production, distribution, and exchange of food,
- access to food, including affordability, distribution, sufficient and safe, and preference (e.g., culturally appropriate),
- utilization, nutritional value, social and cultural value, and food safety, and
- sustained/long-term food stability.

Beyond the above bullets, agriculture based on annual crops or perennial plants will add a new layer and must provide feedstock for new biorefineries. New investigations reveal the potential role of such crops as the source of a very wide range of products, including biofuels (e.g., butanol and hydrogen), biomaterials, and food or animal feed. [12] The combination of sustainable agriculture and biorefineries could play a key role in supplying many products from the most abundant renewable biomass, but this poses new issues as well.[13] There are consequences of the increasing production of biofuels. The cultivation of dedicated biomass is likely to affect bioenergy potentials, global food prices, and water scarcity, and that therefore integrated policies for energy, land-use, and water management will be needed. This is a clear example of "FEW nexus" thinking.

Case studies from the agrifood sector show how acting on nexus challenges can add value through mitigating risk, reducing cost, and raising productivity.[14] Measures like reductions in energy and water consumption save costs and also reduce environmental impacts. They can often be achieved by systems thinking and better engineering design. If the demand for biofuel crops significantly increases in response to concerns about global warming, then the current global trend toward a diet higher in meat will lead to severe competition for agricultural land and a crisis for natural ecosystems.[15]

From a global perspective food security would exist when all people have physical, social and economic access to sufficient, safe, nutritious food to meet dietary needs and food preferences.[16] This security also increases the demand for food and energy which requires sufficient water. (The production and processing of food are the largest users of fresh water, consuming an average of 70% of the human use of fresh water.) Modern agriculture and food production requires large amounts of energy which also affects the water resources, disruption of groundwater recharge, and water quality. Modern agriculture uses approximately 30% of total global energy. Fuels are required for land preparation, fertilizer production, irrigation and sowing, harvesting, transportation of crops, and processing and distribution of foods.[17]

8.1.5   CIRCULARITY IN THE FEW NEXUS

An assessment of agricultural raw material supply chains for food production points out that the choice of sustainability tends to reflect the viewpoint of those who select them and that "sustainability" is not quite the same as "vulnerability" or security. [18] Circularity will bring them into better alignment. Thus, the implications for the FEW nexus of different policy choices will be somewhat different if security is the criterion rather than sustainability, and different again if only environmental footprint is considered (or any other restricted set of indicators).

An interesting case study illustrates this need for circularity, and it comes from the fast-moving consumer goods industry, which includes a wide variety of typically nondurable products related to the energy-water-food system, including food and beverages, household care, and personal care products. The industry is characterized by high sales volumes and relatively low prices.[3] A vision for this industry's future in a circular economy, would entail the necessity of a different type of changes in FEW: (a) water rinse and reuse, (b) the cycling of pure materials and food, and (c) the rise of the circular retailer. This is a very different approach based on approaching the FEW nexus through circularity.

In the twenty-first century, climate change is the biggest risk to global food security, and therefore, it represents a severe threat to human health and long-term wellbeing.[3] The FEW nexus will help to improve choices that align with circularity. Climate change is expected to reduce the quantity of food produced. The climate change will cause many regions throughout the world to experience reductions in crop yields and challenges from drought, pests, inadequate water supply, and soil degradation. This in turn will necessitate a better use of existing resources based on how they affect the nexus.

8.2   CHEMISTRY

The depletion of the supply of new energy, water, and natural resources seems inevitable, so the skills and expertise chemists and agricultural chemistry is especially needed. The issue is certainly not that society has run out of challenges for AC in the twenty-first century. In fact, chemistry may now be the most important of the sciences in its potential to impact sustainability by applying the FEW nexus using insights from nature and the four spheres. In short, chemistry can use the techniques involving atoms and elements to work with "everything that involves molecules."[19] This expansion will require fresh intellectual and practical challenges, resulting in creative solutions to sustainability problems. For industrial chemistry, the expansion of scope would open the door to new commercial opportunities, and to future growth.

8.2.1 PAST

Resulting from past and legacy practices, agriculture contributes approximately 10% of GHGs in the United States, which amounted to 650.7 MMT of CO_2 equivalents of greenhouse gases.[20] There are a wide variety of agricultural sources of GHGs, including soil management, enteric fermentation, manure management, fertilization, biomass burning, and CO_2 emissions from electricity generation and farm equipment. Soil management emissions are associated with cropping practices. Included in this area are fertilization, irrigation, tillage, deposition of livestock manure, and other depositions on cropland.[21]

An important way to lower GHG emissions is through Green Chemistry (GC). GC has been a major driver of sustainable development and has had an important impact in recent years. Using the previous discussion as a starting point, GC, sustainability, and Circular Economy (CE) concepts can assist in ensuring the needed changes for reducing GHG. Since the nature of chemistry is to produce intermediate goods that are generally used by other industries, this will also be true for agriculture. Chemistry is found in most agriculture production systems. The chemical industry is able to contribute to a transition towards greater economic, environmental and social sustainability. The focus of GC is the environment, providing tools and processes for the chemical industry to implement Sustainable Chemistry (SC) system and to realize the transition towards sustainability and CE. GC has the tools through which it is possible realize the SC system. In particular, the SC, in a CE system, can be involved in processes of production and recycling of wastes, ensuring more sustainable environmental, economic, and social systems. Furthermore, GC and CE are aligning and highlighting the successful applications of different tools and adopted approaches in a holistic vision.

8.2.2 PRESENT

Taking this vision, we now apply it to existing problems. Sustainable development goals (SDGs) were proposed for harmonious development for future sustainability. [22] The aim for the SDG in agriculture is highly consistent with the idea of circular economy (CE) which can be achieved by applying green chemistry principles (GCPs). The relationship between GCPs and SDGs are linked from the perspective of input chemical selection and reduction, sustainable design/reactions, and safety management. The benefits of CE create a kind of green business model.

By integrating GCPs and SDGs, the specific applications of renewable resources towards sustainable development are introduced. Sustainability must be applied harmoniously through the FEW nexus.

8.2.3 FUTURE

Agriculture is facing an enormous sustainability challenge: It must ensure that enough high-quality food is available to meet the needs of a continually growing population and do so in ways minimally impacting the environment. Current and future agronomic production of food, feed, fuel and fiber requires innovative solutions for existing and future challenges, such as effects resulting from climate climate change, resistance to pests, increased regulatory demands, renewable raw materials or requirements resulting from food chain partnerships. Modern agricultural chemistry has to support farmers to manage these tasks. Today, the so-called 'side effects' of agrochemicals regarding yield and quality are gaining more importance. Agrochemical companies with a strong research and development focus will have the opportunity to shape the future of agriculture by delivering innovative integrated solutions.

8.3 AGRICULTURAL CHEMISTRY ROLE

The goals for agriculture are specific to achieve sustainable food production by 2050, there will need to be a 75% reduction in the current yield gap, more efficient use of nitrogen and phosphorus fertilizers, and recycling of phosphorus to reduce eutrophication and water use. These will need to be done in alignment with the FEW nexus. In addition, there will need to be implementation of means to mitigate agricultural GHG, adaptation of land management practices to shift agriculture, and forest management to become positive carbon sinks.[23] These align agriculture with the necessary environmental goals.

Multiple approaches will be required to meet the increasing demand for food production, including water management, improved use of fertilizers, and identifying crops that are tolerant of environmental changes. Crops developed with drought tolerance offer promise; however, they must undergo safety testing and environmental safety assessments similar to the first generation of genetically modified (GM) crops with insect and herbicide resistance. The environmental safety of GM crops with drought, heat, or osmotic stress tolerance may be more complex because they will be grown in different environments. To ensure that no unforeseen hazards occur in the plants, they must be tested under both irrigated and water-limited conditions. Drought tolerance could enhance traits that increase the reproductive and vegetative growth and competitive ability of plants under selective conditions, such as drought conditions.

The World Bank[24] anticipates that climate change represents potential risks to agricultural productivity that could drive more than 100 million people back into poverty by 2030. To contribute to global food security, there are currently efforts to develop new varieties of crops that are tolerant to drought using conventional breeding as well as the tools of modern biotechnology; however, few have reached the market. There is a need to develop globally harmonized protocols on how to best assess these effects of GM crop varieties with drought tolerance, in which one is careful with the amount of detail required for environmental risk assessment research. As drought-tolerant crop varieties developed by conventional breeding methods are actively being introduced as well, knowledge of these could be used for comparisons. Harmonized protocols and procedures are going to be required to facilitate the development and introduction of drought-tolerant crops based on an adequate safety assessment for human and animal consumption and with environmental safety.[25]

8.3.1 WATER AVAILABILITY AND SCARCITY

AC will need to address the issue of water supplies. Water shortages are being experienced all over the world including the United States.[26] California is facing severe drought and various states have had disagreements relating to the supply of water. Internationally, water shortages in Africa, Asia, and Latin America are well known. In the Middle East, Iran is facing a water shortage potentially so serious that officials are making contingency plans for rationing in the greater Teheran area, home to 22 million, and also in other major cities around the country. AC will be most effective in how it uses water and how it accommodates agriculture to use less water.

8.3.2 WATER RECLAMATION

Water is considered used if it is taken and returned to the same source (instream use), such as in thermoelectric plants that use water for cooling and are by far the largest users of water. While used water is returned to the system for downstream uses, it has usually been degraded in some way, mainly as a result of thermal or chemical pollution, and the natural flow has been altered, which does not factor into an assessment if only the quantity of water is considered. Chemists have an invaluable role to play in keeping the water usable. Water is consumed when it is removed completely from the system, such as by evaporation or consumption by crops or humans. When assessing water use and water availability, all these factors must be considered as well as spatiotemporal considerations, making precise determination of water availability very difficult.

Water contamination reduces its availability. We need to maintain water quality by monitoring water resources for various known and unknown contaminants rigorously and regularly from point and nonpoint source pollution.[27] Wastewater can originate from many places: households, industries, commercial developments, road runoff, etc. As diverse as the sources of wastewater are, so too are their potential composition and contaminants. The composition of wastewater affects not only the treatment processes applied but also their source recovery opportunities. Understanding wastewater profiles and their abundance at different stages is a first step in recognizing appropriate opportunities for pretreatment.

Wastewater contains pollutants/contaminants that have to be removed and/or reduced before the water is returned to a surface water source (river, ocean, bay, lake, etc.) or to groundwater. As part of sustainable intensification agriculture (discussed in the next chapter), waters resulting from regular agricultural activities should be purified and returned to streams as pristine as they were found.

8.3.3 WATER QUALITY

The amount of testing necessary for reclaimed water should relate to how it is going to be recycled for use. For example, if it is recycled into a surface water supply, its quality after purification should match or exceed the requirements of the surface water to which it is being added. Similar rules may be followed for mixing with groundwater. Recycled wastewater for drinking must meet potable water requirements, with the added assurance by ultratrace analysis that no toxic contaminants are present.[28]

The range and potential areas for reclamation of impure water is a significant task for chemists and agricultural chemists. Evaluating the performance of wastewater treatment represents a challenging and complex task as it usually involves engineering, environmental and economic factors. Key performance indicators, such as energy consumption, pollutant removal, global warming potential, and wastewater treatment fees are critical. In terms of engineering performance, both the membrane bioreactors and constructed wetlands are stable, effective and reliable during their operating periods. When the environmental impacts of wastewater treatment technologies are compared via a life cycle assessment, the ecological technologies showed superior performance, in terms of environmental impacts, especially for the global warming potential and eutrophication potential. In contrast, ecological technologies had higher unit land use due to their large area. In overall, both the membrane bioreactors and constructed wetlands show excellent overall performance. Wastewater treatment plants are typical case studies for addressing the interactions of water and energy elements.[29]

8.3.4 IMPACT OF CONTAMINATED WATER ON FOOD

The results of external contamination of water must be monitored by AC. Urban agriculture plays an important role in sustainable food supply.[30] The contamination levels found are often high. In order to treat it, or better to prevent it, green and sustainable chemistries are needed.

High concentrations of arsenic (exceeding 1000 mg L -1) have been reported in shallow tube wells (STWs).[31] Green leafy vegetables act as arsenic accumulators, with arum (kochu), gourd leaf, Amaranthus, and Ipomea (kalmi) topping the list. Rice shows the presence of As(III), dimethyl arsenic acid, and As(V); greater than 80% is in the inorganic form. More than 85% of the arsenic in rice is bioavailable. A highly increased accumulation of arsenic in vegetables grown with arsenic-contaminated water compared with those grown with uncontaminated water. It is important to note that arsenic at lower levels is found in all soils, including American farm fields. The fertile soils fanning out across the Mississippi River floodplain are up to five times as high in arsenic as in other parts of Louisiana, Mississippi, and Arkansas, according to studies done by the USGS.[32]

Cadmium in rice is a well-known phenomenon. Other metal-contaminated rices are "arsenic rice," "mercury rice," and "lead rice." It appears that the main sources of cadmium pollution are emissions from smelting plants. For five metals (arsenic, cadmium, lead, manganese, and zinc), strong correlations of concentrations in uncultivated soils indicate a common source, suggesting that emissions from these plants using this metal may be a major contributor to elevated concentrations of these five metals in uncultivated soils in this area. The fields are ringed by factories and irrigated with water tainted by industrial waste. A recent study by USGS found that some of the synthetic compounds found in a wheat field sprayed with biosolids are bisphenol A, hexahydrohexamethyl-cyclopent- 2-benzopyran, nonlylphenol ethoxylates, triclosan, and warfarin.[32]

8.3.5   FOOD AND BIOFUELS

There are two types of liquid biofuels: biodiesel and ethanol. Biodiesel is produced primarily from triacylglycerol, a lipid. Ethanol is made by using yeast to ferment sugar extracted from sugarcane and sugar beets, or from the starch of grains and cassava. The current production of ethanol requires much less land per unit of biofuel than does biodiesel. In the US in 2011, fermentation of corn-derived sugars yielded about 49 billion liters of ethanol, which was used mostly as a fuel additive. This accounted for 38% of the country's land used to grow corn for grain. Ethanol makers depend on glucoamylase enzymes to break down starch from corn and other crops, into sugar for fermenting into ethanol. Producing fuels from a nonedible biomass such as corncobs and cornstalks promises to be a large opportunity for enzyme manufacturers.[33] Scientists are developing enzymes that can degrade cellulose into sugars to produce fuel. One way to increase biofuel production is to gain benefits from the leftover plant materials that are currently being discarded. In addition to taking advantage of these existing sources of biomass, a shift toward cultivation of plants that naturally produce more biomass per ha, using less fertilizer is required. Some countries are using biomass to provide high levels of fuel. The adoption of flex-fuel vehicles allows the use of a wide range of mixtures of ethanol and gasoline, depending on the relative prices.

8.3.6   RENEWABLE SOURCES OF ENERGY

The developments of Green Chemistry (GC) and AC will open paths to new energy sources. We need to use energy sources that have minimal impact on our environment. Some of the renewable sources of energy are tidal and wave, solar, wind, and biofuels. Wind has enjoyed some success in the areas that are suitable for its use. The United States is the top producer in wind power today. China and

Germany are second and third respectively. However, Germany is the largest producer of solar energy. The use of biofuels is growing but will continue only if they can be produced by minimal use of corn. Biofuel producers say that they can compete if the price of oil remains above $70 a barrel and gasoline stations offer higher blends of ethanol. The South Dakota-based ethanol producer Poet, in partnership with the Dutch company Royal DSM with its start-up Emmetsburg, Iowa, plant, will produce up to 25 million gallons a year, making use of corncobs, husks, and leaves. DuPont's biggest plant will make ethanol from corn waste and will have a capacity of 30 million gallons a year. Abengoa, S. A. is planning to produce 25 million gallons of biofuel a year and has already started a 21-MW electricity plant at the site powered by biomass. It has developed a proprietary enzyme to mix with cornstalks and wheat straw to produce sugars that will then be fermented and distilled to produce cellulosic ethanol. The more efficient process can increase yields and decrease costs. What is more, because cellulosic ethanol relies on the waste products of corn rather than corn itself, it does not raise demand for corn or raise corn prices.[34]

New work on microalgae strains for aviation fuel production will further solidify the move away from nonrenewable resources for fuels. Assessment involves evaluating 17 candidate microalgae strains. The results show that unmodified biofuel from the most suitable strain could not meet all jet fuel standards. These results highlight the need for a broad action plan including improvement in the processing or modification of biofuel produced from microalgae and revision of the current jet fuel standards to facilitate the introduction of microalgae-based biofuel for the aviation industry.[35]

8.4   FOOD SECURITY WITH FEW NEXUS

As we have seen, the focus on the FEW nexus draws upon the lessons from the four spheres. Further, the lessons learned by AC will help sustainability. (See Figure 8.3) The optimum production conditions for corn, wheat, rice, and potatoes are projected to move to higher latitudes. It remains to be seen how these changes will be implemented because of soil, water, and soil conditions. It is highly likely that there will be needs for genetic alteration of crops to be optimally produced in the higher latitudes. The impacts of climate change on crop productivity may have consequences for food availability in some regions of the world. The stability of food systems in many parts of the world may be at risk because the consequences of climate change can result in short-term variability of food supplies, which will exacerbate food insecurity, particularly in areas currently vulnerable to hunger and undernutrition.[5, 8]

Food security and caloric availability are not the only considerations between agriculture and health. The range of products produced and available in a region is reflected in the composition of diets. Diet composition has major consequences on the health of populations regionally and globally. It has been found that the greatest number of deaths, worldwide and in most regions, including developing countries, can be attributed to dietary risk factors associated with imbalanced diets, such as

FIGURE 8.3   Food security. (See www.shutterstock.com/image-photo/closeup-wheat-spi
kes-agriculture-black-white-2021082614)

those low in fruits and vegetables and high in red and processed meat.[5] The
increasing importance of dietary risk factors represents a general trend away from
communicable diseases associated with undernutrition and poor sanitation to non-
communicable diseases associated with high bodyweight and unbalanced diets.
[36] These results increase the growing importance of developing our production
of sustainable agricultural products that meet the dietary requirement of the
population being served. This underscores the imperative for chemists and AC to
work with agriculture in this area.

In the twenty-first century, climate change is the biggest risk to global food
security, and therefore, it represents a severe threat to human health and long-
term wellbeing. Climate change causes changes in land and water temperatures,
precipitation, and the occurrence of heat waves, floods, droughts, and fires. There
are also indirect effects because of ecological and social disruptions, such as crop
failures, shifting patterns of disease vectors, and displacement of people.[37]

Pathways have been suggested to reach a primary goal that seeks to sustainably
manage and mitigate the contributions that a food system makes to climate change
while developing food production methods that replenish rather than deplete
biodiversity and related ecosystems.[38]

8.4.1   Employing sustainable production methods

In terms of water resource availability, 10 countries used more than 40% of their
water resources for irrigation and were therefore defined as suffering critical water

scarcity [27]. Moreover, a threat is presented by salinization and pollution of water courses and bodies and degradation of water related ecosystems. This is not the only resource whose limited availability is critical for increasing agricultural production. Phosphorus concentration is important in the production of chemical fertilizers.

Another area of chemical fertilizers are nitrogen based, the main criticality associated to them is the highly energy intensive process associated with industrial nitrogen fixation (in the US 70% of greenhouse gas emissions in corn production are related to nitrogen fertilizer).[39] Furthermore, excessive use of chemical fertilizers can cause water pollution through runoff.

Land represents another critical resource. Stiff competition exists for its use because of other human activities (like urbanization and cultivation of crops for biofuels) and where land is available it may no longer be productive because of unsustainable land management (which leads to desertification, salinization, soil erosion, and other consequences) or simply because land banks that must exist for the protection of biodiversity and ecosystems services (such as carbon storage) must be given priority. The FEW approaches can assist through informing policies and regulations that promote the implementation of more efficient production technologies. Some examples are solutions for water conservation (like rainwater harvesting) and efficient water use technologies (on time water delivery and micro irrigation), increased fertilizer use efficiency (through more precise application of fertilizers, nitrogen fixing, use of compost), increased yields to input ratio, and reduced carbon intensity of fuel inputs (by using alternative sources for energy production such as wind and solar power or anaerobic digestion). Notwithstanding these requirements, existing policies created with a silo approach have traditionally focused only on food security, while heavily subsidising water and energy requirements for food production.[40]

8.4.2   CHANGING DIETS

There is a vast range of literature that focuses on finding synergies between a shift towards healthier diets and environmentally friendly ones. While this might not always be the case, a number of parallel solutions for dietary shifts that would lower our impact on the environment such as: the consumption of seasonal products, seeking a balance between energy intake and expenditure and a lower consumption of products such as coffee, tea, cocoa and alcohol that usually come with a high environmental burden and are not necessary from a nutritional perspective. European countries are gradually taking action towards the Farm to Fork Strategy, embracing a Life Cycle Assessment (LCA) perspective to promote the sustainability of food production and consumption.[41]

Modern dietary guidelines need to reflect our best current nutrition guidelines, but they also need to reflect environmental considerations. Our food choices can make significant variation in the contribution to GHGs. The Food and Agriculture Organization (FAO) defines sustainable diets as nutritionally adequate, safe, healthy, culturally acceptable, economically affordable diets that have little environmental impact. Reductions in meat consumption and energy intake were identified as primary factors for reducing diet-related greenhouse gas emissions. High nutritional quality was not necessarily associated with affordability or lower environmental impact. Hence, when identifying sustainable diets, each dimension needs to be assessed by relevant indicators. Finally, some nonvegetarian self-selected diets consumed by a substantial fraction of the population showed good compatibility with the nutritional, environmental, affordability, and acceptability dimensions. Altogether, there is a scarcity of standardized nationally representative data for food prices and environmental indicators and suggest that diet sustainability might be increased without drastic dietary changes.[42]

Our food/dietary choices have major implications in our health, but they also cause major ramifications for the state of the environment. Food and delivery is responsible for more than 25% of GHG emissions.[43] At least 70% of agricultural GHGs are directly or indirectly associated with livestock production.[1] As a result, the choice of consuming large amounts of animal protein is a significant contributor to climate change. We are confronted with a worldwide obesity epidemic, which is associated with high consumption of red and processed meat and low consumption of fruits and vegetables. It is also an important consideration that, as societies become more affluent, the desire for meat production is likely to increase.

The emergence of food security as a key policy issue in developed nations is important in conjunction with the need to reduce greenhouse gas emissions and the implementation of Environmental Management Systems in primary industries. Biotechnological interventions such as biorefinery platforms that produce chemicals and fuels provide opportunities to improve the security and environmental sustainability criteria increasingly sought after by governments. Notably, biotechnology companies are beginning to use Environmental Management Systems employed by other industries to advocate the benefits of green technologies that employ GM, industrial enzymes and bio-materials. Management systems such as Life Cycle Analysis are providing a powerful means to measure benefits and augment change in the biotechnology sector. An important conclusion is that biotechnologies are likely to offer increasingly high impact options for sustainability and security criteria required for food and fuel supply.[44]

8.4.3 REDUCING FOOD LOSS AND WASTE

It has been estimated that throughout the global food chain approximately 30% of food produced for human consumption is lost or wasted.[45] The stages of the food system that experience most wastage can vary significantly when comparing developing and developed countries. In developing countries most of the food loss occurs in the field as a consequence of pests and pathogens and at post-harvest stages, as a consequence of poor infrastructure, technical limitations in harvesting techniques, storage and cooling technologies, packaging and lack of connection to markets.[46] Differently, in the developed world most of the waste occurs at the retail, food service and household level.

The impact of food waste and losses on the environment, in terms of the resources involved in the production, processing, transport and consumption stages have been highlighted, showing the impact of food wastage on climate change, biodiversity, water and land at the global level.[47] Food loss and waste reduction can make a major contribution to making food systems more sustainable.[48]

8.5 SUSTAINABLE SOLUTIONS

The solutions to the water, energy, and food crisis must be carefully integrated to meet the needs to produce resources for the 9 billion inhabitants in the next 40 years. This integration should be both centered around the FEW nexus and the four spheres. Without harmonious solutions the entire system will fail. Sustainable food production requires optimizing health, safety, quality, and consumer appeal of sufficient foods to sustain the growing world population. Agricultural food production has helped sustain this growth by increasing food production. If the world is to continue to meet the food needs of the anticipated population growth, this rate of increased production must continue for the next 25 years. In addition to increasing population, improved nutrition and health care are extending life expectancy, further increasing the need for more high-quality food. These increases must continue in the face of limited energy supplies, limited water supplies, and increased global warming, which results in lower production using current practices.

To meet the growing need for food production, we must accomplish the following goals:

- production of more food on reduced quality land using less water;
- reducing production energy required;
- producing sufficient water in the proper locations at a low cost;
- improving globalization of transport for energy and water resources for food;
- harmonized globalization of compliance regulations.

In conclusion, an integrated approach to dealing with population growth, environmental change, and resource management are critical to our long-term wellbeing. The FEW nexus is the intersection of food, energy, and water, three interdependent components that, together, are the lifeblood of the Earth. By 2025, there will be nine billion people on Earth. Unless we are able to thoroughly understand the connection between all three, global efforts to meet the needs of people on earth will fail. In this sense, the FEW nexus affects everybody, from government, to industry, to academia, to citizens across the globe.

The FEW nexus becomes the "tip of the spear" for the AC approach. The human population will continue to increase in part because of a sustainable nutritious food supply. Better nutrition and improved medicine will result in longer lifespans. Minimizing our negative impacts, such as GHG production, will help mitigate climate change. Modifications in our diet, such as consuming more plant-based protein, will contribute to better environmental stewardship from agriculture. These improvements will occur in the context of the FEW nexus. Understanding and facilitating changes in the nexus will be critical to sustainably accomplish these goals. Sustainability will be accomplished by multiple approaches and incremental steps. It is the role of the scientific community to develop effective technologies, communicate, advocate, educate, and lead the implementation of the new knowledge to the public and political leaders.

There is a tremendous opportunity for food studies to build technology that will deliver foods that will meet the needs for protein foods that will also appeal to the tastes of consumers. Food science also needs to continuously establish improved technology to reduce food waste from production to the dinner plate.[49] The nexus of food, energy, and water is and will continue to need major research and political and communications for the scientific community. We must find clear and consumer friendly communications to explain the utilization of modern technology in food production. Solutions to these issues must also include sustainably produced, safe, nutritious, satisfying, and wholesome foods.

REFERENCES

1. Springmann, M., et al., *Analysis and valuation of the health and climate change cobenefits of dietary change.* Proceedings of the National Academy of Sciences of the United States of America, 2016. **113**(15): pp. 4146–4151.

2. Rockström, J., et al., *A safe operating space for humanity.* Nature, 2009. **461**(7263): pp. 472–475.

3. Keairns, D.L., R.C. Darton, and A. Irabien, *The Energy-Water-Food Nexus.* Annual Review of Chemical and Biomolecular Engineering, 2016. **7**: pp. 239–262.

4. Lawford, R.G., *A design for a data and information service to address the knowledge needs of the Water-Energy-Food (W-E-F) Nexus and strategies to facilitate its implementation.* Frontiers in Environmental Science, 2019. **7**(APR): pp. 1–11.

5. Finley, J.W., *Evolution and future needs of food chemistry in a changing world.* Journal of Agricultural and Food Chemistry, 2020. **68**(46): pp. 12956–12971.

6. Liu, J., et al., *Systems integration for global sustainability.* Science, 2015. **347**(6225): p. 963.

7. Biros, C., C. Rossi, and A. Talbot, *Translating the International Panel on climate change reports: standardisation of terminology in synthesis reports from 1990 to 2014.* Perspectives: Studies in Translation Theory and Practice, 2021. **29**(2): pp. 231–244.

8. Kylili, A., et al., *Adoption of a holistic framework for innovative sustainable renewable energy development: a case study.* Energy Sources, Part A: Recovery, Utilization and Environmental Effects, 2021.

9. Voulvoulis, N., *Water and sanitation provision in a low carbon society: The need for a systems approach.* Journal of Renewable and Sustainable Energy, 2012. **4**(4).

10. Elshall, A.S., et al., *Groundwater sustainability: A review of the interactions between science and policy.* Environmental Research Letters, 2020. **15**(9).

11. Leck, H., et al., *Tracing the Water-Energy-Food Nexus: description, theory and practice.* Geography Compass, 2015. **9**(8): pp. 445–460.

12. Chen, H.G. and Y.H.P. Zhang, *New biorefineries and sustainable agriculture: Increased food, biofuels, and ecosystem security.* Renewable and Sustainable Energy Reviews, 2015. **47**: pp. 117–132.

13. Popp, J., et al., *The effect of bioenergy expansion: Food, energy, and environment.* Renewable and Sustainable Energy Reviews, 2014. **32**: pp. 559–578.

14. Anastasiadis, F., et al., *The role of traceability in end-to-end circular agri-food supply chains.* Industrial Marketing Management, 2022. **104**: pp. 196–211.

15. Powell, T.W.R. and T.M. Lenton, *Future carbon dioxide removal via biomass energy constrained by agricultural efficiency and dietary trends.* Energy and Environmental Science, 2012. **5**(8): pp. 8116–8133.

16. *4th International Conference on Materials Science and Information Technology, MSIT 2014*, in *4th International Conference on Materials Science and Information Technology, MSIT 2014.* 2014, Trans Tech Publications Ltd: Tianjin.

17. Rosegrant, M.W., C. Ringler, and T. Zhu, *Water for agriculture: Maintaining food security under growing scarcity*, in *Annual Review of Environment and Resources.* 2009. pp. 205–222.

18. Springer, N.P., et al., *Sustainable sourcing of global agricultural raw materials: Assessing gaps in key impact and vulnerability issues and indicators.* PLoS ONE, 2015. **10**(6), pp. 1–22.

19. Whitesides, G.M., *Reinventing chemistry.* Angewandte Chemie – International Edition, 2015. **54**(11): pp. 3196–3209.

20. Basu, S., et al., *Estimating US fossil fuel CO2 emissions from measurements of 14C in atmospheric CO2.* Proceedings of the National Academy of Sciences of the United States of America, 2020. **117**(24): pp. 13300–13307.

21. Desai, M. and R.P. Harvey, *Inventory of U.S. Greenhouse Gas Emissions and Sinks: 1990–2015.* Federal Register, 2017. **82**(30): pp. 10767.

22. Aravindaraj, K. and P. Rajan Chinna, *A systematic literature review of integration of industry 4.0 and warehouse management to achieve Sustainable Development Goals (SDGs).* Cleaner Logistics and Supply Chain, 2022. **5**: pp. 1–12.

23. Willett, W., et al., *Food in the Anthropocene: the EAT–Lancet Commission on healthy diets from sustainable food systems.* The Lancet, 2019. **393**(10170): pp. 447–492.

24. Mulholland, E., F. Rogan, and B.P. Ó Gallachóir, *From technology pathways to policy roadmaps to enabling measures – A multi-model approach.* Energy, 2017. **138**: pp. 1030–1041.

25. Mullins, E., et al., *Evaluation of existing guidelines for their adequacy for the food and feed risk assessment of genetically modified plants obtained through synthetic biology.* EFSA Journal, 2022. **20**(7): pp. 1–25.

26. Mann, A., *Food sovereignty and the politics of food scarcity*, in *Global Resource Scarcity: Catalyst for Conflict or Cooperation?* 2017. Taylor and Francis. pp. 131–145.

27. Mohanavelu, A., S. Shrivastava, and S.R. Naganna, *Streambed pollution: A comprehensive review of its sources, eco-hydro-geo-chemical impacts, assessment, and mitigation strategies.* Chemosphere, 2022. **300**: pp. 1–17.

28. Popko, R.R. *Zero liquid discharge project treats waste stream for potable use: Sustainable water management conference.* in *Sustainable Water Management Conference 2013.* 2013. Nashville, TN: American Water Works Association.

29. Su, X., et al., *Systematic approach to evaluating environmental and ecological technologies for wastewater treatment.* Chemosphere, 2019. **218**: pp. 778–792.

30. Dala-Paula, B.M., et al., *Cadmium, copper and lead levels in different cultivars of lettuce and soil from urban agriculture.* Environmental Pollution, 2018. **242**: pp. 383–389.

31. van Halem, D., et al., *Subsurface iron and arsenic removal for shallow tube well drinking water supply in rural Bangladesh.* Water Research, 2010. **44**(19): pp. 5761–5769.

32. Musa, W., J. Ahmad, and C.J. Lamangantjo, *Bioactive compounds in tombili seeds and tubile roots as the alternative for synthetic pesticide to protect wheats from insects and pests.* International Journal of ChemTech Research, 2016. **9**(4): pp. 604–615.

33. Usman, M., S. Cheng, and J.S. Cross, *Biomass feedstocks for liquid biofuels production in hawaii & tropical islands: A review.* International Journal of Renewable Energy Development, 2022. **11**(1): pp. 111–132.

34. Dumortier, J., M. Carriquiry, and A. Elobeid, *Where does all the biofuel go? Fuel efficiency gains and its effects on global agricultural production.* Energy Policy, 2021. **148**: pp. 1–11.

35. Mofijur, M., et al., *Selection of microalgae strains for sustainable production of aviation biofuel.* Bioresource Technology, 2022. **345**: pp. 1–8.

36. Lozano, R., et al., *Global and regional mortality from 235 causes of death for 20 age groups in 1990 and 2010: A systematic analysis for the Global Burden of Disease Study 2010.* The Lancet, 2012. **380**(9859): pp. 2095–2128.

37. Smith, K.R., et al., *Human health: Impacts, adaptation, and co-benefits,* in *Climate Change 2014 Impacts, Adaptation and Vulnerability: Part A: Global and Sectoral Aspects.* 2015. Cambridge: Cambridge University Press. pp. 709–754.

38. Fujimori, S., et al., *Land-based climate change mitigation measures can affect agricultural markets and food security.* Nature Food, 2022. **3**(2): pp. 110–121.

39. Mandrini, G., et al., *Exploring trade-offs between profit, yield, and the environmental footprint of potential nitrogen fertilizer regulations in the US Midwest.* Frontiers in Plant Science, 2022. **13**: pp. 1–16.

40. Lewis, A.C., C.P. Khedun, and R.A. Kaiser, *Assessing residential outdoor water conservation potential using landscape water budgets.* Journal of Water Resources Planning and Management, 2022. **148**(6).

41. Paris, J.M.G., et al., *Changing dietary patterns is necessary to improve the sustainability of Western diets from a One Health perspective.* Science of the Total Environment, 2022. **811**: pp. 1–17.

42. Perignon, M., et al., *Improving diet sustainability through evolution of food choices: Review of epidemiological studies on the environmental impact of diets.* Nutrition Reviews, 2017. **75**(1): pp. 2–17.

43. Qin, Y. and A. Horvath, *Contribution of food loss to greenhouse gas assessment of high-value agricultural produce: California production, U.S. consumption.* Environmental Research Letters, 2021. **16**(1): pp. 1–13.

44. Martindale, W., *Carbon, food and fuel security–will biotechnology solve this irreconcilable trinity?* Biotechnology and Genetic Engineering Reviews, 2010. **27**(1): pp. 115–134.

45. Palmisano, G.O., et al., *Food losses and waste in the context of sustainable food and nutrition security,* in *Food Security and Nutrition.* 2020. Elsevier. pp. 235–255.

46. Lopez Barrera, E. and T. Hertel, *Global food waste across the income spectrum: Implications for food prices, production and resource use.* Food Policy, 2020.
47. Basri, M.S.M., et al., *Progress in the valorization of fruit and vegetable wastes: Active packaging, biocomposites, by-products, and innovative technologies used for bioactive compound extraction.* Polymers, 2021. **13**(20).
48. Cattaneo, A., G. Federighi, and S. Vaz, *The environmental impact of reducing food loss and waste: A critical assessment.* Food Policy, 2021. **98**: pp. 1–16.
49. Finley, J.W. and J.N. Seiber, *The nexus of food, energy, and water.* Journal of Agricultural and Food Chemistry, 2014. **62**(27): pp. 6255–6262.

9 Sustainable intensive agriculture

Intensive agriculture is the most typical method of soil cultivation and the key source of food worldwide. (See Figure 9.1) It relies on reaping high yields with strong and often extreme land exploitation and often extreme inputs. The main benefits of intensive farming include sufficient food supplies at affordable prices.

The goal of sustainable agriculture is to meet society's food and textile needs in the present without compromising the ability of future generations to meet their own needs. Practitioners of sustainable agriculture seek to integrate three main objectives into their work: a healthy environment, economic profitability, and social and economic equity. Sustainability links social issues, in both the sciences and politics. Environmental, economic, and scientific dimensions are actively being pursued. The consideration of the social dimension in agriculture is still rather uncommon.[1] Global sustainability is increasingly understood as a prerequisite to attain and maintain human development at all scales, from local farming communities to cities, nations, and the world.[2,3]

A simple definition of sustainable agriculture suggested by UNESCO is "production practices and systems which are environmentally sound and economically and socially viable." Therefore, "corresponding agricultural cropping systems and management practices would aim at high product quality standards and have an intergenerational time horizon."[4] In this chapter we will merge this with intensive agriculture to form Sustainable Intensive Agriculture (SIA).

9.1 BACKGROUND

Given the prevalence of Intensive Agriculture (IA) and the current global situation, a paradigm shift towards Sustainable Intensive Agriculture (SIA) is needed. It must integrate the dual and interdependent goals of using sustainable practices to meet rising human needs while contributing to resilience and sustainability across scales. Both are required to sustain the future viability of agriculture. This paradigm shift could move agriculture from its current role as the world's single largest driver of global environmental change, to becoming a critical component of a transition to

FIGURE 9.1 Intensive agriculture. (See www.shutterstock.com/image-photo/valley-croa tia-intensive-agriculture-1161894373)

agricultural sustainability that can operate within the biophysical safe operating space on Earth.

This paradigm for sustainable intensification must be defined and translated into an operational framework for agricultural development. This must now be defined – at all scales – in the context of rapidly rising global environmental changes in the Anthropocene, while focusing on eradicating poverty and hunger and contributing to human wellbeing.[5] These pressures result from human actions on the environment, which are causing rising global environmental risks and for the first time constitute the largest driver of planetary change.

Agriculture is at the heart of this difficult situation. It is the world's single largest driver of global environmental change[6] and, at the same time, is most affected by these changes.[7] As productivity expands to meet the needs due to the population increase and shifting diets, many individuals remain hungry, while others suffer obesity, and significant amounts of food are wasted. This paradoxical situation is unsustainable and unethical. The problems of food affordability, food utilization, and food waste make this dilemma more challenging.[8]

9.1.1 PARADIGM

A global food revolution based on this new paradigm for sustainable intensive agricultural development is required. Such a change would enable the achievement of the twin objectives of feeding humanity and living within boundaries of biophysical processes that define the safe operating space of a stable and resilient Earth system.[9] These boundaries ultimately determine the limits of sustainability. This thinking provides a strong scientific justification for a shift from our current paradigm of agriculture focusing on productivity first and sustainability as a question of reducing environmental impacts, to a paradigm where sustainability constitutes the core strategy for agricultural development.

Defining SIA,[10] its evolution, and its role in addressing global food security is a necessary first step.[11] SIA is largely focused on how to enhance agricultural productivity while reducing its environmental impacts.[12] The major task remains how to produce more food while consuming fewer resources. Sustainable intensification seeks to increase agricultural output while keeping the ecological footprint as small as possible. This is a useful and important feature of sustainable agriculture, particularly as mainstream agriculture development still concentrates on productivity and places limited focus on sustainability. Its success will determine planetary survival. It remains focused on avenues for resource efficiency, based on assumptions that efficiency in water and fertilizer use represents the avenues towards sustainable agriculture. Simultaneously it will ensure that intensive production delivers the quantities of food needed.

Equally important, such a comprehensive sustainability paradigm, which not only minimizes environmental impacts but also uses sustainability as the strategy to raise productivity, improve livelihoods, and build resilience and Earth system stability, must meet the dramatic rise in food requirements from a world population of nearly 10 billion by 2050, which most likely will reach 11 billion by the end of the century.[13]

9.1.2 NECESSITY FOR A PARADIGM SHIFT

The volume and scale of global food and agricultural production puts this enterprise in a pivotal position. The future role of agriculture in terms of global sustainability requires a paradigm shift that aims at repositioning world agriculture. Currently, it is the world's single largest driver of global environmental change. It can become a key contributor of a global transition to a sustainable world that functions within a safe operating space on Earth.[14]

The case for intensification of agriculture has been well documented in the literature, starting from a need for increased production, through high-yielding crops, increased irrigation, mechanization, and the role of chemicals that increase production levels.[15] Now, from an environmental perspective, there are concerns for the huge amount of forests which could be converted into farm land, unquantifiable amount of ecosystem services saved, and of massive amounts of CO_2 prevented from being released into the atmosphere.[16] Intensification must be linked to sustainability to reduce consequences of the production increases.

The needs of society (lifestyle and culture) for agricultural products in order to maintain current lifestyles highlight the necessity of SIA. One of the most critical ones, fuels, is an illuminating example. The use of fossil fuels for transportation represents one of the largest anthropogenic contributions to greenhouse gases (GHGs), and the need for clean, renewable, alternative fuel sources has become a global priority. Biofuels production has grown exponentially over the past 30 years, as biofuels essentially contribute no additional GHG in their carbon life cycle and can be grown in almost any farming region in the world. However, the rapid expansion of the biofuel market has revealed weaknesses and areas of potential concern. While the studies and developments are on-going, there is significant variability in the GHG reduction capability of different biofuel production methods in relation to fossil fuel usage. It has even been reported that some of the fuels actually result in net increases of GHG emissions. The area is still maturing and there will continue to be new plants and species that will be possible sources of biofuels.[17] Judicious choice of biofuel sources will also be guided by responsible land and fertilizer use.[18] There will also be the conflict between food for consumption and food for products/processes.

To accomplish this, there must be a shift toward using sustainable principles as the guiding reference for generating productivity enhancements in the twenty-first century. This will fundamentally require research and development in increasing agricultural output by capitalizing on ecological processes in agroecosystems.[19] An agroecosystem is a subset of a conventional ecosystem. Its core is the human activity of agriculture, but is not restricted to the immediate site of agricultural activity (e.g. the farm), but rather includes the region that is impacted by this activity. This can be achieved by managing farmers' fields, watersheds, landscapes, and regions using strategies and practices that maintain biophysical stability and uphold critical feedbacks, such as moisture feedback from forests generating downwind rainfall[20] and carbon sinks in soils and biomass.[21]

Together, these challenges – the social dimension of meeting rapidly rising food requirements and the ecological dimension of building agricultural resilience and Earth system stability – form a social – ecological framework for sustainable intensification of world agriculture.[22]

9.1.3 CRITERIA

Recognizing the central role agriculture plays in determining and regulating Earth's resilience, and the sustainability criteria for agriculture, there is a strong case for adopting sustainable intensification of agriculture as the strategy to meet twin objectives for people and the planet. Are these two objectives mutually compatible? They are if the achievements are guided by principles of green and sustainable science. This chapter lays the foundation for this assertion.

Agriculture is the key to attaining the UN Sustainable Development Goals of eradicating hunger and securing sufficient food for a growing world population, which may require an increase in global food production of between 60 and 110% in a world of rising global environmental risks.[23] Agriculture is also the direct livelihood of 2.5 billion smallholder farmers, and the resilience of these livelihoods to rising shocks and stresses is currently gravely under-addressed.[24]

TABLE 9.1

Six Planetary Boundary Processes Affecting SIA[5]

Boundary Processes	Implication for SIA
Climate change	• 50–80% reduction of CO_2 emissions from energy use by 2050 (compared to 1990); • Transform agriculture from world's single largest carbon source to major carbon sink in soils
Land-use change	• Drastically reduce expansion of agricultural land
Global freshwater use	• 50 % increase in water productivity by 2030
Biosphere integrity	• Zero loss of biodiversity in agricultural landscapes • Adopt watershed and catchment management practices that build ecological landscape resilience
Nitrogen/phosphorous cycles	• Close nutrient loops; • No increase in overall P use, • Raise N and P use in developing countries; reduce in developed countries
Novel entities	• Minimize leakage of agricultural chemicals

Table 9.1 presents the major planetary boundary processes that must be addressed for SIA to successfully contribute to global sustainability.

The challenge to meet the criteria is made even more daunting by the need not only to produce more edible and non-edible food, but also to manage the entire food supply chain much more efficiently, reducing waste which has reached unacceptable proportions (estimated at 30%) along with promoting better distribution, access, and nutrition.[25]

The SIA challenge can be distilled into two coexistent yet independent goals. The criteria and approach for a paradigm shift towards sustainable intensification of agriculture can be summarized into the goals of using sustainable practices to meet rising human needs while contributing to resilience and sustainability of landscapes, the biosphere, and the ecosystem. Together, these social–ecological goals pose an unprecedented challenge for the global food system in the twenty-first century. As will be seen, the best path forward is to adopt a paradigm of sustainable intensive agriculture, with a dual purpose of (i) enabling a step-change in productivity and resilience and (ii) averting unacceptable global environmental risks.

9.1.4 DEFINITION OF SUSTAINABLE INTENSIVE AGRICULTURE (SIA)

The social–ecological goals laid out above allows a succinct definition of sustainable intensive agriculture (SIA): adopting practices along the entire value chain of the global food system that meet rising needs for nutritious and healthy food through practices that build social–ecological resilience and enhance natural capital within the safe operating space of the Earth system.[26]

The above definition allows the addition of a new dimension to sustainable agricultural development. Included in agricultural development must be a reliable methodology for managing natural capital to ensure long-term productivity and social–ecological resilience at field, watershed, and regional scales. The management will ensure that SIA will remain within the planetary boundaries outlined in Table 9.1. This approach will utilize existing research and promote further work. SIA will need to draw upon new greener science and technologies that will provide the means to achieve goal #1, while being guided by further land-use planning and management of natural capital in both agro-ecosystems and natural ecosystems across scales to achieve goal #2. This approach will incorporate a balance between the capacity to deal with shocks and stress with the need to preserve the Earth's ecosystem (in the Anthropocene).[4]

At its foundation, the new paradigm recognizes that the biophysical boundaries of the earth's ecosystem impose a hierarchy of criteria for the attainment and maintenance of sustainability: sustainability is not a relative concept or an act of balancing competing claims. There are absolute biophysical limits. It is only within such biophysically defined boundaries, as far as our current scientific knowledge shows, we stand a high probability of avoiding irreversible shifts in environmental conditions.[27]

Finally, only by defining development within such technically defined criteria or boundaries, can social and economic trade-offs can be assessed, and the earth achieve sustainability.[28]

9.2 INTENSIVE SUSTAINABLE PRACTICES FOR HUMAN NEEDS

Is it possible to identify current agricultural practices to transform them into SIA? The literature provides information on how to prioritize climate-smart and

TABLE 9.2
Co Designing SIA Practices[30]

Phase	Purpose
exploration of the initial situation	• identifies local stakeholders potentially affected by the process, existing farming systems, and specific constraints to the implementation of SIA
co-definition of an innovation platform	• defines the practice and how it favors the sustainability
shared diagnosis	• defines the main challenges needed to be solved before applying it
identification and assessment of incorporation challenges	• assess the potential performances of solutions toward SIA
Experimentation	• tests the prioritized solutions on-farm
assessment of the new SIA process	• validates the ability of the process to reach its initial objectives[31]
strategies for scaling up/out,	• addresses the scaling of the SIA process. For each phase, specific tools or methodologies are used: focus groups, social network analysis, theory of change, life-cycle assessment, and on-farm experiments

intensive agricultural options, but relatively few examples exist on how to re-design climate-smart farming systems with exiting practices. In Table 9.2 we present a process whereby this might happen. This will lead to where and how these may be identified today. Seven phases are suggested to illustrate a process of identifying practices for potential SIA and then taking steps for implementation at scale.[29]

9.2.1 SOURCES OF PRACTICES

The sources of sustainable practices range across all areas of agricultural development. Soil tillage systems, water resource management, crop and nutrient management, livestock practices, integrated landscape management, pest management, and management of ecosystem services are all sources of ready examples. Sustainable intensification agriculture will rely on the acceptance and adoption of many new practices and technologies. Factors such as the demographics of food production producers, land tenure, and labor and credit availability, as well as producers' concepts of what it means to be a "good farmer", should inform the conceptualization and development of new practices and technologies to increase the likelihood of their adoption.[32]

Starting with existing practices, it is possible to identify candidates for SIA. Recent efforts in defining SIA[6,10,33] provide an emerging framework built around the simple principle whereby yields are increased without adverse environmental impact and without the cultivation of more land.[34] These definitions are either not concrete enough or only partial. World agriculture must now also meet social needs and fulfill sustainability criteria that enable food and all other agricultural ecosystem services to be generated within a safe operating space of a stable and resilient Earth system, which in turn can be defined from Earth system science applying the planetary boundary framework (Table 9.1). This is a comprehensive definition of sustainable intensification of agriculture in the Anthropocene.

Appropriate land for food production, however, is a finite resource and hence further expansion could compromise development within Earth's safe operating space (approximately 25% of anthropogenic emissions of greenhouse gases are sequestered on land, of which all occurs in terrestrial non- cultivated ecosystems). [35] If business-as-usual prevails, the expected range of cropland expansion would overshoot the preliminary estimate of the 'safe operating space' well before 2050. [36] Technology innovation is the way to fulfill the food demand of the future ahead, even though climate change and agriculture practices can hamper the further deterioration of the farmlands.[37]

9.2.2 BIOPHYSICAL SAFE SPACE

Past societies have emerged and collapsed with numerous causes precipitating their breakdown. Leading explanations have often pointed to environmental/ecological, social and cultural, or economic factors, with cascading and synergistic effects between them. Issues occur with the exceeding of the carrying capacity by a mixture of overpopulation, resource depletion, diseases or famine, and environmental degradation through the loss of soil fertility, deforestation, and damage to ecosystem integrity. As a result, humanity's resilience capacity will diminish, and the collapse of modern civilization becomes a plausible reality. Valuable lessons that emerge from understanding the waxing and waning of past societies and how the factors involved in their downfall should temper both our carelessness and optimism regarding our civilization's future.[38] It has been proposed that the 'safe operating space' for food production frames the problem and describes the interconnected forces of population growth, consumption growth, environmental change, and food security.[39] The definition of a biophysical safe operating space of the Earth system, within which it has a high likelihood of remaining in a stable state, emerges from the advancements in Earth system science over the past decades, providing evidence of interactions, feedbacks, and thresholds among environmental processes that regulate the Earth system,[31] where the human activities constitute the largest driver of change on Earth.

Agriculture is a primary driver of global change since it is the single largest contributor to the rising environmental risks of the Anthropocene.[40] It is also in the Anthropocene that the challenge of feeding humanity needs to be resolved. The biophysical limits of human life on planet are Earth of paramount importance in the Anthropocene. The planetary boundaries (PBs) framework has emerged as a strong guardrail concept, even though it cannot steer the development of absolute sustainability assessments and realistic policies. PBs must be linked to human consumption as the main socio-economic driver and that planetary concerns can only be addressed through a holistic perspective that encompasses global tele-connections. Future SIA must be informed by PBs.[41]

9.2.3 TRANSFORMING THE ANTHROPOCENE

Human activity is leaving a pervasive and persistent signature on Earth. This new geologic time unit, the Anthropocene, has many characteristics. The ones most germane to the disciplines of agriculture and agricultural chemistry are concerns about dwindling fossil resources, modified carbon, nitrogen, and phosphorus cycles, species invasions worldwide and accelerating rates of extinction.[42] In response, a rapid world transformation to global sustainability is increasingly acknowledged as necessary to enable human development within a functioning and healthy environment.

In this new epoch, humanity faces the challenge of how to transform agriculture that feeds the world, into a force that contributes to eradicate poverty, and contributes to a stable planet. Given the decisive role of world agriculture on human development and on Earth system processes, sustainable intensive agriculture (SIA) is the only strategy that can deliver adequate food quantities with productivity enhancements to meet rising food needs and enable an Earth system operating within planetary boundaries.[5]

Intensive agriculture has been the strategy for meeting food production goals. [43] This must now become sustainable intensification in order to tackle the challenge of improving the health and livelihoods of the 2.5 billion smallholder farmers, which is the other half of the dual goals mentioned earlier. While the obvious benefits to applying SIA will appear in ecological dimension, there will be numerous subtle benefits as well. An improved nutrition status will reduce disease burdens, increase income, improve life expectancies, and provide a host of additional benefits to families and communities. These benefits are also essential drivers of sustainable development. This will ultimately reduce emissions of greenhouse gases and resource footprints.[44]

9.2.4 SUSTAINABLE EXAMPLES WITH INTENSIVE AGRICULTURE (IA)

Calling out and recognizing sustainability in IA activities is helpful. A few will be highlighted here. Commercializing biobased products and biofuels provides significant economic and environmental benefits. There has been substantial public and private research investment, which has significantly improved production efficiency. This is a mechanism by which the SIA will advance. Project financing is the largest barrier to commercialization. Uniting investment capital with technical experience could drive global commercialization. Cooperation among countries could lead to accelerated commercialization. As this happens, better management will be required to guide production of increased quantities. Biobased products have been growing in importance for about two decades. While biofuels have received significant public support including tax credits, blending mandates, and state incentives, biobased products do not have mandates or economic incentives.[45]

Worldwide sustainable management practices of land, water, and biodiversity in agro-ecosystems that increase productivity will be central to the success of SIA. A key part of the effort to long-term SIA requires safeguarding not only local (on-farm) productivity through sustainable practices, but also ecological functions across scales, from watershed, to basin, region, and Earth system scales, to avoid, loss of rainfall during future growing seasons. The major issues related to indiscriminate land use are related to topsoil depletion, groundwater contamination, plant disease outbreaks, air pollution and greenhouse gas emissions. Currently, global vision focused on the environmental impact and use of eco-friendly strategies are increasing. The design of new agroecosystems and food systems are fundamental to more sustainability in soil management systems by improving the release of advanced ecosystems services for farmers. Sustainable agriculture utilizes natural renewable resources in the best way due to their minimizing harmful impacts on the agroecosystems. Many good agricultural practices that growers may use to promote soil quality and soil health by minimizing water use and soil pollution on farms are available from past years.[46]

Furthermore, due to rising risks of water shocks at the local scale (e.g., droughts, floods, and dry spells) it is increasingly important to manage water. Spatial planning strategies are required to safeguard multi-functional landscapes, with a diverse set of ecosystems that can dampen the effects of storm-floods and maximize subsurface flows of water rather than erosive surface runoff. Wetlands, meandering rivers, forests, and landscape mosaics are important natural capital assets that build resilience. Moreover, watershed and river basin management is required to safeguard rainfall.[47] Agroecosystems are important for food production and conservation of biodiversity while continuing to provide several ecosystem services within the landscape. Approaches used for water supply need improvement. Climate smart agricultural practices can improve bio-geochemical interactions within landscape and decrease competition of natural resources between humans and other components of agroecosystems.

SIA must use strategic land-use planning to maintain and improve the supplies and flows involving water, nutrients, energy, carbon, and biodiversity across landscape of natural, semi-natural, and agricultural land uses. In some areas, increases in yield will be compatible with environmental improvements. In other areas of concern to agriculture, yield reductions or land reallocation will be needed to ensure sustainability and deliver benefits such as biodiversity conservation, carbon storage, flood protection, and recreation. An overall increase in production does not mean that yields should increase everywhere or at any cost: the challenge is context and location specific. As sustainable development is becoming more important to ensure the economic success and social well-being of any government, the efficient use and protection of natural resources has increased in importance. The variation in adoption of water sustainability programs in municipalities across the U.S. is hypothesized to rely on three key factors: environmental condition, form of government, and fiscal condition.[48] Use of any SIA activity will most likely be likewise influenced.

9.3 SUSTAINABLE PRACTICES FOR THE FOUR SPHERES

The path to achieving SIA is by realigning our total agricultural system. This requires nothing less than a planetary food revolution which, for the foreseeable future, will largely be driven by the 500 million small farms that comprise a major part of the system and which provide most of the food supply in some of the world's most social–ecologically vulnerable regions. Putting this in perspective, approximately 40% of the world's terrestrial surface has been transformed for use in agricultural areas (crop, fiber, biofuel, and livestock production systems).[49]

The prerequisites for accomplishing the dual goals of SIA will involve working with nature, managing resources, and building twenty-first-century productivity. (See Figure 9.2) Nature has provided many resources, albeit most are finite and in limited supply. It becomes the challenge for agriculture and agricultural chemistry to sustainably use and manage them in the twenty-first century.

9.3.1 WORKING WITH NATURE

Strategies towards agricultural intensification must also include ecological intensification as used in research, development, policy, and the industry,

FIGURE 9.2   Sustainability within the four spheres. (See www.shutterstock.com/image-vector/planet-earth-animals-people-black-silhouette-1656335275)

particularly with respect to the balance between agriculture and nature. Ecological intensification advocates landscape approaches that make smart use of the natural functionalities that ecosystems offer. The aim is to design multifunctional agroecosystems that are both sustained by nature and sustainable in their nature. [50] For example, land provides a range of critical services for humanity (including the provision of food, water and energy). It also provides many services that are often socially valuable but may not have a market value. It is therefore critical to explore how to optimize land use (and if necessary, limit demand), so societies can continue to benefit from all services into the future. Unlike the energy or the transport sectors, however, there is limited understanding or consensus over what 'optimal' land use might look like (from a science perspective), or how to bring it about (from a governance perspective).[51]

Working with nature will require ambitious and absolute targets for sustainability. [52] These may be difficult, but the benefits will far outweigh the cost. In short, the targets are,

- net zero emissions of greenhouse gases,
- very low or zero expansion of agriculture into remaining natural ecosystems, while restoring others providing vital ecosystem services,
- zero loss of biodiversity,

- drastic reduction in excessive use of nitrogen (N) and phosphorus (P) (recycling nutrient flows), and
- major improvement in water productivity and safeguarding of environmental water flows.

Nature has also demonstrated how to accomplish this through circularity (next chapter). The transition from a linear to a circular agriculture allows the measurement of how the degree of circularity of products and systems can promote sustainability. In a linear economy, raw materials are taken from nature and transformed into final products, which are subsequently used and become waste. On the contrary, a circular agriculture is an economic model that is restorative by intent and design.[53]

9.3.2 MANAGING SUSTAINABLE INCREASES

From a production perspective, production increases through SIA must include the following guidelines:

- efficient use of resources by combining locally relevant crop and animal genetic improvement (minimize inputs and close nutrient, carbon, and water cycles),
- employ practices that build landscape-scale resilience by sustaining ecosystem functions and services, such as water flows and biodiversity, and
- actively connect thinking, planning, and practice across scales to fully grasp field to biome and global interactions in the Anthropocene.

This must include improved and more equitable access to knowledge and resources, including land tenure, common property, and markets.[54]

The global agenda for sustainable development includes the alleviation of poverty and hunger by developing sustainable agriculture and food systems. Intensive farming systems and its variations, such as sustainable intensification or ecological intensification, are currently being promoted as technologies that can improve agricultural productivity and reduce environmental impacts. A new bottom-up paradigm is based on three indicators that are fundamental to achieve the environmental, economic, and social sustainability of agriculture and food systems. These are divided into technical, geographic, and social indicators.[55]

Building on previous work[11,50,56] a paradigm shift towards SIA translates to some key operational strategies (Table 9.3).

9.3.3 TWENTY-FIRST CENTURY PRODUCTION

During the twenty-first century, SIA will benefit tremendously from agricultural chemistry and improved agricultural practices. SIA ultimately aims to achieve hunger reduction through biodiversity conservation that secures ecological

TABLE 9.3
SIA Operational Strategies

Strategy	Effect
Plan and implement farm-level practices	Maximize farm productivity
Integrate ecosystem-based strategies with practical farm practices.	Productive and resilient farming
Develop system-based farming practices	Integrate land, water, nutrient, and crop management.
Utilize crop varieties with a high ratio of productivity to use of externally and internally derived inputs.	Enhance productivity
Adopt circular approaches to managing natural resources	Greater efficiency and sustainability
Harness agroecological processes	Promote nutrient cycling, biological nitrogen fixation, and allelopathy.
Assist farmers in overcoming immediate SIA adoption barriers	Making the ecological approach profitable.
Build robust institutions of small farmers, led especially by women	Enable an equitable interface with both markets and government.

functions in agricultural landscapes. It will require well-informed regional and targeted solutions[57] drawing upon the strengths of both ecosystem preservation and ecological stewardship[58] across local, regional, and basin scales. This activity is compatible with optimization methods that will allocate the use of land most efficiently, while sustaining the agroecological systems. To ensure this is possible, heterogeneity, resilience, and ecological interactions between farmed and unfarmed areas must be encouraged. This will become the hallmark of twenty-first-century agriculture.

9.4   THE WAY FORWARD

A transformation to SIA is justified both by necessity (to promote global sustainability, which is necessary for long-term agricultural viability) and by opportunity (to use sustainable practices to bring the world away from ecological tipping points). (See Figure 9.3)

Adopting a paradigm for sustainable intensification within planetary boundaries is a major challenge for research and development that will require new approaches in how and why research for development is formulated, managed, and executed. Pursuing SIA will require approaches that integrate social and natural sciences, in solution-oriented knowledge generation that couples academic and practical knowledge through co-design and co-development of research. The implementation of SIA must include the political economy in which food is traded, how prices are determined, and the business economy along the value chain from field to consumer.

FIGURE 9.3 SIA requires a balance with nature. (See www.shutterstock.com/image-photo/giant-balancing-boulder-rock-called-kummakivi-2018661269)

A lasting paradigm shift will require the ability to place research into policy and enable large-scale change. Political systems and institutional trust will need to work with the many stake-holders in the food system, all of whom will be required to make compromises. While SIA needs to be central to the way we produce food in the future, it also needs to be integrated within a nexus of strategies aimed at achieving food system sustainability, in the broadest sense of the phrase.[6]

9.4.1 Transformation, co-design, and learning

SIA is a new, evolving concept, and its meaning and objectives subject to debate. Sustainable intensification is only part of what is needed to improve food system viability and sustainability. Both agricultural sustainability and food security have multiple social, ethical, and environmental dimensions. Achieving a sustainable, health-enhancing food system for all will require more than just changes in agricultural production. Equally radical agendas will need to be pursued to reduce resource-intensive consumption and waste and to improve all forms of equity.

The transformation must include a mechanism for management for sustainable crop production intensification. This will provide a vision of sustainable food and agriculture that allows access by all to nutritious food with ecosystem-focused natural resources management.[59] The shift demands a new framework for research and development. Major productivity enhancements are required, and the major efforts will come through sustainable intensification of agricultural practices for livelihoods that build farm, community, and biosphere resilience. New research and development will inspire and influence all domains involved in agricultural development.

The promise and hope that SIA, in conjunction with the developments of green chemistry and agricultural chemistry, can succeed is big. Knowledge from different domains, ranging from agricultural chemistry to natural agriculture and ecology to business development must all work together. To be successful, sustainable intensive agriculture must deliver food, livelihoods, and resilience, while contributing to development within Earth's safe operating space. Investing in this would necessitate large R&D. These are not the routine approaches to agricultural development. They will be system-integrating and innovative ventures, and urgently required to meet global sustainability needs.

REFERENCES

1. Janker, J., S. Mann, and S. Rist, *Social sustainability in agriculture – A system-based framework.* Journal of Rural Studies, 2019. **65**: pp. 32–42.
2. Carolan, M., *The sociology of food and agriculture.* The Sociology of Food and Agriculture. 2021: Taylor and Francis. 1–278.
3. Folke, C., et al., *Adaptive governance of social-ecological systems,* in *Annual Review of Environment and Resources.* 2005. pp. 441–473.
4. Hashemi, S.Z., et al., *Assessing agro-environmental sustainability of intensive agricultural systems.* Science of the Total Environment, 2022. **831**: pp. 1–11.
5. Rockström, J., et al., *Sustainable intensification of agriculture for human prosperity and global sustainability.* Ambio, 2017. **46**(1): pp. 4–17.
6. Charles, H., H. Godfray, and T. Garnett, *Food security and sustainable intensification.* Philosophical Transactions of the Royal Society B: Biological Sciences, 2014. **369**(1639): pp. 1–10.
7. Barros, V.R., et al., *Climate change 2014 impacts, adaptation, and vulnerability Part B: Regional aspects: Working group ii contribution to the fifth assessment report of the intergovernmental panel on climate change.* Climate Change 2014: Impacts, Adaptation and Vulnerability: Part B: Regional Aspects: Working Group II Contribution to the Fifth Assessment Report of the Intergovernmental Panel on Climate Change. 2014. Cambridge: Cambridge University Press. 1–1820.
8. Russel, N., *Economics of feeding the hungry: Sustainable intensification and sustainable food security.* Economics of Feeding the Hungry: Sustainable Intensification and Sustainable Food Security. 2017. Taylor and Francis. 1–117.
9. Steffen, W., et al., *Trajectories of the Earth System in the Anthropocene.* Proceedings of the National Academy of Sciences of the United States of America, 2018. **115**(33): pp. 8252–8259.
10. Garnett, T., et al., *Sustainable intensification in agriculture: Premises and policies.* Science, 2013. **341**(6141): pp. 33–34.
11. van Noordwijk, M. and L. Brussaard, *Minimizing the ecological footprint of food: Closing yield and efficiency gaps simultaneously?* Current Opinion in Environmental Sustainability, 2014. **8**: pp. 62–70.
12. Kuyper, T.W. and P.C. Struik, *Epilogue: global food security, rhetoric, and the sustainable intensification debate.* Current Opinion in Environmental Sustainability, 2014. **8**: pp. 71–79.

13. Gerland, P., et al., *World population stabilization unlikely this century*. Science, 2014. **346**(6206): pp. 234–237.
14. Giller, K.E., et al., *The future of farming: Who will produce our food?* Food Security, 2021. **13**(5): pp. 1073–1099.
15. Mazoyer, M., L. Roudart, and I.A. Mayaki, *Report on world development, 2008, World Bank. Agriculture for development. Summary and comments*. Mondes en Developpement, 2008. **36**(3): pp. 117–136.
16. Burney, J.A., S.J. Davis, and D.B. Lobell, *Greenhouse gas mitigation by agricultural intensification*. Proceedings of the National Academy of Sciences of the United States of America, 2010. **107**(26): pp. 12052–12057.
17. Bessou, C., et al., *Biofuels, greenhouse gases and climate change. A review.* Agronomy for Sustainable Development, 2011. **31**(1): pp. 1–79.
18. Kerdan, I.G., et al., *Modelling future agricultural mechanisation of major crops in china: An assessment of energy demand, land use and emissions*. Energies, 2021. **13**(24): pp. 1–31.
19. Struik, P.C., et al., *Deconstructing and unpacking scientific controversies in intensification and sustainability: Why the tensions in concepts and values?* Current Opinion in Environmental Sustainability, 2014. **8**: pp. 80–88.
20. Gordon, L.J., G.D. Peterson, and E.M. Bennett, *Agricultural modifications of hydrological flows create ecological surprises*. Trends in Ecology and Evolution, 2008. **23**(4): pp. 211–219.
21. Le Quéré, C., et al., *Global Carbon Budget 2015*. Earth System Science Data, 2015. **7**(2): pp. 349–396.
22. Jackson, L.E., et al., *Social-ecological and regional adaptation of agrobiodiversity management across a global set of research regions*. Global Environmental Change, 2012. **22**(3): pp. 623–639.
23. Pardey, P.G., et al., *A bounds analysis of world food futures: global agriculture through to 2050*. Australian Journal of Agricultural and Resource Economics, 2014. **58**(4): pp. 571–589.
24. Carr, E.R., *Resilient livelihoods in an era of global transformation*. Global Environmental Change, 2020. **64**: pp. 1–8.
25. Edenhofer, O., et al., *Renewable energy sources and climate change mitigation: Special report of the intergovernmental panel on climate change*. Renewable Energy Sources and Climate Change Mitigation: Special Report of the Intergovernmental Panel on Climate Change. 2011. Cambridge: Cambridge University Press. 1–1075.
26. Davis, K.F., S. Downs, and J.A. Gephart, *Towards food supply chain resilience to environmental shocks*. Nature Food, 2021. **2**(1): pp. 54–65.
27. McKenzie, F.C. and J. Williams, *Sustainable food production: constraints, challenges and choices by 2050*. Food Security, 2015. **7**(2): pp. 221–233.
28. Meacham, M., et al., *Advancing research on ecosystem service bundles for comparative assessments and synthesis*. Ecosystems and People, 2022. **18**(1): pp. 99–111.
29. Andrieu, N., et al., *Co-designing climate-smart farming systems with local stakeholders: a methodological framework for achieving large-scale change*. Frontiers in Sustainable Food Systems, 2019. **3**: pp. 1–19.
30. Waddock, S., *Stewardship of the future: Large system change and company stewardship*, in *Corporate Stewardship: Achieving Sustainable Effectiveness*. 2017, Taylor and Francis. pp. 36–54.

31. Lenton, T.M., et al., *Tipping elements in the Earth's climate system*. Proceedings of the National Academy of Sciences of the United States of America, 2008. **105**(6): pp. 1786–1793.

32. Campbell, A. and A.E.H. King, *Choosing Sustainability: Decision Making and Sustainable Practice Adoption with Examples from U.S. Great Plains Cattle Grazing Systems*. Animals, 2022. **12**(3), pp. 1–13.

33. Rockström, J. and L. Karlberg, *The quadruple squeeze: Defining the safe operating space for freshwater use to achieve a triply green revolution in the anthropocene*. Ambio, 2010. **39**(3): pp. 257–265.

34. Kröbel, R., et al., *Making farming more sustainable by helping farmers to decide rather than telling them what to do*. Environmental Research Letters, 2021. **16**(5), pp. 1–15.

35. Luisetti, T., et al., *Climate action requires new accounting guidance and governance frameworks to manage carbon in shelf seas*. Nature Communications, 2020. **11**(1): pp. 1–10.

36. Creutzig, F., et al., *Assessing human and environmental pressures of global land-use change 2000–2010*. Global Sustainability, 2019. **2**, e1, pp. 1–17.

37. Yaqoob, N., et al., *The effects of agriculture productivity, land intensification, on sustainable economic growth: A panel analysis from Bangladesh, India, and Pakistan Economies*. Environmental Science and Pollution Research, 2022.

38. Abegão, J.L.R., *The limits of sustainability: lessons from past societal collapse and transformation, for a civilization currently defying humanity's safe operating space*, in *World Sustainability Series*. 2022. Springer Science and Business Media Deutschland GmbH. pp. 439–454.

39. Rockström, J., et al., *A safe operating space for humanity*. Nature, 2009. **461**(7263): pp. 472–475.

40. Henderson-Sellers, A. and K. McGuffie, *The future of the world's climate*. The Future of the World's Climate. 2012. Elsevier.

41. Li, M., et al., *The role of planetary boundaries in assessing absolute environmental sustainability across scales*. Environment International, 2021. **152**: pp. 1–13.

42. Waters, C.N., et al., *The Anthropocene is functionally and stratigraphically distinct from the Holocene*. Science, 2016. **351**(6269), p. 137.

43. Gaffney, J., et al., *Science-based intensive agriculture: Sustainability, food security, and the role of technology*. Global Food Security, 2019. **23**: pp. 236–244.

44. Tilman, D. and M. Clark, *Global diets link environmental sustainability and human health*. Nature, 2014. **515**(7528): pp. 518–522.

45. Snyder, S.W., *An introduction to commercializing biobased products: Opportunities, challenges, benefits, and risks*, in *RSC Green Chemistry*, S.W. Snyder, Editor. 2016, Royal Society of Chemistry. pp. 1–7.

46. de Corato, U., *Towards new soil management strategies for improving soil quality and ecosystem services in sustainable agriculture: Editorial overview*. Sustainability (Switzerland), 2020. **12**(22): pp. 1–5.

47. Kimaro, J., *A review on managing agroecosystems for improved water use efficiency in the face of changing climate in Tanzania*. Advances in Meteorology, 2019. **2019**: pp. 1–12.

48. Kwon, S.W. and D.B. Bailey, *Examining the variation in local water sustainability practices*. Social Science Journal, 2019. **56**(1): pp. 107–117.

49. Ramankutty, N., et al., *Farming the planet: 1. Geographic distribution of global agricultural lands in the year 2000.* Global Biogeochemical Cycles, 2008. **22**(1), pp. 1–19.

50. Tittonell, P., *Ecological intensification of agriculture-sustainable by nature.* Current Opinion in Environmental Sustainability, 2014. **8**: pp. 53–61.

51. Benton, T.G., et al., *Designing sustainable landuse in a 1.5 °C world: the complexities of projecting multiple ecosystem services from land.* Current Opinion in Environmental Sustainability, 2018. **31**: pp. 88–95.

52. Newell, P., M. Twena, and F. Daley, *Scaling behaviour change for a 1.5-degree world: Challenges and opportunities.* Global Sustainability, 2021. **4**: pp. 1–13.

53. Rocchi, L., et al., *Measuring circularity: an application of modified Material Circularity Indicator to agricultural systems.* Agricultural and Food Economics, 2021. **9**(1).

54. Ait Issad, H., R. Aoudjit, and J.J.P.C. Rodrigues, *A comprehensive review of Data Mining techniques in smart agriculture.* Engineering in Agriculture, Environment and Food, 2019. **12**(4): pp. 511–525.

55. Sandhu, H., *Bottom-up transformation of agriculture and food systems.* Sustainability (Switzerland), 2021. **13**(4): pp. 1–13.

56. Pretty, J., *Agri-culture: Reconnecting people, land and nature.* Agri-Culture: Reconnecting People, Land and Nature. 2013. Taylor and Francis Inc. 1–261.

57. Tscharntke, T., et al., *Global food security, biodiversity conservation and the future of agricultural intensification.* Biological Conservation, 2012. **151**(1): pp. 53–59.

58. Law, E.A., et al., *Better land-use allocation outperforms land sparing and land sharing approaches to conservation in Central Kalimantan, Indonesia.* Biological Conservation, 2015. **186**: pp. 276–286.

59. Kristensen, D.K., C. Kjeldsen, and M.H. Thorsøe, *Enabling sustainable agro-food futures: exploring fault lines and synergies between the integrated territorial paradigm, rural eco-economy and circular economy.* Journal of Agricultural and Environmental Ethics, 2016. **29**(5): pp. 749–765.

10 Circularity
Environmental, chemical, agricultural

The primary intent of the circular economy (CE) is to close material loops, leading to enhanced environmental performance in business and scientific endeavors.[2] The circular economy is gaining increasing interest as a potential pathway towards global sustainability. Instead of a linear and take-use-disposal pattern, resources are used, recovered and renewed in internal cycles as long as possible, creating further maximum value (see Figure 10.1).[1] Natural systems can be used as models to provide better insight to benefits of circularity. The circular economy[3], can be described as: "an industrial system that is restorative or regenerative by intention and design. It replaces the end-of-life concept with restoration, shifts towards the use of renewable energy, eliminates the use of toxic chemicals, which impair reuse and return to the biosphere, and aims for the elimination of waste through the superior design of materials, products, systems and business models".[1]

10.1 THE NATURE OF CIRCULARITY

In the current context of resource scarcity, global climate change, environmental degradation, and increasing food demand, the circular economy represents a promising strategy for supporting sustainable, restorative, and regenerative agriculture. As we will see in this chapter, the value of CE, when applied to agriculture, is immense. In addition, the principles of CE will be shown to be adapted to the practice of agriculture, and CE strategies for agricultural activity can be defined.[4]

The benefits of CE can be found in its broad application and development in many fields, such as chemistry and agricultural chemistry. The intended outcomes are not always achieved, but the structure provides a tangible path to sustainability within those fields. CE has been one of the most transformational tendencies for the past years. It is now appearing as a global trend, affecting macro, meso and microenvironments, ranging from governments, global organizations (such as the UN), the whole private sector, science, to final consumers and individuals. CE should provide fields, organizations and science the tools to change their actions and business model.[5]

DOI: 10.1201/9781003157991-10

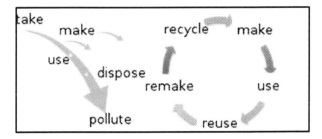

FIGURE 10.1   Linear economy (left) and circular economy (right).[1]

The principles behind the CE are receiving increasing global attention and are being increasingly realized as a methodology to reach a sustainable balance in our world.[6] If its implementation is to improve environmental performance of society, the actions must result from scientific evidence and further contribute to the value of CE. The societal needs and functions offer a promising meso-level link to bridge the micro and macro-levels for assessment, monitoring and setting thresholds in CE.[7]

10.1.1   Circular economy

The basic ideas and reasoning behind economic circularity are not novel. They are the practical extension of closed-looping. When applied to more traditional business and scientific activities, it can become an essential model for sustainability and will result in fundamental changes. It provides an alternative to the current production and consumption model that has guided industrial development for over a century. The importance of the topic is confirmed worldwide. The challenge will be to develop innovative technologies and tools for the efficient use of resources which will lead to sustainable production.[8] CE requires the effective and efficient utilization of resources from the ecosystem, which results in environmental and economic performance optimization. The development of CE initiatives, incorporating the fields of environmental science and chemistry into agriculture and agricultural chemistry will be instrumental in the achievement of agricultural sustainability in the twenty-first century.[9]

Circular economy and the circularity models are necessary for sustainable development in the twenty-first century. These models promote the minimization of emissions, reliance on renewable energy, and elimination of waste based on the utilization of closed-loop systems and the reuse of materials and resources. The incorporation of circular economy practices in resource-consuming agricultural systems is a vital step toward correcting the environmental issues of the currently linear systems.[10] The bioeconomy (renewable resources facet of circular economy) promotes the production of renewable biological resources (i.e., biomass) that transform into nutrients, bio-based products, and bioenergy. These become the new resources for the agriculture/food production industry. The use of recycled agro-industrial wastewater in agricultural activities (e.g., irrigation) can further foster the circularity of the bio-based systems. Agricultural chemistry will facilitate this incorporation by providing research, analyses, and guidance in the following areas:

- minimizing the use of natural resources (e.g., water, energy),
- decreasing the use of chemical fertilizers,
- utilizing bio-based materials (e.g., agricultural/livestock residues), and
- reusing wastewater from agrifood operations.

This work will result in the increased understanding of how better to cooperate with nature and utilize any direct or indirect interactions within the water-energy-nutrients nexus. This use of the circular economy in agriculture can become basis for developing circular bio-based business models, while ensuring sustainability within the agri-food industry.

10.1.2 Transitioning to a circular economy

The transition to a circular paradigm and circular agriculture will require several steps. The first step is the realization and acceptance the current system is inefficient, and it will not achieve (in its current form) desired sustainability. This can be highlighted by quantifying the physical losses and show them as profits and losses in the Agri-Food chain. A summary of physical losses in an Agri-Food chain context should help governments and business leaders to make decisions regarding the reduction, elimination or transformation of physical losses to achieve a more sustainable business strategy.[11]

Awareness and acknowledgment of the issues with the current agri-food system model, should lead to examining the potential benefit of food waste prevention that should be addressed in the context of the emerging circular economy.[12] The topics we raised in earlier chapters now reappear with potential solutions. The two broad areas are food waste and food productivity. The germane topics related to food waste are in food surplus, waste, and losses. These areas are crucial for the sustainability of future food waste management systems, especially in the context of circular economy. Six distinct categories of food waste can be developed that get to the heart of the issue. The six categories are:

- edible,
- naturally inedible (pits),
- industrial residue,
- inedible due to natural causes (pests),
- inedible due to ineffective management and
- not accounted for.

Based on this, surplus food and new ways for material recycling can be identified, in order to highlight the importance of the future food waste biorefineries in the circular bioeconomy. Enhanced, sustainable food production is important. These two topics now set a path for moving toward agricultural circularity.

10.1.3 CIRCULARITY GOALS

A strategy needed to achieve the adoption of the circular agriculture paradigm is provided by reviewing the development of the EU-bioeconomy, 2014–2020. [13] This provides a vision of the emerging Circular Bio-based Economy (CBE). It is necessary to visualize the full potential of all types of sustainable biomass sources, crop residues, industrial side-streams, and wastes. New value is realized by transforming them into value-added products. The potential products from this effort consist of a wide spectrum of value-added products, addressing societal and consumer needs:

- Food and feed,
- bio-based chemicals,
- materials,
- health-promoting products, and
- bio-based fuels.

The functional components of CBE include

- biotechnology,
- microbial production,
- enzyme technology,
- green chemistry,
- integrated physical/chemical processing,
- policies, conducive framework conditions, and
- public private partnerships.

Key operations of CBE include

- biomass supply,
- biorefineries,
- value chain clusters,
- rural development,
- farmers,
- foresters and mariners.

Environmental sustainability is critical in all of these products, components, and operations. However, there still is an urgent need for climate change mitigation, adaptation, and stopping biodiversity loss. Improved conditions can allow expansion into biomass feedstocks, and a broadening of bio-based product portfolio. The products will be guided by including higher-value, health-promoting products for humans, animals, plants, and soil. This effort also showed that industrial needs and academic research involvement are important. Impact assessment is included in this effort, as a way to achieve continual improvement. The key quality of this circularity goal is that biological resources are found worldwide, and that a circular bio-based economy can be implemented globally, while stimulating rural development.

10.2 ENVIRONMENTAL CIRCULARITY – A NECESSARY FIRST STEP

One of the necessary steps to accomplish the temperature goal of the Paris Agreement, which is key for agricultural sustainability, is the utilization of environmental circularity. It is important to be realistic and optimistic about how much CE strategies can contribute to climate change mitigation. Interestingly, six sectors (industry, waste, energy, buildings, transport, and agriculture) all potentially can contribute to this goal. These sectors are not mutually exclusive, and all directly/indirectly impact agriculture. Few studies contribute insights on how the CE can support mitigation, so that more work needs to be done in identifying actual tools and processes. It has been shown in previous literature, that the lowest gains in mitigation are to be expected in agriculture.[14] This may change as sustainable agriculture matures.

10.2.1 INTRODUCTION

The shift away from fossil fuels is needed to reduce CO_2 emissions. This requires the use of renewable carbon and energy sources, including biomass in the bioeconomy. This becomes the main entry point for agriculture and agricultural chemistry to impact climate change in the environment. The bioeconomy will have an increasingly significant role in the global economy with the growth of traditionally bio-based sectors. In the future, the energy, transportation, and chemical sectors will increasingly play a role in the bioeconomy, especially as agriculture produces biomass as an industrial feedstock. Looking at the total demand and considering the sustainability limits of biomass production leads to the conclusion that the expected demand for all industries that could process biomass exceeds the sustainably available capacity. This situation should be addressed in two distinct ways: Use the skills in the organic chemical sector within the bioeconomy and increase recycling of wastes and residues including CO_2 to enhance the circular bioeconomy.[15]

The CE paradigm will contribute to environmental sustainability. Understanding the circular economy as resilient and sustainable, principles can guide the operations within the circular economy under the environmental sustainability paradigm. There is a high variability of approaches performed by the governments to reach the circular economy. Moreover, some circular economy strategies reveal shortcomings in their capacity to generate real changes. Further research is needed to connect more essential elements to enhance a socially fair development that preserves the environment.[16]

10.2.2 LINKING ENVIRONMENTAL AND AGRICULTURAL CIRCULARITY

The global challenge of feeding an ever-increasing world population in a sustainable manner provides the goal for agricultural circularity. This effort is directing scientists' attention towards nutritious and sustainable foods whose production have low impacts on environment while supporting both the economy and society. One way this may be accomplished is by addressing food wastes in all their forms and manifestations. This effort becomes a way to address environmental circularity. Nutrients are found in wastes, and nutrient circularity is an important circular economy practice. Since nutrients, indeed all food, result from the use of natural resources, the circularization of nutrients will directly be an example of the circularization of natural resources. Further, the valorization of food waste as precious biomass to grow new food and feed contribute to the goal of feeding our world. These foods can become part of people diets in many areas of the world. The food waste biorefinery viability becomes a concrete response by agricultural chemistry on the issue of food sustainability.[17]

Agriculture has damaged the environmental landscape through human actions. The agro-food system has grown through intensified use of synthetic fertilizers, territorial specialization, and integration in global food and feed markets. This rapid expansion has led to increased nitrogen (N) losses to aquatic environments and the atmosphere, which, despite increasing environmental regulations, continues to harm ecosystems and human well-being. Agricultural chemistry can reduce environmental impacts and demonstrate the possibility to feed the projected global population by cutting in half the current level of environmental N losses through the CE. This activity links agricultural circularity to environmental circularity to create a more sustainable system.[18]

10.2.3 AREAS OF ENVIRONMENTAL CIRCULARITY APPLICATIONS

Several examples can be used to illustrate the importance of environmental circularity that lead to agricultural circularity. Most water usage follows a linear model, which increasingly causes cumulative emissions of pollutants, waste stocks, and causes the irreversible deterioration of water and other resources. Moving towards a circular model in the water sector will require a change in the configuration of use of water. One way is through converting to nature-managed water systems. A holistic and practical methodology is needed that integrates existing tools, methods, and incorporating existing and/or newly-developed indicators.[19] A clear example occurs when cities face increasing water pressure and supply issues, jeopardizing the balance between growth and sustainable water resource use. Green, resilient, smart, circular, blue, water sensitive, or water-wise city concepts are increasingly part of the design of strategies to redesign how cities use water. The question is how they consider water circularity and employ water circularity principles. Water circularity-related challenges are still a small niche of concern, but will grow in importance as the population increases. Water circularity principles will need to be considered a critical component in future city design.[20]

A second example is from the construction industry, and how the manufacture of its materials affects the environment. The production of construction materials consumes significant quantities of raw materials and utilizes a significant amount of energy. Construction materials do contribute to sustainable building management. However, their environmental impact has to be measured based on raw materials' input, the production process followed, and the energy sources used in the future are needed ways that increase use of the recycling material used as an input material to the production process, improve material circularity indicator. This increased material circularity indicator means more closed loops, less linear flows, and a decrease of the raw materials used. Circular economy can be used to minimize waste, reduce environmental impact, and more efficiently use resources towards sustainability and economic efficiency in the built space and construction industries.[21]

The last example is one of measuring circularity. Measuring the circularity of resources is essential to assessing the performance of a circular economy. Indicators are needed that will provide the metrics to assess circularity.[22] In a circular economy, the material and product's value maintenance implies providing a unique way for their multiple transformations in an economy. Preserving the value of a material and product in the economy for as long as possible is the result of quality and completeness of the process steps. A metric, or combination of metrics, should measure the retention of value of the products within the entire closed loop.[23] This metric will be useful in gaging the utility of programs and tools that aim to accomplish environmental circularity.

10.3   CHEMICAL CIRCULARITY

Chemistry is the science that manipulates and understands materials. The earth is running out of traditional resources needed for manufacturing materials such as chemicals, minerals, and petroleum. Thus, these components are available only at increasing economic and environmental costs. As an important contribution to a sustainable future, chemistry and its products must be adapted to a circular economy (CE)—a system aimed at eliminating waste, circulating and recycling products, and saving resources and the environment. In doing this, the focus becomes chemical circularity (CC). (See Figure 10.2)

By beginning with waste within the agricultural and chemical sectors, including industries that produce large volumes of chemical or metal-rich wastes (such as agri-food, textiles, electronics, plastics, metals, and alloys), we gain a better understanding of chemistry and its links to resources with sinks. Reflecting on the systemic role of chemistry will not contribute much to sustainability unless the

FIGURE 10.2   Chemical circularity. (See www.shutterstock.com/image-photo/concepts-chemistry-researcher-working-laboratory-430172755)

tools and knowledge chemistry provides are grounded in circularity. The chemistry necessary for a CE will come to fruition only through a new attitude toward chemistry education, chemical research and engineering, and product design.[24] As an important contributor to a sustainable future, chemistry and its products must be adapted to a circular economy.[25] Nearly 140,000 industrial chemicals are marketed worldwide, and new chemicals are becoming more complex (e.g., stereochemistry, functional groups) (2). Products of the chemical and allied industries contain mixtures of elements and molecules, and these products and their constituents are found everywhere, including in waste products, soils, water, air, plants, food, animals, and the human body.[26]

Initially, chemistry will contribute to the goal of circularity through the inclusion of green/sustainable chemistry principles. These principles will drive the changes in the resource-dependent chemical industry. In the present state of environmental challenges, changes in the chemical industry using sustainable product stewardship are critical. The drivers that enable the circularity and the methodological approach/tools that help to practice circularity in the chemical industry are useful and broadly applicable. The CC helps to understand and address the sustainability issues of the chemical processes/products.[27]

Within a CE, products should be used as long as possible until the end of their lives. Chemistry can help modern circularity thinking to include the design of products with adapted lifetime, reusability, ease of repair, and recycling ability – all made with renewable resources.[28] However, greater success will have to come from changes at the product-design level, led by scientists who strive to decipher, at the atomic and molecular levels, how chemical products and their underpinning synthetic chemistry fit into a CE. With today's diverse and interconnected chemical, material, and product flows, manufacturers must learn what can and should be circulated and recycled and what can and should not.[29]

Understanding the current drivers, potentialities and challenges related to the role of chemical sciences in a circular economy is of fundamental importance. The creation, development and use of green chemicals derived from renewable materials can be seen as more than simple opportunities in research and innovation. The latest trends related to green chemical products, processes and services concerning eco-design and solution approaches should focus on ways to extend circularity through sustainability in chemistry. Emphases on establishing new relationships with goods, materials, energy and, mostly, long-term cooperation and integration models among all partners involved will demonstrate this dynamic growth.

10.3.1 CIRCULAR CHEMISTRY TO ENABLE A CIRCULAR ECONOMY

Circular economy is considered a new opportunity to build a more sustainable world from both the social and the economic point of view.[30] It has become obvious that the linear route of production, in which scarce resources are consumed and their value-added products are degraded to waste, is a root cause of several impending global crises such as climate change and diminished biodiversity. Within the framework of sustainability, low-carbon chemistry is gaining importance and consequently, infusion of circular economy (CE) concepts will play a vital role to encompass the green/sustainable chemistry principles more comprehensively to the resource-dependent chemical industry. In the present state of environmental challenges, retrofitting the chemical industry with sustainable product stewardship is significantly important.[27]

In a CE, products should be used as long as possible until the end of their lives. The incorporation and absorption of Green Chemistry is a major step toward sustainability. Green chemistry has been a major driver of sustainable development and increased in importance in recent years. There still need to be more studies identifying possible developments for future research. Green Chemistry (GC), Sustainability and Circular Economy (CE) concepts are related to each other and this section shows this relationship.

Since the nature of chemistry is to produce intermediate goods that are generally used by other industries, chemistry involves most of production systems which affect global sustainability. Chemical industry is able to contribute to a strong transition towards a greater economic, environmental and social sustainability. Even if the main focus of GC is the environment, GC is getting closer to circularity and sustainability by providing the main tools for chemical industry to implement Sustainable Chemistry (SC) system and to realize the transition towards sustainability and CE.

GC is the tool through which it is possible realize the SC system. In turn, agricultural chemistry (AC) is a major contributing effort to sustainable agricultural chemistry. CC is a major contributing effort to in particular, the SC, in a CE system. It can be involved in processes of production and recycling, ensuring more sustainable environmental, economic and social systems. Furthermore, results show how GC and CE are getting closer to each other highlighting the ongoing alignment of purposes among different tools and adopted approaches in a holistic vision.[31]

The Green Chemistry 12 guiding principles (Table 10.1) focus on the direct sustainability assessment of chemical reactions and are perfectly suited for the optimization of linear production routes. The developments towards a circular economy, however, require a re-evaluation of what defines a sustainable chemical process, and needs to take into account the "triple bottom line" of people, planet and profit. Notably, innovative chemistry designed with sustainability in mind is only effective when translated into economically viable applications.

Green chemistry principles (GCP) have been comprehensively deployed in industrial management, governmental policy, educational practice, and technology development around the world. Circular economy always aims to balance the economic growth, resource sustainability, and environmental protection. The circular economy concept should be integrated into GCP to effectively achieve

---

TABLE 10.1
The Twelve Principles of Green Chemistry (GC)[32]

Principle	Explanation
1. Prevent waste.	It is better to prevent waste than to treat or clean up waste after it has been created.
2. Maximize atom economy.	Chemical processes should maximize incorporation of all materials used into the final product.
3. Less hazardous synthesis.	Chemical processes should avoid using or producing substances toxic to humans and the environment.
4. Design benign chemicals.	Chemicals should be designed to achieve their function while minimizing their toxicity.
5. Use safer solvents and auxiliaries.	Auxiliary substances should be rendered redundant wherever possible and harmless when used.
6. Increase energy efficiency.	Energy requirements of chemical processes should be minimized.
7. Use renewable feedstocks.	A raw material or feedstock should be renewable rather than depleting.
8. Reduce chemical derivatives.	Reduce generation of derivatives, since such chemical steps require more reagents and produce additional waste.
9. Use catalytic (not stoichiometric) conditions	Using catalysts is preferable compared to using reagents in stoichiometric amounts.
10. Design for degradation.	Chemical products should be designed to deteriorate after fulfilling a function without persisting in the environment.
11. Real-time analysis for pollution prevention.	Analytical methods allow in-process monitoring and control prior to the formation of hazardous substances.
12. Minimize potential for accidents.	Substances used in a chemical process should be chosen to minimize the potential for accidents.

practical methodologies. To achieve an effective pathway on GCP practices towards a circular economy, the following strategies are helpful:

- strengthening of cross-departmental collaboration,
- developing cleaner production to green products,
- building integrated chemical management system,
- implementing green chemistry education, and
- establishing business model linked to waste-to-resource supply chain.[33]

As it deals with materials, formulations, and transformations, chemistry provides key functions for sustainability. Chemistry can contribute to the development of a circular economy, becoming circular chemistry (CC). There are, likewise, 12 principles for a 'circular chemistry' (Table 10.2). In this way, a framework analogous to that of green chemistry will result. These can facilitate the transition to a circular economy. This approach aims to make chemical processes truly circular by expanding the scope of sustainability from process optimization to the entire lifecycle of chemical products. It strengthens resource efficiency across chemical value chains and highlights the need to develop novel chemical reactions to reuse and recycle chemicals, which in turn enables development towards a closed-loop, waste-free chemical industry.[34]

In order for a circular economy to keep products, components and materials at their highest utility and value at all times, the processes must be handled at the molecular level. Chemistry is crucial for achieving this. Chemists understand their role in designing and developing indispensable materials and technologies, but also simultaneously recognize the potentially detrimental effects that this may have on the environment; chemists are becoming increasingly aware that each step must be designed or reassessed with sustainability in mind.[35]

A key strategy to achieve material circularity is through minimizing waste. This serves as an essential tool to secure sustainability of the planetary environment. Elaborations of the 3R waste hierarchy (reduce, reuse, recycle), plus circular economy, 'zero waste' and zero discharge movements, are pathways that meet the sustainability challenge. Requisite technical solutions require major inputs from the chemical sciences to deliver the material basis of sustainability. These must address not only the concerns with limited resources, but also the totality of consequences connected to massive material and product flows. Closing the open loops, energy efficient practices, and maximal/efficient utilization of residual materials adherently play a key role. Circulating the molecules refers to recycling and reuse and is always critical in achieving the required properties or alternatively, finding secondary uses which ultimately needs new chemistry. Closing the loops is the path to achieve sustainability in a comprehensive way. [27]

10.3.2 INTEGRATING CHEMISTRY INTO A CIRCULAR ECONOMY

Chemistry integrates into the CE by developing new resource alternatives. The use of renewable bioresources for organic molecules has a long history and is increasingly practiced in industrial chemistry. Agro-industrial or forestry waste products are ready source of materials. They are complex and difficult to separate. Thus, the new challenge lies in transforming waste into specific products with defined properties and degrees of complexity, which meet current needs. Additional complications arise from the buildup of unwanted chemicals in products during use, natural aging, and the recycling process itself, as well as the connected flows of materials and products. (For example, mechanical recycling (remolding) results in polluted and lower-quality recyclates, thus preventing reuse in the same application).[29] This is an area in which the skills of chemists will be invaluable

Understanding the current drivers, potential opportunities, and challenges related to the role of chemical sciences in a circular economy is critical when bioresources are taken into account. The creation, development and use of green chemicals derived from renewable materials can be seen as more than simple opportunities in research and innovation. Green chemical products, processes and services concerning eco-design and solution approaches will be essential.[1] This will be developed later with discussions on the biorefinery.

In Table 10.3 (Integrating chemistry into CE) are listed all the ways in which chemistry integrates with CE to become circular chemistry (CC). These

TABLE 10.2
The Twelve Principles of Circular Chemistry (CC)[35]

Principle	Explanation
1. Collect and use waste.	Waste is a valuable resource that should be transformed into marketable products.
2. Maximize atom circulation.	Circular processes should aim to maximize the utility of all atoms in existing molecules.
3. Optimize resource efficiency.	Resource conservation should be targeted, promoting reuse and preserving finite feedstocks.
4. Strive for energy persistence.	Energy efficiency should be maximized.
5. Enhance process efficiency.	Innovations should continuously improve in- and post-process reuse and recycling, preferably on-site.
6. No out-of-plant toxicity.	Chemical processes should not release any toxic compounds into the environment
7. Target optimal design.	Design should be based on the highest end-of-life options, accounting for separation, purification and degradation.
8. Assess sustainability.	Environmental assessments (typified by the LCA) should become prevalent to identify inefficiencies in chemical processes.
9. Apply ladder of circularity	The end-of life options for a product should strive for the highest possibilities on the ladder of circularity.
10. Sell service, not product.	Producers should employ service-based business models such as chemical leasing, promoting efficiency over production rate.
11. Reject lock-in.	Business and regulatory environment should be flexible to allow the implementation of innovations.
12. Unify industry and provide coherent policy framework.	The industry and policy should be unified to create an optimal environment to enable circularity in chemical processes.

considerations should factor into decisions regarding chemical selection and chemical process selections. Ultimately, these will provide reduced waste streams and enhanced sustainability.

10.3.3 EXAMPLES

There are many ways that illustrate how chemistry is becoming circular. Each could become almost a book on its own. We will highlight some work, and show how it moves toward agriculture. Current textile production and processing practices provide materials with desirable performance properties, such as stretch and moisture management, but these processes are also contributors to global greenhouse gas emissions, microplastic pollution, and toxic wastewater. Fortunately, green alternatives to current textile fibers that support a transition to a sustainable, circular materials economy are beginning to emerge. Bioengineering of fibers at the nano-, micro-, and macroscale provides several avenues to improve both the environmental impacts and technical performance of textile materials. This results from efforts to bioengineer fibers and textiles from the biopolymer components to biofabrication schemes. These include the genetic engineering of microorganisms for biofabrication, green chemistry processing of raw materials, and green manufacturing techniques. Sustainable resources for textiles can come from biomass. The biotextile production can also include the utilization of waste streams to both improve the circularity and commercial viability of the processes.[36]

Plastics have revolutionized modern life, but have created a global waste crisis driven by our reliance and demand for low-cost, disposable materials. The development and use of plastics have made plastic a "societal necessity." Chemistry, through biomass, is developing viable alternatives. While chemical recycling and upcycling of polymers may enable circularity through separation strategies, chemistries that promote closed-loop recycling inherent to macromolecular design, and transformative processes that shift the life-cycle landscape are necessary. Polymer upcycling schemes may enable lower-energy pathways and minimal environmental impacts compared with traditional mechanical and chemical recycling. The emergence of industrial adoption of recycling and upcycling approaches is encouraging, solidifying the critical role for these strategies in addressing the fate of plastics and driving advances in next-generation materials design.[28]

While the plastics recycling industry has made many advances in its relatively short life, there are still many technical and societal hurdles to be overcome. Each stage along the path, from design of packaging and materials of construction to sortation, recycling, and reprocessing are occasions for innovation. The most relevant issues are introduced to provide a starting point for research across all fields of polymer science to aid in reducing the environmental impact of plastic packaging waste.[36] This will include design of plastics with finite environmental lifetimes.

A sustainable future depends on the availability of high-grade metals – both common (e.g., aluminum and copper) and specialty (noble, rare earth) ones – sourced

TABLE 10.3
Integrating Chemistry into CE[29]

Principle		Explanation
• Keep molecular complexity to the minimum.	→	Required for the desired performance, including end of life
• Design products for recycling,	→	Including all additives and other components of the product.
• Reduce and simplify diversity and Dynamics.	→	Use fewer chemicals overall (both number and quantity), design for less resource intensity, and adapt innovation speed of products to adaptation speed of recycling.
• Avoid complex products.	→	Avoid multiple components, materials.
• Minimize use of product components	→	Seek easily separated and recycled products (e.g., solvents, metals).
• Design products	→	Design for capture and recycling for complete fast mineralization at the end of their lives
• Prevent raw materials overuse	→	Reduced use and efficient recovery and recycling of materials
• Avoid entropic losses and transfers	→	Reduce dissipation of metals, energy).
• Avoid rebound effects	→	Using less carbon often means higher demand for metals
• Be responsible for/develop ownership of your product	→	Occurs throughout complete life cycle, including recycling.
• Ensure traceability	→	Use of product digital passports (e.g., composition of products, components, and processes).
• Develop and apply circular metrics	→	Giving credit to the use of by-products.
• Change traditional chemical practices	→	Change "bigger-faster" into "optimal adapted-better-safer" and change ownership to rent, lease, and share business models.
• Keep processes as simple as possible	→	Minimum number of steps, auxiliaries, energy, and unit operations (e.g., separations, purification).
• Design processes for optimal material recovery	→	Recovery of auxiliaries, unused substrates, and unintended by-products.

from virgin ores or from recycling. Copper, for example, is indispensable for many products, such as wiring, wind turbines, electric motors, information technology, generators, sensors, and electronic devices. Since the beginning of the twentieth century, copper production has grown more than 3000%, with a predicted market deficit of 600,000 metric tons by 2021.[37] New copper ores frequently fall short of the desired quality; thus, upgrading requires more energy and resources. Extraction of ores from deep mines creates a large environmental footprint, as this process generates more waste, mobilizes more toxic elements (e.g., arsenic), and is more likely to require access to protected lands of indigenous peoples, risking social disturbance. In order to continually mine virgin ores or recycle the metals will require the use of energy and generation of increased pollution, which will impact agriculture in the twenty-first century.

10.3.4 AFTERTHOUGHTS ON CIRCULAR CHEMISTRY

Within our linear economy, green chemistry has allowed the optimization of chemical processes, leading to less environmentally demanding chemistry practices. In doing so, it has laid the groundwork for an environmentally friendly culture in the chemical discipline. Now, further steps need to be taken towards sustainability. With the development of a circular economy, a set of principles for sustainable chemistry practice that is analogous to those of green chemistry highlights necessary actions we need to take. Circular chemistry will indeed become the necessary path forward.[35]

Circular chemistry offers a holistic systems approach: by making chemical processes truly circular, products can (ideally) be repurposed near-indefinitely, with energy as the only input. The chemical sector has the opportunity to take a leading role in combatting scarcity and environmental crises as a result of ineffective waste management, as the development of novel chemical reactions to reuse molecules and materials will lead to a closed-loop chemical industry. Life cycle thinking and circularity will reinvent chemistry,[38] and will provide the basic principles for developing novel chemical products and processes that use waste as resource. This will contribute to realizing the circular economy and addressing the United Nations Sustainable Development Goals.[39]

10.4 AGRICULTURAL CHEMISTRY CIRCULARITY

Agriculture must ensure that enough high-quality food is available to meet the needs of a continually growing population and it must do so sustainably. Further, it must provide necessary biomass for future sustainability goals. Current and future agronomic production of food, feed, fuel and fiber requires innovative solutions for existing and future challenges, such as climate change, resistance to pests, increased regulatory demands, renewable raw materials or requirements resulting from food chain partnerships. Modern agricultural chemistry has to support farmers to accomplish these tasks. Today, the so-called 'side effects' of agrochemicals regarding yield and quality are gaining more importance.[40] The modern agrochemical industry has to support farmers to manage these diversified tasks in accordance with further understanding of the crosslinked biological system and generation of innovation. Currently, several modern active ingredients are already matching these expectations, and further ones will need to follow in order to fulfil these ambitious criteria. The growth of agricultural circularity must include the chemical circularity to meet its goals.

The necessary shift away from fossil fuels needed to reduce CO_2 emissions requires the use of renewable carbon and energy sources, including biomass in the circular bioeconomy. The bioeconomy has a significant share in the EU economy with traditionally bio-based sectors. In the future, the energy, transportation, and chemical sectors will have additional high expectations of the bioeconomy, especially for agriculture and forestry to produce biomass as an industrial feedstock. The availability of feedstocks often only look at individual applications. Looking at the total demand and considering the sustainability limits of biomass production leads to the conclusion that the expected demand for all industries that could process biomass exceeds the sustainably available capacity. To mitigate this conflict between feedstock demand and availability, it is proposed that the organic chemical sector be fully integrated into the bioeconomy and this includes agricultural chemistry. In addition, recycling of wastes and residues including CO_2 should lead to a circular bioeconomy. Assessing the demand and use of biomass will result in the bioeconomy.[15]

10.4.1 INTRODUCTION

Recent advances in the application of circular bioeconomy technologies for converting agricultural wastewater to value-added resources verifies the requisite path forward. The properties and applications of the value-added products from agricultural wastewater reveal the enormous potential for this resource. Various types of agricultural wastewater can be included. Different types of circular technologies for recovery of humic substances (e.g., humin, humic acids and fulvic acids) and nutrients (e.g., nitrogen and phosphorus) from the agricultural wastewater are needed. Advanced technologies, such as chemical precipitation, membrane separation and electrokinetic separation highlight the contributions from agricultural chemistry. The environmental benefits of the circular technologies compared to conventional wastewater treatment processes have important implications for its utility in the circular technologies for agricultural wastewater. [41]

Global food production is under pressure to produce more from limited resources, with further expectations to reduce waste and pollution and improve social outcomes. Circular economy principles aim to minimize waste and pollution, reduce the use of non-renewable resources and increase the lifespan of products and materials. Waste reduction should begin with systems design, while recycling should be at the bottom of the hierarchy. On-farm resource use efficiency has been widely studied, but there are also opportunities to repurpose waste and integrate systems. The use of organic waste products as fertilizer and supplementary feed occurs to some extent, but they present both opportunities and challenges. More farm waste recycling opportunities are becoming available, with new products available from waste processing. Circular strategies in agriculture require more analysis to determine economic, social, cultural and environmental outcomes.[42]

10.4.2 DIMENSIONS AND MODELS FOR AGRICULTURAL CIRCULARITY

Circular and green economy considerations are relevant issues in the agricultural sector and drive the academic and policy decisions. The environmental and social problems raised by the economic development of the global economy are deeply involved in it. Circular and green economy affect the agri-food systems and their supply chains. Dominant research areas include "Circular Economy," "Green Economy," "Food Waste," and "Environmental Impacts." Future research should focus on the nexus "food waste and environmental impacts," and emphasizes the need to adopt a multidisciplinary approach to investigate the complex nexus between the food waste and the environment.[43]

The current agri-food supply chain is plagued by different problems such as food loss and waste generation along the supply chain, but the circular economy offers possibilities to enhance and optimize the production and consumption to move to a sustainable paradigm. The circular economy can be a successful approach to intervene and moderate the impacts generated in the agri-food sector. It offers actions and solutions to readmit wastes and by-products in the productive chain. This requires an examination of the circular economy model in the agri-food sector, with particular relevance to the reuse and valorization of wastes and by-products. This topic is of particular relevance in the scientific community, and the concept is continuously evolving. Europe plays a leading role in the research, thanks to the involvement of the Member States, policy makers and stakeholders. Nevertheless, some aspects such as the development of a new economic circular model and some limitations of the current policies deserve further investigation, especially in the agri-food industry.[44]

Achieving sustainable socio-economic development requires approaches that enhance resource use efficiencies and must address current cross-sectoral challenges in an integrated manner. Existing evidence suggests an urgent need for polycentric and transformative approaches. The potential of improved water and energy use efficiency and linking them to food production within the context of a circular economy will be most helpful. Identifying successes, opportunities, challenges, and pathways towards a circular economy facilitated developing a conceptual framework to guide strategic policy formulations towards a more sustainable economy. The initiatives try to attain the Sustainable Development Goals (SDGs) by mitigating negative environmental impacts due to waste and pollution. There is a need to enhance transformational change as a catalyst for employment creation and the attainment of a green economy while reducing waste. Transformative approaches have been identified to provide pathways towards global climate targets and protection of the environment from further degradation. They are a catalyst to achieve SDG 12 on ensuring sustainable consumption and production patterns.[45]

A major effort in developing "circular agriculture" is through the major goal of eliminating waste by appropriate planning. Biochar has recently gained popularity in the environmental sector as a versatile material for waste reduction and increasing the efficacy of the circular economy. It has demonstrated possibilities towards environmental impact, battling climate change, and creating an efficient circular economy model. Research on biochar's benefits is still widely ongoing. This is interrelated to the biochar's inherent properties that are deeply impacted by certain variables like feedstock types and treatment conditions. The conversion of waste into biochar and its application in different regions has implications for its use in circularity. Waste materials can be upcycled to make biochar and then used towards betterment of the environment. Biochar can be produced from different wastes and can be used in agriculture, wastewater treatment, anaerobic digestion and various other sectors thereby proving its multidimensional role towards protection of the environment and successfully building up a circular economy based environmental management model. This finally ends up closing the loop thereby demonstrating an actual circular economy.[46]

10.4.3 EXAMPLES OF AGRICULTURAL CIRCULARITY

There are many examples of agricultural circularity. These demonstrate current activities, but also suggest future trends. The generation of agroindustrial byproducts is rising fast worldwide. The production of bioethanol, and the processing of oil palm, cassava, and milk are industrial activities that, in 2019, generated huge amounts of wastewaters. Thus, it is urgent to reduce the environmental impact of these effluents through new integrated processes applying biorefinery and circular economy concepts to produce energy or new products. Some of the most important agro-industrial wastes, including their physicochemical composition, worldwide average production, and possible environmental impacts can be processed and valorized in the biorefinery. In addition, some alternatives, focusing mainly on energy savings and the possibilities of generating value-added products are key results.[43]

What can be learned from the quality of a cup of coffee? It is an example of global food production. It can also be viewed as an example of the role agricultural chemistry circularity can play. Coffee quality is primarily determined by the type and variety of green beans chosen and the roasting regime used. Furthermore, green coffee beans are not only the starting point for the production of all coffee beverages but also are a major source of revenue for many sub-tropical countries. Green bean quality is directly related to its biochemical composition which is influenced by genetic and environmental factors. Post-harvest, on-farm processing methods are now particularly recognised as being influential to bean chemistry and final cup quality. In addition, the potential of coffee waste biomass in a biobased economy context for the delivery of useful bioactives is becoming a topic of growing relevance within the coffee industry. There is a need for more intensive and united effort to build up our knowledge both of green bean composition and also how perturbations in genetic and environmental factors impact bean chemistry, crop sustainability and ultimately, cup quality.[44] This scenario is important for every crop in the global agrifood industry in order to align them in sustainable agriculture.

In response the global challenge of feeding an ever-increasing world population, agricultural chemists are researching nutritious and sustainable foods whose production should have low impacts on environment, economy and society. When the input feedstock can be waste nutrients, such products become even greener. Nutrients circularity is a form of the circular economy practice. Valorization of food waste as precious biomass to grow new food and feed in the form of nutrients is a major accomplishment. Mushrooms, microalgae and insects have been part of people diets in many areas of the world and they are often used for cultivation and breeding of waste nutrients recovery. Proof of such food waste biorefinery viability is already given by several research efforts featuring the main traits of a suitable growing medium: optimal pool of nutrients and optimal pH. However, much work still needs to be done in order to assess the optimal growth and cultivation conditions and the health security of the harvested/bred edibles.[17]

To begin to address the problem of plastics in our world, agricultural waste might be a solution. Recent strategies are available for upcycling agri-food losses and waste (FLW) into functional bioplastics and advanced materials. This valorization of food residuals and waste contribute positively to the food-energy-waste nexus. Low value or underutilized biomass, biocolloids, water-soluble biopolymers, polymerizable monomers, and nutrients are introduced as feasible building blocks for biotechnological conversion into bioplastics. The latter can be incorporated in multifunctional packaging, biomedical devices, sensors, actuators, and energy conversion and storage devices, contributing to the valorization efforts within the circular bioeconomy. Agricultural chemists will provide the means to effectively synthesize, deconstruct and reassemble or engineer FLW-derived monomeric, polymeric, and colloidal building blocks. Multifunctional bioplastics with differing structural, chemical, physical properties depending upon the accessibility of FLW precursors. Processing techniques are analyzed within the fields of polymer chemistry and physics. The prospects of FLW streams and biomass surplus is expected to lead to next-generation bioplastics and advanced materials.[47]

Likewise, the manufacturing and construction industries are benefitting from biomass. Biocomposites, being environmentally-friendly alternatives to synthetic composites, are gaining increasing demand for various applications. Biocomposite development should be integrated within a circular economy (CE) model ensuring a sustainable production that is simultaneously innocuous towards the environment. Adoption of the CE concept in biocomposite development is both logical and desirable. Biocomposites have desirable properties, environmental and economic impacts. An LCA of biocomposites show lower greenhouse gas emissions and carbon footprints. In addition, the opportunities and challenges pertaining to the implementation of CE have been discussed in detail. Recycling and utilisation of bio-based constituents were identified as the critical factors in embracing CE. Therefore, the development of innovative recycling technologies and an enhanced use of novel biocomposite constituents could lead to a reduction in material waste and environmental footprints. The circularity of biocomposites look positive and this will stimulate further research in enhancing the sustainability of these polymeric materials.[48]

The bioeconomy must also deal with waste. Bio-wastes from different agro-based industries are increasing at a rapid rate with the growing human population's demand for the products. The industries procure raw materials largely from agriculture, finish it producing the required major product, and generating huge amounts of bio-wastes which are mostly disposed unscientifically. This creates serious environmental problems and loss of resources and nutrients. Traditional bio-wastes disposal possess several weaknesses which again result in negative impact in the eco-system. Anaerobic digestion, composting, co-composting, and vermicomposting are currently given importance due to the improved and modified methods that can produce enhanced transformation of bio-wastes into suitable soil amendments. The advanced and modified methods like biochar assisted composting and vermicomposting is growing with the updated knowledge in the field. More work needs to be done to couple effective and efficient methods to utilize industry generated bio-wastes for circularity between agriculture – industrial sectors to promote sustainability.[49]

10.5   CONCLUSION

The Circular Bio-based Economy (CBE), in order to achieve its full success, must release the full potential of all types of sustainably sourced biomass, crop residues, industrial side-streams, and wastes by transforming them into value-added products. (See Figure 10.3) The resulting product portfolio will consist of a

FIGURE 10.3   Circular bio-based economy. (See www.shutterstock.com/image-vector/earth-hands-environment-concept-black-white-475584787)

wide spectrum of value-added products, addressing societal and consumer needs. Food and feed, bio-based chemicals, materials, health-promoting products; and bio-based fuels. The steps will involve the full scientific portfolio in agricultural chemistry, including biotechnology, microbial production, enzyme technology, green chemistry, and integrated physical/chemical processing. Drivers of CBE will include a huge range of resources. These range from biomass supply, biorefineries, value chain clusters, rural development, small-medium-large farmers. There is an urgent need for climate change mitigation and adaptation, and stopping biodiversity loss (taken up in Chapter 11). Improved framework conditions can be drivers but also obstacles if not updated to the era of circularity. Parallel to this, diversification of industrial segments and types of funding instruments developed is needed, reflecting industrial needs and academic research involvement.[13]

REFERENCES

1. Zuin, V.G., *Circularity in green chemical products, processes and services: Innovative routes based on integrated eco-design and solution systems.* Current Opinion in Green and Sustainable Chemistry, 2016. **2**: pp. 40–44.
2. Castro, C.G., et al., *The rebound effect of circular economy: Definitions, mechanisms and a research agenda.* Journal of Cleaner Production, 2022. **345**: pp. 1–13.
3. Hobson, K., *Closing the loop or squaring the circle? Locating generative spaces for the circular economy.* Progress in Human Geography, 2016. **40**(1): pp. 88–104.
4. Velasco-Muñoz, J.F., et al., *Circular economy implementation in the agricultural sector: Definition, strategies and indicators.* Resources, Conservation and Recycling, 2021. **170**: 1–15.
5. Nobre, G.C. and E. Tavares, *The quest for a circular economy final definition: A scientific perspective.* Journal of Cleaner Production, 2021. **314**: pp. 1–13.
6. Rigamonti, L. and E. Mancini, *Life cycle assessment and circularity indicators.* International Journal of Life Cycle Assessment, 2021. **26**(10): pp. 1937–1942.
7. Harris, S., M. Martin, and D. Diener, *Circularity for circularity's sake? Scoping review of assessment methods for environmental performance in the circular economy.* Sustainable Production and Consumption, 2021. **26**: pp. 172–186.
8. De Felice, F. and A. Petrillo, *Green transition: The frontier of the digicircular economy evidenced from a systematic literature review.* Sustainability (Switzerland), 2021. **13**(19): pp. 1–26.
9. Alhawari, O., et al., *Insights from circular economy literature: A review of extant definitions and unravelling paths to future research.* Sustainability (Switzerland), 2021. **13**(2): pp. 1–22.
10. Rodias, E., et al., *Water-energy-nutrients synergies in the agrifood sector: A circular economy framework.* Energies, 2021. **14**(1): pp. 1–17.
11. Cravero, R.A., M.M. Capobianco-Uriarte, and M.P. Casado-Belmonte, *Rethinking the physical losses definition in agri-food chains from eco-efficiency to circular economy,* in *Environmental Footprints and Eco-Design of Products and Processes.* 2021. Springer. pp. 93–117.
12. Teigiserova, D.A., L. Hamelin, and M. Thomsen, *Towards transparent valorization of food surplus, waste and loss: Clarifying definitions, food waste hierarchy, and role in the circular economy.* Science of the Total Environment, 2020. **706**: pp. 1–13.

13. Lange, L., et al., *Developing a sustainable and circular bio-based economy in EU: by partnering across sectors, upscaling and using new knowledge faster, and for the benefit of climate, environment & biodiversity, and people & business.* Frontiers in Bioengineering and Biotechnology, 2021. **8**: 1–16.

14. Cantzler, J., et al., *Saving resources and the climate? A systematic review of the circular economy and its mitigation potential.* Environmental Research Letters, 2020. **15**(12), pp. 1–19.

15. Kircher, M., *Economic trends in the transition into a circular bioeconomy.* Journal of Risk and Financial Management, 2022. **15**(2): pp. 1–24.

16. Suárez-Eiroa, B., E. Fernández, and G. Méndez, *Integration of the circular economy paradigm under the just and safe operating space narrative: Twelve operational principles based on circularity, sustainability and resilience.* Journal of Cleaner Production, 2021. **322**: pp. 1–13.

17. Girotto, F. and L. Piazza, *Food waste bioconversion into new food: A mini-review on nutrients circularity in the production of mushrooms, microalgae and insects.* Waste Management and Research, 2022. **40**(1): pp. 47–53.

18. Billen, G., et al., *Reshaping the European agro-food system and closing its nitrogen cycle: The potential of combining dietary change, agroecology, and circularity.* One Earth, 2021. **4**(6): pp. 839–850.

19. Nika, C.E., et al., *Nature-based solutions as enablers of circularity in water systems: A review on assessment methodologies, tools and indicators.* Water Research, 2020. **187**: pp. 1–12.

20. Miranda, A.C., et al., *Assessing the inclusion of water circularity principles in environment-related city concepts using a bibliometric analysis.* Water (Switzerland), 2022. **14**(11): pp. 1–23.

21. Giama, E., M. Mamaloukakis, and A.M. Papadopoulos. *Circularity in production process as a tool to reduce energy, environmental impacts and operational cost: The case of insulation materials.* in *4th International Conference on Smart and Sustainable Technologies, SpliTech 2019.* 2019. Institute of Electrical and Electronics Engineers Inc.

22. Cobo, S., A. Dominguez-Ramos, and A. Irabien, *Trade-offs between nutrient circularity and environmental impacts in the management of organic waste.* Environmental Science and Technology, 2018. **52**(19): pp. 10923–10933.

23. Shevchenko, T. and Y. Danko, *Progress towards a circular economy: New metric for circularity measurement based on segmentation of resource cycle.* International Journal of Environment and Waste Management, 2021. **28**(2): pp. 240–262.

24. Kümmerer, K., *Sustainable chemistry: a future guiding principle.* Angewandte Chemie – International Edition, 2017. **56**(52): pp. 16420–16421.

25. Christmann, P., *Mineral resource governance in the 21st century and a sustainable European Union.* Mineral Economics, 2021. **34**(2): pp. 187–208.

26. Barra, R.O., *The 2019 global environment outlook and global chemicals outlook: challenges for environmental toxicology and chemistry in Latin America.* Current Opinion in Green and Sustainable Chemistry, 2020. **25**: pp. 1–4.

27. Mohan, S.V. and R. Katakojwala, *The circular chemistry conceptual framework: A way forward to sustainability in industry 4.0.* Current Opinion in Green and Sustainable Chemistry, 2021. **28**: pp. 1–9.

28. Clark, J.H., *Green biorefinery technologies based on waste biomass.* Green Chemistry, 2019. **21**(6): pp. 1168–1170.

29. Kümmerer, K., J.H. Clark, and V.G. Zuin, *Rethinking chemistry for a circular economy.* Science, 2020. **367**(6476): pp. 369–370.
30. Salviulo, G., et al., *Enabling circular economy: the overlooked role of inorganic materials chemistry.* Chemistry – A European Journal, 2021. **27**(22): pp. 6676–6695.
31. Silvestri, C., et al., *Green chemistry contribution towards more equitable global sustainability and greater circular economy: A systematic literature review.* Journal of Cleaner Production, 2021. **294**: pp. 1–24.
32. Anastas, P. and N. Eghbali, *Green chemistry: principles and practice.* Chemical Society Reviews, 2010. **39**(1): pp. 301–312.
33. Chen, T.L., et al., *Implementation of green chemistry principles in circular economy system towards sustainable development goals: Challenges and perspectives.* Science of the Total Environment, 2020. **716**: pp. 1–16.
34. Bote Alonso, I., M.V. Sánchez-Rivero, and B. Montalbán Pozas, *Mapping sustainability and circular economy in cities: Methodological framework from europe to the Spanish case.* Journal of Cleaner Production, 2022. **357**: pp. 1–25.
35. Keijer, T., V. Bakker, and J.C. Slootweg, *Circular chemistry to enable a circular economy.* Nature Chemistry, 2019. **11**(3): pp. 190–195.
36. Billiet, S. and S.R. Trenor, *100th anniversary of macromolecular science viewpoint: Needs for plastics packaging circularity.* ACS Macro Letters, 2020. **9**(9): pp. 1376–1390.
37. Valenta, R.K., et al., *Re-thinking complex orebodies: Consequences for the future world supply of copper.* Journal of Cleaner Production, 2019. **220**: pp. 816–826.
38. Whitesides, G.M., *Reinventing chemistry.* Angewandte Chemie – International Edition, 2015. **54**(11): pp. 3196–3209.
39. *Transforming our world: the 2030 Agenda for SustainableDevelopment. Sustainable Development Goals Knowledge Platform* [2015]; Available from: https://go.nature.com/2FWSseG
40. Jeschke, P., *Progress of modern agricultural chemistry and future prospects.* Pest Management Science, 2016. **72**(3): pp. 433–455.
41. Mehta, N., et al., *Advances in circular bioeconomy technologies: From agricultural wastewater to value-added resources.* Environments – MDPI, 2021. **8**(3): pp. 1–23.
42. Burggraaf, V.T., et al., *Application of circular economy principles to new Zealand pastoral farming systems.* Journal of New Zealand Grasslands, 2020. **82**: pp. 53–59.
43. Martinez-Burgos, W.J., et al., *Agro-industrial wastewater in a circular economy: Characteristics, impacts and applications for bioenergy and biochemicals.* Bioresource Technology, 2021. **341**: pp. 1–11.
44. Chiaraluce, G., D. Bentivoglio, and A. Finco, *Circular economy for a sustainable agri-food supply chain: A review for current trends and future pathways.* Sustainability (Switzerland), 2021. **13**(16): pp. 1–21.
45. Naidoo, D., et al., *Transitional pathways towards achieving a circular economy in the water, energy, and food sectors.* Sustainability (Switzerland), 2021. **13**(17): pp. 1–15.
46. Singh, E., et al., *Circular economy-based environmental management using biochar: Driving towards sustainability.* Process Safety and Environmental Protection, 2022. **163**: pp. 585–600.
47. Otoni, C.G., et al., *The food–materials nexus: next generation bioplastics and advanced materials from agri-food residues.* Advanced Materials, 2021. **33**(43): pp. 1–41.

48. Shanmugam, V., et al., *Circular economy in biocomposite development: State-of-the-art, challenges and emerging trends.* Composites Part C: Open Access, 2021. **5**: pp. 1–16.

49. Ravindran, B., et al., *Cleaner production of agriculturally valuable benignant materials from industry generated bio-wastes: A review.* Bioresource Technology, 2021. **320**: pp. 1–12.

11 Smart agriculture through agricultural chemistry

The science and technology developments over the past years have provided tools for application in agriculture and particularly contributing for more sustainable agricultural systems (see Figure 11.1). This technological development is continuously evolving. We are nearing a point where this knowledge must become common practice and be seen to improve agricultural production while leading to sustainability goals. This will become smart agriculture (SA). New opportunities will come from SA and technologies such as the Internet of Things (IoT) are ways to improve agricultural efficiency and use of scarce resources such as soil, water, and energy.[2]

Climate change, with increasing temperatures and atmospheric carbon dioxide levels, constitutes a severe threat to the environment and all living organisms. Severe consequences for the health of crops, affecting both the productivity and quality of raw material destined to the food industry are a clear result. In this chapter we will see that SA, or sometimes called Climate-Smart Agriculture (CSA) will combat this issue through a double goal of reducing environmental impacts (use of pesticides, nitrogen and phosphorus leaching, soil erosion, water depletion and contamination) and improving raw material use and food quality. Organic farming, biofertilizers and to a lesser extent nano-carriers can be seen as potential chemical changes. On the other hand, advanced devices and Precision Agriculture (PA) will allow the agricultural practices to be more profitable, efficient, and to contribute more to reduce pest issues and to increase the quality of agricultural products and food safety. Thus, adoption of technologies through CSA when applied to sustainable farming systems is a challenging and dynamic issue for facing negative trends due to environmental impacts and climate changes.[3]

The challenges facing agriculture also result from the increasing population, the urban sprawl, scarcity of resources, climate change, and waste management. [4] These issues call for innovative solutions. New technological developments may play a relevant and determinant role, namely with novel practices associated with, for example, the vertical/urban farming, seawater and desert farming [3], as well as SA. It is evident that the entire agri-food industry is vulnerable to climate

DOI: 10.1201/9781003157991-11

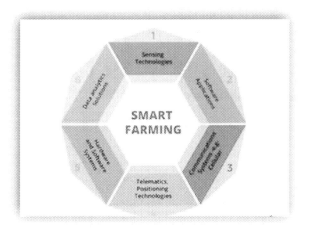

FIGURE 11.1   Scientific components of smart agriculture.[1]

change and variability. As these problems persist, the industry's quest to deliver on set targets of the sustainable development goals will be strongly challenged without appropriate interventions. Adopting SA tools and technologies would seem to be a suitable strategy to achieving food security while also mitigating and adapting to climate-related risks. Among numerous technologies that are found under the umbrella of SA are soil and water conservation technologies and climate information services as highly valued promising options for climate change adaptation and risk management

In this chapter, we will highlight the evolving way agricultural chemistry is practiced in the twenty-first century. This will demonstrate the main relationships between the smart agriculture (modern agricultural practices based on the new technologies, such as the Internet of Things (IoT), that combine scientific research and innovation) and the sustainability (capacity to meet the present needs without compromising the future generations).[5]

11.1   INTRODUCTION

Smart farming is a management approach that seeks to modernize the agricultural industry with the currently available technologies for agricultural systems using digital techniques to deliver sustainable agriculture and ensure food security amid climate change. Smart farming is software-managed and sensor-monitored, and its technologies drive climate-smart agriculture (CSA). The CSA will incorporate robotics, Internet of Things, and remote sensing. These are valuable tools for tracking, monitoring, automating and analyzing agricultural operations. CSA is a key approach for optimizing agricultural production processes in order to address the interlinked challenges of food security and climate change.[6] Smart farming must grow in importance due to the combination of the expanding global population, the increasing demand for higher crop yield, the need to use natural resources efficiently, the rising use and sophistication of information and communication technology, and the increasing need for a climate-smart agriculture.

The innovative technologies developed in the different fields of science (chemistry, nanotechnology, artificial intelligence, genetic modification, etc.) have opened new and challenging possibilities for the agri-food industry. For agriculture, these new approaches are particularly relevant and may bring interesting contributions, considering the specificities of the sector, often dealing with contexts of land abandonment and narrow profit margins.[7] Difficulties will arise in terms of costs, education, and regional applicability.

"Sustainable intensification" (SI) and CSA are highly complementary. SI is an essential means of increasing production volumes while adapting to climate change. This also will result in lower emissions per unit of output. With its emphasis on improving risk management, information flows and local institutions to support adaptive capacity, CSA provides the foundations for incentivizing and enabling intensification. While SI and CSA are crucial for global food and nutritional security, they are only part of the needed solution, which includes broader practices that reduce consumption and waste, build social safety nets, facilitate trade, and enhance diets.[8]

In the twenty-first century, agriculture must find new approaches that are relevant and bring necessary contributions to accomplishing its goals of sustainability. Nonetheless, the question is about the connection with the promotion of sustainability. There will be impacts on the sustainability from the transformations in the farming organization. There will be new opportunities, food supply, arising from the precision agriculture, agricultural intelligence, vertical/urban farming, circular economy, internet of things, and crowd-farming. After reviewing "climate-smart agriculture," we may find that the new and wider concept of "integrated-smart agriculture" is preferred.

Finally, we must keep in mind that agriculture is the backbone of most economies, and the global agricultural sector is dominated by smallholder farming systems. The farming systems are facing constraints such as small land size, lack of resources, and increasing degradation of soil quality that hamper sustainable crop production and food security. The effects of climate change (e.g., frequent occurrence of extreme weather events) exacerbate these problems. Applying appropriate technologies like smart agriculture (SA) can help to resolve the constraints of smallholder farming systems. Newly developed SA practices, such as integrated soil fertility management, water harvesting, and agroforestry can help ameliorate the harsh situation. These practices are commonly related to drought resilience, stability of crop yields, carbon sequestration, greenhouse gas mitigation, and higher household income. However, the adoption of the practices by smallholder farmers is often limited, mainly due to shortage of cropland, land tenure issues, lack of adequate knowledge about SA, slow return on investments, and insufficient policy and implementation schemes.[9]

11.1.1 FOOD SECURITY

The emergence of industrial agriculture in the twentieth century resulted in increased agricultural productivity and efficiency, but the concept of global food security was not envisioned.[10] It is anticipated that the current and anticipated impacts of climate change on the agricultural sector will exacerbate the problem of food insecurity. Climate-smart agriculture appears to be a mechanism that could potentially assist in the attainment of food security. It may also reduce the impacts of climate change and improve the ability of farmers to adapt. Ecosystem management initially should be used for the design and implementation of climate-smart agriculture. By making farming more connected and intelligent, smart agriculture helps reduce overall costs and improve the quality and quantity of products, the sustainability of agriculture for the consumer. Increasing control over production leads to better cost management and waste reduction. The ability to trace anomalies in crop growth, for instance, helps eliminate the risk of losing yields. Additionally, automation boosts efficiency. All of these processes lead directly to food security.[11]

Climate-smart agriculture (CSA) is a methodology for transforming and redirecting agricultural systems to support food security under conditions of climate change. Widespread changes in rainfall and temperature patterns threaten agricultural production. These changes increase the vulnerability of people dependent on agriculture for their livelihoods, which includes most of the world's poor. Climate change disrupts food markets, posing population-wide risks to food supply. The effects can be reduced by increasing the adaptive capacity of farmers and increasing resilience and resource use efficiency in agricultural production systems. CSA facilitates coordinated actions by farmers, researchers, private sector, civil society and policymakers towards climate-resilient pathways through four main action emphases:

- Increasing information and documentation,
- Improving local agricultural practices,
- Promoting application of climate and agricultural policies, and
- Linking climate and agricultural financing.

CSA ideology differs from "business-as-usual" approaches by emphasizing the need to implement flexible, context-specific solutions, using innovative policy and financing actions.[12] An efficient use of the agricultural resources, such as water, soil and energy, is crucial for competitiveness and food and security.[13]

11.1.2 SMART FARMING TECHNOLOGIES

The intelligent farm potentially includes the use of technologies such as shown in Table 11.1.

TABLE 11.1
Components of Smart Agriculture

Component	Description/Application
Sensors	soil scanning and water, light, humidity and temperature management
Telecommunications technologies	advanced networking and GPS
Hardware and software	specialized applications for enabling IoT-based solutions, robotics and automation
Data analytics tools	decision making and prediction. Data collection is a significant part of smart farming as the quantity of data available from crop yields, soil-mapping, climate change, fertilizer applications, weather data, machinery and animal health continues to escalate
Satellites and drones	gathering data around the clock for an entire field. This information is forwarded to IT systems for tracking and analysis to give an "eye in the field" that makes remote monitoring possible

The combination of these technologies increases the volume and quantity of data. This data feeds into a decision support system so that farmers can see what is happening at a more granular level than in the past. For example, by precisely measuring variations within a field and adapting the strategy accordingly, farmers can greatly increase the effectiveness of pesticides and fertilizers and use them more judiciously.

Smart farming systems enables farmers to carefully manage the demand forecast and delivery of goods to market just in time to reduce waste. Precision agriculture facilitates managing the supply of land and formulating the right growing parameters – for example, moisture, fertilizer or material content – to provide production for the right crop that is in demand. Control systems manage sensor input, delivering remote information for supply and decision support, in addition to the automation of machines and equipment for responding to emerging issues and production support.

11.1.3 METRICS FOR THE THREE PILLARS OF CSA

The three pillars of CSA are food security, adaptation, and mitigation. Given the technology improvements and the probable cost to implement CSA, there should also be improvements in the current state of assessing CSA in small-, medium- and large-holder farms. Climate Smart Agriculture assessment tools should cover not only biophysical but also socio-economic aspects of CSA, focusing on household level in all countries. Many tools in CSA are evaluated on the scientific

FIGURE 11.2 Climate smart agriculture. (See www.shutterstock.com/image-vector/smart-farm-modern-colorful-line-design-1864893580)

and technical levels while looking at productivity. (See Figure 11.2) This largely ignores potential social (e.g., food security, gender) and economic (poverty) aspects of the sustainability of intensified production. Climate change adaptation by farmers is the CSA pillar with the weakest contribution toward sustainability within the approaches.[14]

The targets that smart agriculture tries to reach and respond to encompass not only the climate effects, but also other environmental factors as well as social and economic aspects linked with the life of farmers and rural communities all over the world. Hence, the concept of smart agriculture should be expanded to include all these dimensions and an integrated-smart agriculture approach should express the future direction of agriculture.[15]

11.1.4 ENVIRONMENTAL IMPACTS AND CLIMATE CHANGE

Within the agricultural sector, food security is linked to the environmental impacts on the climate change. The climate-smart agriculture (CSA) approach appears to be a promising solution for the sustainability according to the following three actions:

- improve the resilience of the farming sector to the climate change,
- mitigate the greenhouse gas (GHG) emissions from farming activities, and
- guarantee the food security.[16]

The resilience will be the main challenge to deal with the climate change, but the transition to smart solutions is unstoppable. The objectives are to achieve sustainability, resilience, wellbeing, and development with new approaches. The new approaches found within smart agriculture will be a valuable contribution for the farm profitability and financial performance.[7]

11.2 AGRICULTURAL SUSTAINABILITY AND CLIMATE CHANGE

Climate change is of high concern to economists, ecologists, and agriculturalists as agriculture and climate change are closely related. While farmers have to modify their practices due to weather changes, any changes made must also reduce the impact of modern agriculture on climate change. The connection between modern industrial agriculture in its drive to sustainability and the effects on/from climate change demand close attention. Climate change adaptation strategies in agriculture and mitigation of negative effects they may currently produce are of simultaneous importance.

The goals of smart agriculture are intended to promote more sustainability, increasing farming productivity, adapting to the climate change challenges, and reducing the greenhouse gas emissions. For these aims to be realized, many new technologies will be employed. Artificial intelligence may bring interesting contributions, as well as, other new approaches and technologies, such as the Internet of Things (IoT). In the current contexts of climate change, these innovation are crucial to achieve sustainable development goals (SDGs).[17]

In the relationships between the agricultural sector, food security, and the environmental impacts, in light of climate change, the climate-smart agriculture sharpens the focus as a promising solution for the sustainability. It is particularly effective with the following objectives:

- improve the resilience of the farming sector to the climate change,
- mitigate the greenhouse gas (GHG) emissions, and
- guarantee the food security.[16]

The transition to smart solutions is gaining momentum. The new objective is to simultaneously achieve sustainability, resilience, wellbeing, and development.[18]

11.2.1 CURRENT SITUATION

Agriculture has changed dramatically since the end of World War II. Food and fiber productivity has soared due to new technologies, mechanization, increased chemical use, specialization, and government policies that favored maximizing production and reducing food prices. These changes have allowed fewer farmers to produce more food and fiber at lower prices.

Although these developments have had many positive effects and reduced many risks in farming, they also have resulted in significant costs, both environmental and social. (See Table 11.2)

Sustainable agriculture integrates the three main goals of environmental health, economic profitability, and social equity. Due to excessive vulnerability to climate change and its effects on sustainability, greater investments in more adapted institutions and technologies that are susceptible to these effects are needed. Consequently, the way food systems are managed needs to be changed if the goal is to achieve food security and sustainable development more quickly. The nexus "climate-smart agriculture-food systems-sustainable development" can lead to sound ways that could allow rapid transformation of food systems in the context of climate change pressure.[19]

11.2.2 CONSUMERS

The success of global sustainability must come in the broad areas of environment, society and economy. To achieve sustainable development, agriculture is one of the main fields and it is key to address economic, environmental, and ethical problems. Considering this, consumers are increasingly demanding foods produced under sustainable practices and desire to see progress in the process of enhancing food sustainability. Studies have been conducted as to whether consumers wanted sustainability in food production and their willingness to pay for it in different food categories. In general, results showed that consumers thought that a sustainable

---

TABLE 11.2
Cost of Current Farming Situation

Costs of Current Farming

- topsoil depletion
- groundwater contamination
- air pollution
- greenhouse gas emissions
- the decline of family farms
- neglect of the living and working conditions of farm laborers
- new threats to human health and safety due to the spread of new pathogens
- economic concentration in food and agricultural industries
- disintegration of rural communities

product is "environmentally friendly," "healthier," has been grown using "few chemicals" and "have better quality." The main conclusion is that consumers are not fully aware of the importance of sustainability; in general, consumers tend to associate sustainable production with just organic farming and higher quality.[20] This means that more consumer education is needed.

Climate change and food sustainability are two challenges and are highly interlinked. On the one hand, climate change puts pressure on food sustainability. On the other hand, farming significantly contributes to anthropogenic greenhouse gas emissions. Climate-smart agriculture can help both areas, but this also requires greater "buy-in" to the CSA. A key component of this acceptance involves adaptation measures, including technological advancements, improving farming practices, and financial management. To realize success, there are potential actions and trade-offs of climate-smart agricultural measures by producers and consumers. The major challenges to accomplishing the acceptance of CSA are shown in two findings: *(1) The benefits of measures are often site-dependent and differ according to agricultural practices (e.g., fertilizer use), environmental conditions (e.g., carbon sequestration potential), or the production and consumption of specific products (e.g., rice). (2) Climate-smart agricultural measures on the supply side are likely to be insufficient or ineffective if not accompanied by changes in consumer behavior, as climate-smart agriculture will affect the supply of agricultural commodities and require changes on the demand side in response.* This connection between demand and supply require simultaneous policy and market incentives. The link to consumer behavior is often neglected but is an essential component of climate-smart agriculture.[21]

11.2.3 Politics

As Climate Smart Agriculture gains prominence in global agricultural practices, it can provide "triple win" contributions to food productivity, adaptation, and mitigation to climate change. To maintain sustained inertia will require increasing support and promotion by governments. There are three broad areas where help will be needed:

- market policies that emphasize private-sector driven agricultural development in the face of rising input costs and falling commodity prices,
- expansion in diversified livelihood strategies amongst smallholder households as a response to the highly unpredictable biophysical environment and economic climate under which they live, and
- a growing competition for land and other productive resources.

This will must involve a greater effort in the political and economic processes surrounding these three issues.[22]

Political action is potentially useful for promoting climate-smart agriculture. Political ecology theory shows how three inter-related socio-political processes that perpetuate smallholder farmer vulnerability significantly influence climate-smart responses: inequality, unequal power relations and social injustice. CSA needs to work to promote global incorporation at different levels to lessen the vulnerability of smallholder farmers to current and future climate change impacts. Interventions to support climate-smart agriculture should examine local risks, specificities and priorities of smallholder farmers.[23]

To improve the sustainability in the agricultural sector through a smarter agriculture, the institutions (namely the associations and cooperatives) will need to play a stronger role. The adaptability of farmers (especially smaller ones) to the new technological demands might be difficult. The role of associations and cooperatives will be of great help to educate them to the new technological demands. To further this successful commitment of the farmers and to design adjusted policy instruments to promote the adoption of innovative approaches in a sustainable perspective much work is needed. Indeed, the agricultural policies and planning impact significantly the structure of the farms and the evolution of the sector.[24]

International organizations can also help smallholder farmers to manage agricultural systems in response to climate change. Analyzing these organization's priorities and highlighting their knowledge gaps are crucial for designing future pathways of CSA. There is insufficient discussion on the issues relating to governance measures and equality of gender participating in farming activities, within a larger focus on techno-managerial measures of CSA. Research and training related to CSA must offer opportunities for marginalized and disproportionately vulnerable populations to participate and raise their voices and share innovative ideas at different levels of governance.[25]

11.3   CLIMATE SMART AGRICULTURE

Innovative technologies are crucial to achieve sustainable development goals (SDGs) in the face of climate change.[26] Climate-Smart Agriculture (CSA) in the vital agri-food industry has a key role to play in transforming agriculture across the globe and reducing the contribution it makes to climate change globally (Greenhouse Gas emissions). Climate change is a major and unavoidable challenge to global development and deliberations on food security and sustainable farm livelihoods. Starting from the climate change impact on farm livelihoods, vulnerability and adaptation strategies, it will then be possible to evaluate the effects of CSA. This approach to agriculture can curtail widespread detrimental effects of climate change on farm households, farm productivity and farm profitability through adaptation strategies. This approach is important because of the livelihood status of various strata of farmers, along with their vulnerability, resilience capacity and the changing and unpredictability of many climate problems. Climate change is a global process with local visibility; in countries that are dependent on an agrarian economy, a balanced mix of adaptation mechanisms to lower the risk to vulnerable farm livelihoods and re-route them towards smart interventions can be important for climate-proofing agriculture.[27]

11.3.1 OVERVIEW

CSA is an approach involving several technologies that helps transform agri-food systems towards green and climate resilient practices. CSA supports reaching internationally agreed goals such as the SDGs and the Paris Agreement. It aims to tackle three main objectives:

- sustainably increasing agricultural productivity and incomes,
- adapting and building resilience to climate change, and
- reducing and/or removing greenhouse gas emissions, where possible.

CSA attempts to meet the growing demand for food, fiber and fuel, despite the changing climate and fewer opportunities for agricultural expansion onto additional lands. By doing this, it contributes to economic development, poverty reduction and food security. This in turn maintains and enhances productivity and resilience of natural and agricultural ecosystem functions, thus building natural capital. In order to accomplish all of this, it will be necessary to develop adaptation and mitigation approaches in addition to novel technologies.[28]

Climate-smart agriculture is envisioned as a promising approach to feed the growing world population under the challenge of climate change. While the CSA concept encompasses three pillars (productivity, adaptation, and mitigation), interest in institutional aspects has been gradual in CSA, as shown earlier. There has been more attention spent to understand influences on the adoption of CSA options. What are the important components: whether investments in physical infrastructure and actors' interaction, or how historical, political, and social context carry the most weight? Rethinking the approach to promoting CSA technologies by integrating technology packages and institutional enabling factors can provide potential opportunities for effective scaling of CSA options.[29]

11.3.2 SCIENCE IN CSA

For CSA to become widely used and effective as a sustainable solution in Agricultural Chemistry in the twenty-first century, three areas must be addressed. Current gaps in knowledge within CSA for interdisciplinary research and science-based actions fall within three themes:

- farm and food systems,
- landscape and regional issues and
- institutional and policy aspects.

The first two themes comprise crop physiology and genetics, mitigation and adaptation for agriculture, barriers to adoption of CSA practices, climate risk management and energy and biofuels; closely aligned with this are the areas of modelling adaptation and uncertainty, achieving multifunctionality, food systems, and ecosystem services, rural migration from climate change and metrics. The final area comprises designing research that bridges disciplines, integrating stakeholder input to directly link science, action and governance.

In addition to interdisciplinary research among these broad areas, imperatives must include developing

- models that allow adaptation and transformation at either the farm or landscape level;
- capacity approaches to examine multifunctional solutions for agronomic, ecological and socioeconomic challenges;
- studies providing direct evidence and metrics to support behaviors that improve resilience and natural environment;
- reductions in the risk that can present barriers for farmers during adoption of new technology and practices; and
- an understanding of how climate affects the rural labor force, and the stability of food production.

Effective work in CSA will involve all stakeholders, address governance issues, examine uncertainties, incorporate social benefits with technological change, and will establish the means to finance a green development framework.[28]

11.3.3 IMPACTS

Adoption of climate smart agriculture affects sustainability and food security. CSA is an effective strategy to guide and direct agricultural activities on many levels. It directly improves the agricultural populations' well-being for farm households with access to capital, strong social networks and access to integrated food markets. Farmers adopting CSA fare better than non-adopters, although CSA adoption does not fully counterbalance the severe climate pressures. Unfortunately, farmers with poor connections to food markets or inadequate access to the internet benefit less from CSA. This situation calls for an active role for policy makers in encouraging adaptation through CSA adoption by increasing access to capital, improving food market integration and building information sharing among farmers.[30]

Many projections of the broad impact of climate change on the crop production sectors of agriculture are reported in the literature. However, they may be too general to capture the magnitude of impact and to signal the need for adaptation strategies and policy development efforts that are tailored to promoting climate-smart agriculture in many parts of the world. From community to national and regional levels, various strategies and policies are also being taken to guide actions and investment for climate-smart agriculture.[31]

The impacts of climate change in agriculture are not confined to the third world nations. The potential effects of climate change in first world countries also present unprecedented challenges to regional agriculture. A variety of climate-smart agricultural methodologies have been proposed to retain or improve crop yields, reduce agricultural greenhouse gas emissions, retain soil quality and increase climate resilience of agricultural systems. One component that is commonly neglected when assessing the environmental impacts of climate-smart agriculture is the biophysical impacts, where changes in ecosystem fluxes and storage of moisture and energy lead to perturbations in local climate and water availability. [32] This demonstrates once again how the four spheres are interrelated and interdependent.

CSA is an approach to secure the management of landscapes with the purpose of adapting crop production while accounting for the climate changes on the planet. However, the CSA concept is a narrow approach about the current farming methodologies and a broader approach, involving interdisciplinary dimensions, and using recent sustainability metrics is needed.[33] This means that CSA must become dynamic through changing and adapting to new global pressures. It has the capacity to do this, but it will need the resources provided by green and sustainable chemistries.

Smart agriculture with an integrated view targeting sustainability can help address the challenges resulting from the urgent need to feed the growing world population resulting from inequality of population distribution around the globe. Even the asymmetries coming from ultra-high density urban areas as opposed to depopulated rural areas, in addition to the scarcity of natural resources, climate change demands, waste management needs, and circular economy promotion can be addressed by CSA. Artificial intelligence and agricultural digitalization contribute to improve productivity while reducing the input/output ratios, which make the systems effective. Optimization of natural resources and of production factors together with cleaner pest control strategies, move the agricultural practices toward environmental sustainability.[34]

Climate-smart agriculture improves energy/water/soil management and increasing resilience of the agricultural sector, contributing to guaranteed food security and attenuating GHG emissions. This develops a compromise between intensive agricultural production and global warming consequences.[13] Concepts of precision agriculture, smart irrigation and smart fertilization are possible due to IoT and CSA tools. They help to save resources, by for example collecting data from climate, soil, or plants to manage agricultural inputs, while improving productivity and reducing environmental impacts, either related with GHG emissions or the biosystem's ecology. Through controlled and automated farming systems, it is possible to improve competitiveness, with positive socioeconomic impacts, specifically in rural communities. The challenges are particularly relevant for developing countries where food security is at risk. Also, the sector of organic farming greatly benefits from these technologies, since in these farming systems better management is crucial for profitability. Financial commitments from governments and development agencies will be crucial for improving large scale adoption of climate-smart agriculture.[35]

11.3.4 INTENSIFICATION WITHIN THE CONSTRAINTS OF CSA

To achieve sustainable food production in the era of global environmental problems such as climate change, increasing population and natural resource degradation including soil degradation and biodiversity loss, mechanisms to increase the volume of production are needed. Climate change is necessitating changes to agricultural systems. The Green Revolution, while it multiplied agricultural production several fold, it did so at a huge environmental cost including climate change. It jeopardized the ecological integrity of agroecosystems by intensive use of fossil fuels, natural resources, agrochemicals and machinery. Moreover, it threatened the traditional agricultural practices. Agriculture is one of the largest business sectors that provide livelihood and employment to a large number of people, even while contributing to climate change. Therefore, a climate-smart approach to sustainable food production is needed. Traditional agriculture must be transformed, so as to meet the world's needs in context of sustainable food production in changing climate.[36]

The intensification will need to be monitored and evaluated. CSA assessment tools are critical in covering not only biophysical but also socio-economic aspects of CSA. Moving forward, several steps are needed to intensify agricultural practices while remaining within the constraints of CSA:

- It must make better use of recent advances in indicator development for sustainability assessments, including work on quantification of water and land footprints in relation to farm management;
- It must use household level analyses to quantify pathways from productivity toward food security and improved nutrition as well as improving ways of adopting CSA; and
- It must use recent advances in system specific quantification of greenhouse gas emissions by focusing on modeling and empirical work.[14]

11.4　TOOLS FOR CLIMATE SMART AGRICULTURE

Among the technologies available for present-day farmers are the following (see Table 11.3). While these are available in principle, their use is determined by local economies and education.

With the increase of these new technologies the productivity of agricultural and farming activities is also highlighted. This is critical to improving yields and cost-effectiveness. IoT can improve the efficiency of agriculture and farming processes by eliminating human intervention through automation. The fast rise of Internet of Things (IoT)-based tools has changed nearly all life sectors, including business, agriculture, surveillance, etc. These radical developments are upending traditional agricultural practices and presenting new options in the face of various obstacles. IoT aids in collecting data that is useful in the farming sector, such as changes in climatic conditions, soil fertility, amount of water required for crops, irrigation, insect and pest detection, and horticulture. IoT enables farmers

TABLE 11.3
Available Technologies in Smart Agriculture

Category		Representative Technologies
Sensors	⮕	soil, water, light, humidity, temperature management
Software	⮕	specialized software solutions that target specific farm types or applications agnostic IoT platforms
Connectivity	⮕	cellular, Long Range Radio
Location	⮕	GPS, Satellite
Robotics	⮕	Autonomous tractors, processing facilities
Data analytics	⮕	standalone analytics solutions, data pipelines for downstream solutions

to effectively use technology to monitor their farms remotely round the clock. Several sensors, including distributed WSNs (wireless sensor networks), are utilized for agricultural inspection and control, which is very important due to their exact output and utilization. In addition, cameras are utilized to keep an eye on the field from afar.[37]

Agriculture must employ the vast set of tools described in this chapter to meet environmental, economic and social challenges that make the case for transition towards sustainability. Digitization is one of the most significant ongoing transformation processes in global agriculture. Information and Communication Technologies (ICTs) range from traditional communication aids (e.g. telephones, televisions), internet and mobile applications, to Big Data analytics and information systems, Cloud computing, Internet of Things, remote sensing and drones, and artificial intelligence. Different terms have been used to refer to the application of ICTs in agriculture e.g. digital agriculture, e-agriculture, smart agriculture, precision agriculture. Smart agriculture is almost a 'Third Green Revolution', after the plant breeding and genetics revolutions. ICTs can help reduce inefficiencies, increase resource productivity, decrease management costs, and improve traceability and transparency.[38]

11.4.1 CLOUD COMPUTING

Cloud Computing is a well-established architecture for building service-centric systems that do not rely on localized data storage or manipulation. However, ultra-low latency, high bandwidth, security, and real-time analytics are limitations in Cloud Computing when analyzing and providing results for a large amount of data. Fog and Edge Computing offer solutions to the limitations of Cloud Computing. The number of agricultural domain applications that use the combination of Cloud, Fog, and Edge has been increasing in the last few decades.[17] Cloud computing technology and Artificial Intelligence (AI) techniques are becoming significant components in many applications in agriculture. Cloud computing allows the mass storage of various types of data combined with ready availability for descriptive, predictive, and prescriptive analytics. AI techniques, such as deep neural networks, have drawn on this data to achieve, promising disease detection and yield prediction outcomes. Not many agricultural systems have applied cloud computing and AI. This is a significant shortcoming in making it available for farmers to use practically.[39]

11.4.2 ARTIFICIAL INTELLIGENCE

Specifically, the use of artificial intelligence may be crucial for a more efficient use of the resources, but also, for a better disease and pest control, data analysis, productions' management and to fill the gap between farmers and knowledge, allowing in this way higher productivities and competitiveness. The scientific community interlinks the artificial intelligence with sustainability and the attainment of a more circular economy.[40] Potential benefits of the artificial intelligence has been known for decades. The smart agriculture aims to use AI to promote more sustainability, increasing the farming productivity, dealing with the climate change implications and reducing the greenhouse gas emissions. For these aims the artificial intelligence will bring interesting contributions, as well as, other new approaches and technologies, such as the IoT.

With the maturation of artificial intelligence, deep learning is gradually applied to the field of agriculture and plant science. However, the high-quality results of deep learning require the collection and use of massive numbers of samples. In the field of plant science and biology, it is not easy to obtain a large amount of labeled data. The emergence of few-shot learning solves this problem. It imitates the ability of humans' rapid learning and can learn a new task with only a small number of labeled samples, which greatly reduces the time cost and financial resources.[41]

11.4.3 DATA MINING

Precision agriculture is a crucial way to achieve greater yields by utilizing the natural deposits in a diverse environment. The ability of farmers to access and utilize information pertaining to agriculture is paramount. The yield of a crop may vary from year to year depending on the variations in climate, soil parameters and fertilizers used. Automation in the agricultural industry moderates the usage of resources and can increase the quality of food in the post-pandemic world. Agricultural robots have been developed for crop seeding, monitoring, weed control, pest management and harvesting. Physical counting of fruitlets, flowers or fruits at various phases of growth is labor intensive as well as an expensive procedure for crop yield estimation. Remote sensing technologies offer accuracy and reliability in crop yield prediction and estimation. The automation in image analysis with computer vision and deep learning models provides precise field and yield maps. Machine vision and deep learning need to be explored for improving automated precision farming.[42]

Agriculture remains a vital sector for most countries and it represents the main source of food for the population of the world. Smart Agriculture is one of the solutions to deal with the growing demand for food and in Smart Agriculture, the role of information is increasing. Information on weather conditions, soils, diseases, insects, seeds, fertilizers, etc. constitutes an important contribution to the economic and sustainable development of this sector. Smart management consists of collecting, transmitting, selecting and analyzing data. As the amount of agricultural data increases significantly, robust analytical techniques capable of processing and analyzing large amounts of data to obtain more reliable information and much more accurate predictions are essential. Data Mining is expected to play an important role in Smart Agriculture for managing real-time data analysis with massive data.[34]

11.4.4 INTERNET OF THINGS (IoT)

The technological progress in our world opens several opportunities for the different socioeconomic sectors, including the farming sector, especially in smart agriculture. This array of new technologies also brings various challenges and threats that may compromise the sustainability of the development process worldwide. The Internet of Things (IoT) refers to the digital interconnection of everyday objects (e.g., tools, technologies, and any other items found on the internet) with the Internet. This allows the connecting of objects instead of human intervention, allowing the intercommunication of data between sensors and digital controllers, for example. IoT is a set of networks which connect things capable of processing and communicating data between each other. The IoT may contribute to increase the agricultural efficiency in the use of soil, water and energy (some agricultural resources where an efficient management is crucial). It also creates new risks associated with confidentiality and integrity through the removal of human intermediacy due to autonomous machine activity.[2]

The Internet of Things (IoT) and the Internet of Everything (IoE) have had great impact on the farms and are the means to improve the productivity in the use of several agricultural resources, namely the water, through approaches of smart irrigation and precision agriculture.[43] The smart irrigation systems are important for the collection and work on environmental data. The IoT allows the implementation of automated operations with reduced supervision over the whole food chain and during agricultural production, including in greenhouse agriculture control and in diverse farming systems. Where water management is critical, the IoT may contribute significantly for a more balanced use, as well as in the soil health assessment and fertilization management. These contribute to a more competitive agriculture. As potable water becomes one of the scarcest resources, new technologies found in the IoT will be determinant for a more efficient management.[44]

The Internet of Things (IoT) has met requirements for security and reliability in domains like automotive industry, food industry, as well as precision agriculture. The System of Systems (SoS) expands the use of local clouds for the evolution of integration and communication technologies. SoS devices need to ensure Quality of Service (QoS) capabilities including service-oriented management and different QoS characteristics monitoring. Smart applications depend on information quality since they are driven by processes which require communication robustness and enough bandwidth. Interconnectivity and interoperability facilities among different smart devices can be achieved using reliable technologies. Smart IoT devices with wide applicability areas including smart building, smart energy, smart cities, and smart agriculture will be essential to move this technology forward. The advantages of reliable systems are driven by parameters such as transmission speed, latency, security, etc.[45]

11.5   RESOURCES AND ENGINEERING THAT COMPRISE CSA

11.5.1   MONITORING

Climate-smart agriculture (CSA) is a network of technologies and strategies that can sustain agriculture growth in a changing climate. (See Figure 11.3) Researchers are finding ways to collect big data, which are required to clarify local climate change and its impacts on agriculture to pinpoint the farming strategies for the practice of CSA. Numerous approaches including big data analytics, IoT, wireless sensor network (WSN)-based monitoring systems, machine learning, and AI algorithms are an information source to assist in delivering novel insights and explication to the problems. This accumulated data would be employed for monitoring and analyzing the effects of climate change.[29]

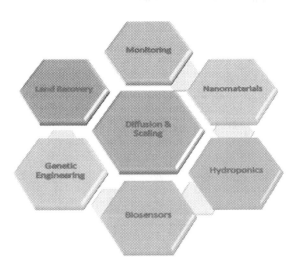

FIGURE 11.3   Functional components in CSA

11.5.2 NANOMATERIALS FOR FERTILIZERS AND PESTICIDES

Fertilizers play an essential role in increasing crop yield, maintaining soil fertility, and providing a steady supply of nutrients for plant requirements. The excessive use of conventional fertilizers can cause environmental problems associated with nutrient loss through volatilization in the atmosphere, leaching to groundwater, surface run-off, and denitrification. To mitigate environmental issues and improve the longevity of fertilizer in soil, controlled release fertilizers (CRFs) have been developed. The application of CRFs can reduce the loss of nutrients, provide higher nutrient use efficiency, and improve soil health. This will simultaneously achieve the goals of climate-smart agricultural (CSA) practices. CRFs can prevent direct exposure of fertilizer granule to soil and prevent loss of nutrients such as nitrate and nitrous oxide emissions and CRFs are less affected by the change in environmental parameters, and that can increase longevity in soil compared to conventional fertilizers. Finally, CRFs can maintain required soil nitrogen levels, increase water retention, reduce GHG emissions, which leads to optimum pH for plant growth, and increase soil organic matter content.[46]

Pesticides have also been used extensively in the field of plant protection to maximize crop yields. However, unmanaged application of pesticides has resulted in severe consequences such as pesticide resistance, risk in human health, soil degradation, and other important global issues. Recently, in addition to the combination of nanotechnology with plant protection strategies which has offered new avenues to mitigate these global issues, there has been a rapid development of new pesticides. Unlike certain conventional pesticides that have been applied inefficiently and lacked targeted control, pesticides delivered by nanotechnologies have optimized formulations, controlled release rate, and minimized or site-specific application. They are receiving increasing attention and are considered as an important part in sustainable and smart agriculture.[47]

In order to fully incorporate nanotechnologies, the only way forward is to embrace smart agricultural practice in a sustainable manner. The use of modern electronics and material science to increase production, without further increasing fertilizer or pesticide input, can be referred to as smart and sustainable agriculture. While scientists have made giant leaps in the field of 'biology at nanoscale' nanoparticles and nanosensors have huge potential in agricultural advancements, if used wisely with proper caution. Nanoparticles can be used for getting higher yield and for crop protection. Nanosensors can contribute to smart farming by growth monitoring, real time detection of pests, and continuous monitoring of local environment.[48]

11.5.3 HYDROPONICS

Soilless cultivation, while seeming contrary to the basic assumption of agriculture, represent a valid opportunity for the agricultural production sector. This will especially be true in areas characterized by severe soil degradation and limited water availability. This agronomic practice potentially offers an environment-friendly agricultural technology and a promising tool in the vision of a general challenge in terms of food security. There are significant questions regarding hydroponics, including what the processes and mechanisms are occurring in the hydroponic solutions that will ensure an adequate nutrient concentration during plant growth. Questions remain about an optimal nutrient acquisition without negatively influencing crop quality (e.g., solubilization/precipitation of nutrients/elements in the hydroponic solution, substrate specificity in the nutrient uptake process, nutrient competition/antagonism and interactions among nutrients). Equally important are new emerging technologies that might improve the management of soilless cropping systems such as the use of nanoparticles and beneficial microorganism like plant growth-promoting rhizobacteria (PGPRs). Finally, what are the tools (multi-element sensors and interpretation algorithms based on machine learning logics to analyze such data) that might be exploited in a smart agriculture approach to monitor the availability of nutrients/elements in the hydroponic solution and to modify its composition in real-time?[49]

11.5.4 Biosensors

The in vivo monitoring of key plant physiology parameters will be a key technology for precision farming. A biomimetic textile-based biosensor, which can be inserted directly into plant tissue is highly desired: the device is able to monitor, in vivo and in real time, variations in the solute content of the plant sap. The biosensor has no detectable effect on the plant's morphology even after six weeks of continuous operation. The continuous monitoring of the sap electrolyte concentration in a growing tomato plant revealed a circadian pattern of variation. The biosensor has the potential to detect the signs of abiotic stress, and therefore might be exploited as a powerful tool to study plant physiology and to increase tomato growth sustainability. Also, it can continuously communicate the plant health status, thus potentially driving the whole farm management in the frame of smart agriculture.[50]

The new biosensor technologies and approaches are, important methodologies to support the farmers in other agronomic practices. In some circumstances the government supports are decisive.

The wireless sensor network, for example, is an interesting tool to collect data about weather, soil, and plant conditions to provide the farmers with information to better manage the pest and disease control and the fertilizers' use [44]. The biosensors are additional techniques to collect information about the condition of the plant and assess its exposure to biotic and abiotic stresses [45].

Intense farming is an activity causing significant damage to vital resources such as lands and water. This is primarily due to unsustainable agricultural practices and the resulting environmental pollution. Furthermore, the increasing world population and the impact of climate change worsen these growing problems. Pioneering technologies for monitoring these challenges include nanostructured (bio)sensors. Nanotechnology-based (bio)sensors, thanks to the exploitation of fascinating properties of functional materials at the nanoscale, can support farmers in delivering fast, accurate, cost-effective, and in field analyses of i) soil humidity, ii) water and soil nutrients/pesticides, and iii) plant pathogens.[51] They can also monitor in real-time the negative effect that intense farming have on environmental sustainability.

11.5.5 GENETIC ENGINEERING

The topic and discussion of genetic engineering will be explored in Chapter 12. One example here will demonstrate its importance. Finger millet (FM; Eleusine coracana (L.) Gaertn.) is becoming increasingly vulnerable to various climate-induced stresses. The genetic and genomic resources will be important for improving the crop in the twenty-first century. Currently, sizable-untapped genetic resources exist that offer promise for FM improvement in response to biotic and abiotic stresses. It will be important to develop and utilize molecular markers, genetic maps, and DNA sequence, but the data are scanty to support the efficient and accelerated delivery of the climate-smart FM varieties. This is partly attributable to the delayed availability of complete genome sequence (CGS) of FM. Following the latest developments in FM genomic research, based on the CGS, a diversity of genomic resources have been reported.[52]

The innovative technologies developed in the different fields of science (nanotechnology, artificial intelligence, genetic modification, etc.) have opened new and exciting possibilities for several activities in different agriculture sectors. These new approaches are particularly relevant and may bring interesting contributions, considering the specificities of the sector, often dealing with contexts of land abandonment and narrow profit margins.[7] In the frame of modern agriculture facing the predicted increase of population and general environmental changes, the securement of high quality food remains a major challenge to deal with. Vegetable crops include a large number of species, characterized by multiple geographical origins, large genetic variability and diverse reproductive features. Due to their nutritional value, they have an important place in human diet. In recent years, many crop genomes have been sequenced permitting the identification of genes and superior alleles associated with desirable traits. Such technologies, however, require efficient transformation protocols as well as extensive genomic resources and accurate knowledge before they can be efficiently exploited in practical breeding programs.[53]

11.5.6 LAND RECOVERY

Due to climate change, more countries will suffer from drought conditions. Such conditions will result from surface water being only available for very limited periods during the year. This exacerbates the food security challenge. The cool but sunny desert makes it possible to produce crops under controlled environments with limited evapotranspiration. However, the high cost of inorganic hydroponic fertilizers makes adoption of hydroponics by smallholder subsistence farmers a great challenge.[54]

In an era of global warming, long-standing challenges for rural populations, including land inequality, poverty and food insecurity, risk being exacerbated by the effects of climate change. Innovative and effective approaches, such as Climate Smart Agriculture (CSA), are essential to alleviate these environmental pressures without hampering efficiency. "Climate Smart Land Reform" (CSLR) contains "a framework" of four driving pillars, namely land redistribution, tenure reform, rural advisory services and markets and infrastructure. The framework opens relevant channels through which land reform can foster CSA adoption and thus contribute to the attainment of sustainable increases in agricultural productivity, climate change adaptation and climate change mitigation. The framework also includes relevant channels through which more 'traditional' objectives of land reformers, including economic, social and political objectives, can be achieved. In turn, the (partial) attainment of such objectives would lead to improvements in agroecological and socioeconomic conditions of rural areas and populations. The CSLR framework represents an innovative way of conceptualising how land reforms can generate beneficial effects not only in terms of equity and efficiency but also of environmental sustainability.[55]

11.5.7 Diffusion

Climate-smart agriculture must be a technologically innovative response to the challenges faced by agriculture due to climate change. Its implementation needs a change of mentality on the part of farmers in the direction of an approach that takes into account how the increase in technologically induced productivity affects climate change. CSA is seeking to overcome the food security problem and develop rural livelihoods while minimizing negative impacts on the environment. However, when synergies exist, the situation of small-scale farmers is often overlooked, and they are unable to implement new practices and technologies. CSA indicates that none of the decisions made by policymakers can be realistic and functional as long as the voice of the farmers influenced by their decisions is not heard.[56]

11.5.8 Scaling

Climate change is an inherent and indispensable component of global development and deliberations on food security and sustainable farm livelihoods. Climate change impacts and effects farm livelihoods, vulnerability and adaptation strategies. Preordained holistic strategies such as CSA can curtail widespread detrimental effects of climate change on farm households, farm productivity and farm profitability through adaptation strategies. The implementation of this approach is driven by the livelihood status of various strata of farmers, along with their vulnerability, resilience capacity and climate knowledge with regards to contrasting rationales of climatic vagaries. Climate change is a global process with local visibility; in countries that are dependent on an agrarian economy, a balanced mix of adaptation mechanisms to de-risk the vulnerable farm livelihoods and re-route them towards smart interventions can be important for climate-proofing agriculture. The above mentioned framework provides a robust future course of action for circumventing adverse climatic impacts[27] by scaling from local to global farming approaches.

11.6 AFTERTHOUGHTS

Climate variability and change is a major source of risk to smallholder farmers globally. Climate related risks are linked to low productivity, food insecurity and poverty. However, the research and development community is widely promoting Climate-Smart Agriculture (CSA) to transform livelihoods under a changing environment. To date, adoption of CSA practices is low across the third world countries despite their demonstrated effectiveness. The low adoption challenge calls for prudent policy and institutional efforts in finding ways to effectively take CSA practices to scale. CSA scaling (upgrading) is the expansion of the adoption of the proven and beneficial CSA practices and/or technologies. The scaling of CSA practices, and technologies is not automatic, there is need for facilitation in terms of conducive policy and institutional actions. Policy strategies are important as they clearly define the rules of the game that will ultimately establish responsibilities in the scaling process by stakeholders. Effective and complementary institutional actions towards scaling can minimize farmer challenges, reduce adoption constraints, and improve sustainability in scaling processes, which can ultimately improve impacts of CSA practices and technologies to society.[57]

Agriculture not only provides the fuel for billions of people but also employment opportunities to a large number of people. The agricultural industries are seeking innovative approaches for improving crop yields because of unpredictable climatic changes, the rapid increase in population growth and food security concerns. Thus, artificial intelligence in agriculture also called "Agriculture Intelligence" is progressively emerging as a part of the industry's technological revolution. Applications of agriculture intelligence such as precision farming, disease detection, and crop phenotyping with the help of numerous tools such as machine learning, deep learning, image processing, artificial neural network, deep learning, convolution neural network, wireless sensor network (WSN) technology, wireless communication, robotics, Internet of Things (IoT), different genetic algorithms, fuzzy logic and computer vision to name a few will all be an integral part of Smart Agriculture.[58]

Climate-Smart Agriculture (CSA) is more responsive to the achievement of Sustainable Development Goals 2 and 13. CSA practices have the potential to alleviate low yields among farmers. Further investigation of the determinants of practice of CSA would inform the efforts to adapt production to the effects of climate change. Information from websites and other data can help to understand the financial, natural, physical, and social capital required to execute the CSA technological practices. The mode of communicating the CSA practices determines its adoption, therefore knowledge of such determinants and that of socioeconomic and institutional factors will shape CSA technological development and diffusion strategies. Understanding of these is essential to tailoring the CSA practices to the farmers' most pressing needs and to the development of the practices that can easily be accessed and adopted by the farmers.[59]

CSA still faces a number of challenges, including lack of clear conceptual understanding, limited enabling policy and financing. The prospects of CSA globally hinge on the capacities of farming households and the region's national institutions to understand the environmental, economic and social challenges in the context of climate change, and consequently to self-mobilize to develop and implement responsive policies at appropriate scales.[60]

REFERENCES

1. Research, B. 2021. Available from: www.beechamresearch.com/downloads/.
2. Anand, P., et al., *IoT vulnerability assessment for sustainable computing: Threats, current solutions, and open challenges.* IEEE Access, 2020. **8**: pp. 168825–168853.
3. Agrimonti, C., M. Lauro, and G. Visioli, *Smart agriculture for food quality: facing climate change in the 21st century.* Critical Reviews in Food Science and Nutrition, 2021. **61**(6): pp. 971–981.
4. Hossain, A., et al., *Next-Generation Climate-Resilient Agricultural Technology in Traditional Farming for Food and Nutritional Safety in the Modern Era of Climate Change,* in *Plant Abiotic Stress Physiology Volume 1: Responses and Adaptations.* 2022. CRC Press. pp. 225–291.
5. Horton, P. and B.P. Horton, *Re-defining sustainability: living in harmony with life on Earth.* One Earth, 2019, **1**(1): pp. 86–94.
6. Adamides, G., *A review of climate-smart agriculture applications in Cyprus.* Atmosphere, 2020. **11**(9): pp. 1–15.
7. Martinho, V.J.P.D. and R.P.F. Guiné, *Integrated-smart agriculture: Contexts and assumptions for a broader concept.* Agronomy, 2021. **11**(8): pp. 1–21.
8. Campbell, B.M., et al., *Sustainable intensification: What is its role in climate smart agriculture?* Current Opinion in Environmental Sustainability, 2014. **8**: pp. 39–43.
9. Zerssa, G., et al., *Challenges of smallholder farming in Ethiopia and opportunities by adopting climate-smart agriculture.* Agriculture (Switzerland), 2021. **11**(3): pp. 1–26.
10. Akamani, K., *An ecosystem-based approach to climate-smart agriculture with some considerations for social equity.* Agronomy, 2021. **11**(8): pp. 1–19.
11. Munaweera, T.I.K., et al., *Modern plant biotechnology as a strategy in addressing climate change and attaining food security.* Agriculture and Food Security, 2022. **11**(1): pp. 1–28.
12. Lipper, L., et al., *Climate-smart agriculture for food security.* Nature Climate Change, 2014. **4**(12): pp. 1068–1072.
13. Agrawal, S. and A. Jain, *Sustainable deployment of solar irrigation pumps: Key determinants and strategies.* Wiley Interdisciplinary Reviews: Energy and Environment, 2019. **8**(2): pp. 1–14.

14. van Wijk, M.T., et al., *Improving assessments of the Three Pillars of Climate Smart Agriculture: current achievements and ideas for the future.* Frontiers in Sustainable Food Systems, 2020. **4**: pp. 1–14.

15. Taylor, M., *Climate-smart agriculture: what is it good for?* Journal of Peasant Studies, 2018. **45**(1): pp. 89–107.

16. Acosta-Alba, I., E. Chia, and N. Andrieu, *The LCA4CSA framework: Using life cycle assessment to strengthen environmental sustainability analysis of climate smart agriculture options at farm and crop system levels.* Agricultural Systems, 2019. **171**: pp. 155–170.

17. Herrero, M., et al., *Articulating the effect of food systems innovation on the Sustainable Development Goals.* The Lancet Planetary Health, 2021. **5**(1): pp. e50–e62.

18. Tripathy, A.S. and D.K. Sharma, *Image processing techniques aiding smart agriculture*, in *Modern Techniques for Agricultural Disease Management and Crop Yield Prediction*. 2019, IGI Global. pp. 23–48.

19. Zougmoré, R.B., P. Läderach, and B.M. Campbell, *Transforming food systems in africa under climate change pressure: Role of climate-smart agriculture.* Sustainability (Switzerland), 2021. **13**(8): pp. 1–17.

20. Sánchez-Bravo, P., et al., *Consumer understanding of sustainability concept in agricultural products.* Food Quality and Preference, 2021. **89**, pp. 1–11.

21. Scherer, L. and P.H. Verburg, *Mapping and linking supply- and demand-side measures in climate-smart agriculture. A review.* Agronomy for Sustainable Development, 2017. **37**(6): pp. 1–17.

22. Shilomboleni, H., *Political economy challenges for climate smart agriculture in Africa.* Agriculture and Human Values, 2020. **37**(4): pp. 1195–1206.

23. Chandra, A., K.E. McNamara, and P. Dargusch, *The relevance of political ecology perspectives for smallholder Climate-Smart Agriculture: A review.* Journal of Political Ecology, 2017. **24**(1): pp. 821–842.

24. Martinho, V.J.P.D., *Exploring the topics of soil pollution and agricultural economics: Highlighting good practices.* Agriculture (Switzerland), 2020. **10**(1): pp. 1–19.

25. Gardezi, M., et al., *Prioritizing climate-smart agriculture: An organizational and temporal review.* Wiley Interdisciplinary Reviews: Climate Change, 2021.

26. Rosenzweig, C., et al., *Climate change responses benefit from a global food system approach.* Nature Food, 2020. **1**(2): pp. 94–97.

27. Das, U. and M.A. Ansari, *The nexus of climate change, sustainable agriculture and farm livelihood: contextualizing climate smart agriculture.* Climate Research, 2021. **84**: pp. 23–40.

28. Steenwerth, K.L., et al., *Climate-smart agriculture global research agenda: Scientific basis for action.* Agriculture and Food Security, 2014. **3**(1): pp. 1–39.

29. Totin, E., et al., *Institutional perspectives of climate-smart agriculture: A systematic literature review.* Sustainability (Switzerland), 2018. **10**(6): pp. 1–20.

30. Bazzana, D., J. Foltz, and Y. Zhang, *Impact of climate smart agriculture on food security: An agent-based analysis.* Food Policy, 2022. **111**: pp. 1–20.

31. Zougmoré, R., et al., *Toward climate-smart agriculture in West Africa: A review of climate change impacts, adaptation strategies and policy developments for the livestock, fishery and crop production sectors.* Agriculture and Food Security, 2016. **5**(1): pp. 1–16.

32. Bagley, J.E., J. Miller, and C.J. Bernacchi, *Biophysical impacts of climate-smart agriculture in the Midwest United States.* Plant Cell and Environment, 2015. **38**(9): pp. 1913–1930.

33. Torquebiau, E., et al., *Identifying climate-smart agriculture research needs.* Cahiers Agricultures, 2018. **27**(2): pp. 1–7.

34. Ait Issad, H., R. Aoudjit, and J.J.P.C. Rodrigues, *A comprehensive review of Data Mining techniques in smart agriculture.* Engineering in Agriculture, Environment and Food, 2019. **12**(4): pp. 511–525.

35. Martinez-Baron, D., et al., *Small-scale farmers in a 1.5°C future: The importance of local social dynamics as an enabling factor for implementation and scaling of climate-smart agriculture.* Current Opinion in Environmental Sustainability, 2018. **31**: pp. 112–119.

36. Singh, R. and G.S. Singh, *Traditional agriculture: a climate-smart approach for sustainable food production.* Energy, Ecology and Environment, 2017. **2**(5): pp. 296–316.

37. Rehman, A., et al., *A Revisit of Internet of Things Technologies for Monitoring and Control Strategies in Smart Agriculture.* Agronomy, 2022. **12**(1): pp. 1–21.

38. El Bilali, H., et al. *Information and communication technologies for smart and sustainable agriculture.* in *30th International Scientific-Experts Conference of Agriculture and Food Industry, AgriConf 2019.* 2020. Springer.

39. Dozono, K., S. Amalathas, and R. Saravanan, *The impact of cloud computing and artificial intelligence in digital agriculture,* in *6th International Congress on Information and Communication Technology, ICICT 2021,* X. Yang, et al., Editors. 2022. Springer Science and Business Media Deutschland GmbH. pp. 557–569.

40. Ruiz-Real, J.L., et al., *A look at the past, present and future research trends of artificial intelligence in agriculture.* Agronomy, 2020. **10**(11): pp. 1–16.

41. Yang, X., et al., *A survey on smart agriculture: development modes, technologies, and security and privacy challenges.* IEEE/CAA Journal of Automatica Sinica, 2021. **8**(2): pp. 273–302.

42. Darwin, B., et al., *Recognition of bloom/yield in crop images using deep learning models for smart agriculture: A review.* Agronomy, 2021. **11**(4): pp. 1–22.

43. Kamienski, C., et al., *Smart water management platform: IoT-based precision irrigation for agriculture.* Sensors (Switzerland), 2019. **19**(2): pp. 1–20.

44. Shafi, U., et al., *Precision agriculture techniques and practices: From considerations to applications.* Sensors (Switzerland), 2019. **19**(17): pp. 1–25.

45. Marcu, I., et al., *Arrowhead technology for digitalization and automation solution: Smart cities and smart agriculture.* Sensors (Switzerland), 2020. **20**(5): pp. 1–27.

46. Jariwala, H., et al., *Controlled release fertilizers (CRFs) for climate-smart agriculture practices: a comprehensive review on release mechanism, materials, methods of preparation, and effect on environmental parameters.* Environmental Science and Pollution Research, 2022.

47. Hou, Q., et al., *Ncs-delivered pesticides: A promising candidate in smart agriculture.* International Journal of Molecular Sciences, 2021. **22**(23): pp. 1–15.

48. Moulick, R.G., et al., *Potential use of nanotechnology in sustainable and 'smart' agriculture: advancements made in the last decade.* Plant Biotechnology Reports, 2020. **14**(5): pp. 505–513.

49. Sambo, P., et al., *Hydroponic solutions for soilless production systems: issues and opportunities in a smart agriculture perspective.* Frontiers in Plant Science, 2019. **10**: pp. 1–17.

50. Coppedè, N., et al., *An in vivo biosensing, biomimetic electrochemical transistor with applications in plant science and precision farming.* Scientific Reports, 2017. **7**(1): pp. 1–9.

51. Antonacci, A., et al., *Nanostructured (Bio)sensors for smart agriculture.* TrAC – Trends in Analytical Chemistry, 2018. **98**: pp. 95–103.

52. Wambi, W., et al., *Genetic and genomic resources for finger millet improvement: opportunities for advancing climate-smart agriculture.* Journal of Crop Improvement, 2021. **35**(2): pp. 204–233.

53. Cardi, T., N. D'Agostino, and P. Tripodi, *Genetic transformation and genomic resources for next-generation precise genome engineering in vegetable crops.* Frontiers in Plant Science, 2017. **8**: pp. 1–15.

54. Mupambwa, H.A., et al., *The unique Namib desert-coastal region and its opportunities for climate smart agriculture: A review.* Cogent Food and Agriculture, 2019. **5**(1): pp. 1–22.

55. Rampa, A., Y. Gadanakis, and G. Rose, *Land reform in the era of global warming—can land reforms help agriculture be climate-smart?* Land, 2020. **9**(12): pp. 1–24.

56. Azadi, H., et al., *Rethinking resilient agriculture: from climate-smart agriculture to vulnerable-smart agriculture.* Journal of Cleaner Production, 2021. **319**: pp. 1–10.

57. Makate, C., *Effective scaling of climate smart agriculture innovations in African smallholder agriculture: A review of approaches, policy and institutional strategy needs.* Environmental Science and Policy, 2019. **96**: pp. 37–51.

58. Pathan, M., et al., *Artificial cognition for applications in smart agriculture: A comprehensive review.* Artificial Intelligence in Agriculture, 2020. **4**: pp. 81–95.

59. Waaswa, A., et al., *Climate-Smart agriculture and potato production in Kenya: review of the determinants of practice.* Climate and Development, 2021.

60. Partey, S.T., et al., *Developing climate-smart agriculture to face climate variability in West Africa: Challenges and lessons learnt.* Journal of Cleaner Production, 2018. **187**: pp. 285–295.

12 Crop protection and agricultural green chemistry

For agriculture to produce sustainable quantities of food, the crops must be protected. Crop protection chemicals must be sustainably produced and used in an environmentally conscious manner. Arguably, the most important challenge facing farmers is crop protection.[1] (See Figure 12.1)

12.1 INTRODUCTION

The modern agrochemical industry is currently confronted with chemical tasks and many challenges.[2] These challenges include consumer preferences, which are focusing more and more on quality, including ethical aspects and a healthier diet, high demand for the raw materials, tougher corporate competition, stringent regulations and the need for continual innovation. Within these challenges is the basic requirement to guarantee the food, feed, fiber, and fuel production are in high yield and quality. The best economical, ecological, and environmental practices for sustainable agriculture must be practiced, all while the individual farmer is attempting to secure a personal livelihood.[3]

Scientific and technological advances occurring today have been opening new possibilities for farmers around the world. The networked digital farm of the future is already making agriculture more efficient and sustainable today. Agriculture is practiced on a massive scale, with one in three persons working in agriculture globally. Agrochemicals protect crops from competition for nutrients by weeds, insect damage, and disease. These chemicals will continue to be necessary agricultural tools, as farmers strive to feed a growing global population. According to the Food and Agriculture Organization of the United Nations, more than 4 million metric tons of agrochemical crop protection chemicals (CPCs) are applied annually across the world. Because of this tremendous usage volume, the potential for significant process safety hazards during the production and use of crop protection chemicals is magnified.[4]

Efforts are being made to improve the types of agrichemicals being used. Current high-throughput field phenotyping methods can be used as a basis for developing and improving techniques useful for precision agriculture.[5] There is the dynamic

DOI: 10.1201/9781003157991-12

FIGURE 12.1 The challenge of crop protection. (See www.shutterstock.com/image-photo/tractor-spraying-pesticides-wheat-fields-1389867185)

interrelationship of the biological system (soil, plant, climate or habitat), combined with current challenges such as climate change, biotic or abiotic stress, soil erosion, the growing world population, energy and workforce.[6] In addition, with limited arable land and a continually growing world population, the available farmland per capita is expected to decrease dramatically.[7] All of these stressors that must be addressed in the efforts of crop protection.

The drive to improve agrichemicals in the twenty-first century can showcase the efforts in green chemistry. In the 20 years since the publication of *Green Chemistry: Theory and Practice* by Anastas and Warner,[8] the 12 principles of green chemistry have provided a framework for the responsible design of new chemicals and chemical processes. These principles are key features that reduce environmental footprints and improve the safety of chemical processes and products. Because of the sheer magnitude of agricultural production, the potential impact of applying the principles of green chemistry to the design and production of crop protection manufacturing processes and products far exceeds that of the pharmaceutical industry. In addition, since crop protection products are applied directly to the environment on food crops, their fate and toxicities are especially relevant to consumers, farmers, and governmental regulatory agencies. The 12 principles of green chemistry provide a useful framework with which to improve the safety and sustainability of agrochemical production and use, especially as we move through the twenty-first century.[9]

Modern trends in the greener synthesis and fabrication of inorganic, organic and coordination compounds, materials, nanomaterials, hybrids and nanocomposites will improve the environmental impact of CPCs. Green chemistry deals with the classic 12 principles, contributing to the sustainability of chemical processes, energy savings, lesser toxicity of reagents and final products, lower damage to the environment and human health, decreasing the risk of global warming, and more rational use of natural resources and agricultural wastes. Non-hazardous solvents including ionic liquids, use of plant extracts, fungi, yeasts, bacteria and viruses can be useful.[10]

   This leads to a fundamental question of what the role of the modern agricultural chemistry in the future will be.[11]

12.2   CLASSES OF CPCS USED IN TWENTY-FIRST-CENTURY AGRICULTURE

12.2.1   NEED FOR INNOVATION

Continual innovation in modern agricultural chemistry is vital, as in the field of crop protection and agricultural stress relief. Without crop protection, yields would be reduced by around 50% as compared to those currently attained with crop protection against pests, weeds, and diseases. However, given the losses due to pests, weeds, and agricultural pathogens, as well as to storage and logistics failure, theoretically attainable yield might be around 170% of current yields. This means that, through innovations in the field of crop protection and stress relief (e.g. without abiotic stress), the current total yield and quantity could be further increased significantly.[12]

   To meet the evolving demands, a continuous supply of novel crop protection agents is needed. For new active ingredients, the market demands are rapidly changing because of emerging resistance in insects, weeds and agricultural pathogens, as well as resistance management, increased regulatory demands, requirements resulting from food chain partnerships, invasive species of insects, weeds and agricultural pathogens and shifts in existing pests. Therefore, there is a strong need for innovative active ingredients with favorable properties such as novel modes of action (MoAs), for crop enhancement effects that could be used to improve plant health.[13]

   Agrochemical companies aim to supply integrated solutions to the market, but major research and development activities are still required to focus on meeting this evolving challenge in agricultural chemistry. The major efforts in this research continue to focus on economically important target-segments of crop protection, including fungicides, herbicides, insecticides and safeners, as well as nematicides.[11]

12.2.2   FUNGICIDES FOR DISEASE CONTROL

Fungicides, herbicides, and insecticides are all pesticides used in plant protection. A fungicide is a specific type of pesticide that controls fungal disease by specifically inhibiting or killing the fungus causing the disease. Not all diseases caused by fungi can be adequately controlled by fungicides. For many years, the basic cellular functions have been primary targets in fungicide discovery research. Only six MoAs dominate this market segment. These MoAs in agricultural market products include around 30% demethylation inhibitor fungicides (DMIs), which have been used to manage a wide range of turfgrass diseases for many decades.

The loss of agrochemicals due to consumer attitudes and perceptions, changing grower needs and everchanging regulatory requirements is higher than the number of active ingredients being introduced into the market. Therefore, the development of new agrochemicals is essential. Especially needed are those that can provide improved efficacy and favorable environmental profiles. In this context, introduction of halogen atoms into a molecule is an important tool to influence its physico-chemical properties. Since 2010, around 81% of the launched agrochemicals contain halogen atoms.[13]

The use of synthetic fungicides to control fungal diseases has been growing. However, limitations due to eco-toxicological risks are a major hurdle. Therefore, it is necessary to replace or integrate high risk chemicals with safer compounds for human health and environment. Consequently, research on the selection, evaluation, characterization, and use of biocontrol agents (BCAs) has increased since 2000. BCA chemicals are still scarce when compared to the growing demand for their use in sustainable agricultural management. To foster development and utilization of new effective bioformulates, there is a need to optimize BCA activity, to share knowledge on their formulation processes and to simplify the registration procedures. Studies based on new fungicides and molecules can significantly contribute to the achievement of such objectives.[14]

12.2.3 HERBICIDES FOR WEED CONTROL

Herbicides, also commonly known as weedkillers, are substances used to control undesired plants and weeds. Selective herbicides control specific weed species, while leaving the desired crop relatively unharmed, while non-selective herbicides (sometimes called total weedkillers in commercial products) can be used to clear waste ground, industrial and construction sites, railways and railway embankments as they kill all plant material with which they come into contact. High-yield farming production systems are based on efficient herbicides as well as on integrated cultivation systems. The global trend towards simplification of crop rotation will continue, as illustrated by corn and soy bean farming in the United States and wheat and oilseed production in the European Union. Conservation tillage will also to increase, and the importance of herbicide resistance will grow significantly. It can be assumed that the number of available herbicides for the farmer will further decline because of difficult regulatory requirements. Three herbicide classes represent approximately 50% of the world market and the consolidation process in agricultural chemistry has resulted in only a few remaining companies with dedicated and broad herbicide research capability.

To avoid significant problems for agriculture, a new herbicide technology is urgently required. For herbicide market products, plant-specific approaches are the most important. Unfortunately, no major MoA has been introduced into the marketplace for more than two decades.[15] Today, only six mechanisms of action dominate, and these account for around 80% of the herbicide market. These MoAs include

- 5-enolpyruvylshikimate-3-phosphate synthase inhibitor (glyphosate),
- acetolactate synthase (sulfonyl ureas),
- very-long-chain fatty acid elongase (chlorinated acetanilides), and
- photosystem II (triazines).

The presence of herbicide-resistant weeds, particularly in major field crops, is a widespread problem, however, and a significant challenge for global food security. [16] Unfortunately, more than 60% of the global herbicide market is represented by products with MoAs that already today have serious resistance issues.

Weeds are among the most dominant adversaries of agricultural and nutritional crops causing major yield loss, ranging between 10% to 98% of total crop yield, which may vary by crop, or region or even by crops within the same field. This loss in crop yield results when parasitic weeds attach themselves to another plant, their 'host', and draw nutrients from it causing huge damage to the host crop and consequently huge economic loss. *Orobanche spp.*, *Striga spp.*, and *Cuscuta spp.* are the most common parasitic agricultural weeds with economic consequences in many areas of the world. Applying control methods selective enough for killing these parasitic weeds without causing crop damage is difficult as the application of chemical herbicides which causes soil and water contamination and adverse effects to beneficial organisms and hence loss in the nutritional benefits of the cultivated crop. In some cases chemical compounds which result from herbicide degradation process may continue to be significantly toxic to health and environment.[17]

12.2.4 SAFENERS FOR WEED CONTROL

Herbicide safeners are chemical compounds used in combination with herbicides to make them "safer"; that is, to reduce the effect of the herbicide on crop plants, and to improve selectivity between crop plants vs. weed species being targeted by the herbicide. Herbicide safeners can be used to pretreat crop seeds prior to planting, or they can be sprayed on plants as a mixture with the herbicide.[18]

It has been known for some time that broad-spectrum herbicides can be combined with a safener for crop protection and efficient weed management. The safener induces the degradation of the active ingredient only in the crop, but not in the weed. Full crop selectivity of some highly effective herbicides can be obtained by using safeners, which increase the degradation or detoxification of the herbicide in the corresponding agricultural crop.[19] This effect is connected with the increased expression of genes coding for enzymes responsible for degradation in the crop, such as cytochrome-P450-dependent monooxygenases, glutathione S-transferases and transporters. Since 2009, the new corn safener cyprosulfamide from Bayer CropScience has been combined with the herbicide thiencarbazone-methyl in the herbicidal product Adengo®. The high crop tolerance to two or three highly active ingredients in one commercial product is ensured by this novel safener cyprosulfamide, which can protect corn against herbicide damage via root uptake and via leaf uptake. In addition, it was found that cyprosulfamide, alone or with abscisic acid protected plants (e.g. rice) from salinity stress and induced vigorous growth, including the formation of new tillers and early flowering. Safeners for dicotyledenous crops such as soybean, canola and sugar beet have not been identified.[19,20]

12.2.5 INSECTICIDES FOR PEST CONTROL

Humans have used insecticides for centuries. The spectrum and potency of available insecticidal substances has greatly expanded since the industrial revolution, resulting in widespread use and unforeseen levels of synthetic chemicals in the environment. Concerns about the toxic effects of these new chemicals on non-target species became public soon after their appearance, which eventually led to the restrictions on their use. Since that time, new, more environmentally friendly insecticides have been developed, based on naturally occurring chemicals, such as pyrethroids (derivatives of pyrethrin), neonicotinoids (derivatives of nicotine), and insecticides based on the neem tree vegetable oil (*Azadirachta indica*), predominantly azadirachtin. Although these new substances are more selective toward pest insects, they can still target other unintended organisms. Neonicotinoids, for example, have been implicated in the decline of the bee population worldwide.[21]

Molecular targets play important roles in agrochemical discovery, and particularly in herbicide research. Numerous pesticides target the key proteins in pathogens, insects, or plants. Investigating ligand-binding pockets and/or active sites in the proteins' structures is usually the first step in designing new green pesticides. Thus, molecular target structures are extremely important for the discovery and development of green, sustainable pesticides. New molecular target studies for fungicidal, bactericidal, insecticidal, herbicidal, and plant growth-regulator targets currently used in agrochemical research are critical. The data will be helpful in pesticide design and the discovery of new green pesticides.[22]

Insecticide resistance is a long-standing problem affecting the efficacy and utility of crop protection compounds. Insecticide resistance also impacts the ability and willingness of companies around the world to invest in new crop protection compounds and traits. The Insecticide Resistance Action Committee (IRAC) was formed in 1984 to provide a coordinated response by the crop protection industry to the problem of insecticide resistance. Since its inception, participation in IRAC has grown from a few agrochemical companies in Europe and the US to a much larger group of companies with global representation and an active presence involving an even wider array of companies in more than 20 countries. The focus of IRAC has also evolved from that of defining and documenting cases of insecticide resistance to a pro-active role in addressing insecticide resistance management providing an array of informational and educational tools (videos, posters, pamphlets) on insect pests, bioassay methods, insecticide mode of action and resistance management, all publicly available. A key tool developed by IRAC is the Insecticide MoA Classification Scheme, which has evolved from a relatively simple acaricide classification started in 1998 to the far broader scheme that now includes biologics as well as insecticides and acaricides. The IRAC MoA Classification Scheme coupled with expanding use of MoA labeling on insecticide and acaricide product labels provides a straightforward means to implement the management of insect resistance.[23]

Interactions among pesticides at common molecular targets combined with detoxification systems often determine their effectiveness and safety. Compounds with the same mode of action or target are candidates for cross resistance and restrictions in their recommended uses. Discovery research is therefore focused on new mechanisms and modes of action. Interactions in detoxification systems also provide cross resistance and synergistic and safener mechanisms As shown with serine hydrolases and inhibitors, cytochrome P450 and insecticide synergists, and glutathione S-transferases and herbicide safeners. Secondary targets are also considered for inhibitors of serine hydrolases, aldehyde dehydrogenases, and transporters. The mechanistic aspects of interactions, depends on potency, exposure, ratios, and timing. The benefits of pesticide interactions are the additional levels of chemical control to achieve desired effects. The risks are the unpredictable interactions of complex interconnected biological systems.[24]

For effective insecticide products, neuronal and muscle targets are most important. Only three MoAs dominate, accounting for most of this market segment. These MoAs as opposed to agricultural market products include nicotinic acetylcholinereceptor (nAChR) competitive modulators such as theneonicotinoids, sodium channel modulators (SoCh) such as the pyrethroids and acetylcholinesterase (AChE) inhibitors organophosphates and carbamates.[25]

The removal of contamination caused by application of pesticides is a major concern. Constructed wetland (CW) is a popular sustainable best management practice for treating different wastewaters. Recent findings on the occurrence of pollutants other than nutrients that occur in agricultural field runoff and wastewater from animal facilities, including pesticides, insecticides, veterinary medicine, and antimicrobial-resistant genes in the agricultural runoff water, their removal by different wetlands (surface flow, subsurface horizontal flow, subsurface vertical flow, and hybrid), show removal mechanisms, and illustrate the factors that affect the removal. The information can be used to highlight the current research gaps and needs for resilient and sustainable treatment systems. Factors, including contaminant property, aeration, type, and design of CWs, hydraulic parameters, substrate medium, and vegetation, impact the removal performance of the CWs. Hydraulic loading and hydraulic retention were found to be important for the removal of agricultural pollutants from wetlands. The pollutants in agricultural wastewater, excluding nutrients and sediment, and their treatment utilizing different nature-based solutions, such as wetlands, requires more investigation.

More long-term research in the actual field utilizing environmentally relevant concentrations to seek actual impacts of weather, plants, substrates, hydrology, and other design parameters is needed.[26] This is a vital component for sustainable agricultural chemistry in the twenty-first century.

12.2.6 Nematicides

Plant-parasitic nematodes (PPN) have caused huge economic losses to agriculture worldwide and seriously threaten the sustainable development of modern agriculture. Chemical nematicides are still the most effective means to manage nematodes. However, the long-term use of organophosphorus and carbamate nematicides has led to a lack of field control efficacy and increased nematode resistance. To meet the huge market demand and slow the growth of resistance, new nematicides are needed through chemical Research and Development. The rational design and synthesis of new chemical scaffolds to screen for new nematicides is still a difficult task. Research progress on nematicidal compounds in the past decade, highlighting the structure-activity relationship and mechanism of action, has yielded some new nematicidal active fragments.[27]

PPNs such as the root-knot nematode (*Meloidogyne spp.*), cyst nematodes (*Globodera spp.* and *Heterodera spp.*) and migratory nematodes (*Radopholus spp.*, *Pratylenchus spp.* and *Helicotylenchus spp.*) infest many important agricultural crops, such as soybean, coffee, potato, sugarbeet, corn, banana, etc., and they are responsible for approximately 12% of world crop production losses. Besides damaging roots, they can also transmit viruses and make plants more vulnerable to attack by bacterial and fungal pathogens in the soil. PPNs have been controlled through extensive application of the fumigant nematicide methyl bromide, which is now restricted owing to its ozone-depleting properties. [28] Currently available older active ingredients used as nematicides such as the acetylcholinesterase (AChE) inhibitor organophosphates (OPs) and carbamates have unfavorable toxicological and ecotoxicological profiles, are applied at high application rates but are gradually being withdrawn from further use. Therefore, the search for innovative solutions useful for integrated nematode management is very important.[11]

The use of nematicides is an effective way of controlling PPNs. However, the long-term use of traditional organophosphorus and carbamate chemical nematicides can lead to increased nematode resistance. With the increasing awareness of the necessity for the protection of the environment and human health, highly toxic nematicides no longer meet the developmental requirements of modern agriculture. Recently, many studies have been undertaken on the isolation and nematicidal activity of natural products against PPNs and *Caenorhabditis elegans*. As an important model nematode, *C. elegans* plays a vital reference role in studying plant-parasitic nematodes regarding nematicidal activity, metabolic mechanism, and modes of action and target.[29]

12.2.7 Managing microbes

Research on the ability to manage microbes as a long-term crop protection strategy is important. Soil microbes have critical influence on the productivity and sustainability of agricultural ecosystems, yet the magnitude and direction to which management practices affect the soil microbial community remain unclear. Conventional grain cropping, organic grain cropping, and grain cropping-pasture rotation, all have effects on the soil microbial community structure and putative gene abundances of Nitrogen (N) transformations. A better assessment of the soil microbial community in terms of variation scale and regulatory importance of management intensity vs. plant type is needed. Farming systems significantly affect the biodiversity of soil fungi but not bacteria. Bacterial and fungal communities in the above three cropping systems suggest that management practices play minor roles in shaping the soil microbial community compared to plant type (i.e., woody vs. herbaceous plants). However, management practices prominently regulated habitat-specific taxa. Data indicates that while moderately affecting the overall structure of the soil microbial community, management practices, particularly fertilization and the source of N (synthetic vs. organic), were important in regulating the presence and abundance of habitat-specific species. Farming management moderately shapes the soil microbial community structure and promotes habitat-specific species.[30] This is a further example of working with nature to achieve a sustainable agricultural chemistry in the twenty-first century.

12.3 PRINCIPLES OF GREEN CHEMISTRY APPLIED IN CROP PROTECTION

Since CPCs are essential in the practice of agriculture, their manufacture and use must align with current environmental demands. The sustainability of agriculture is driven by the efforts in agricultural chemistry (AC). AC is itself becoming green through the practice of green chemistry. Systematic consideration of the 12

FIGURE 12.2 Green principles to guide agricultural chemistry. (See www.shutterstock.com/image-photo/image-chemical-flasks-stylized-ball-hourglass-1029329263)

principles of green chemistry as applied in agriculture and AC (See Figure 12.2) provides a useful framework with which to improve the safety and sustainability of agrochemical production and use.[9] Green chemistry includes metrics and methodologies, such as life cycle analysis, that provide a more comprehensive, quantitative assessment of chemical processes. These processes produce the crop protection chemicals. Unfortunately, complete details of commercial manufacturing processes are typically not disclosed in scientific publications or patents, and much information is kept as trade secrets. Because of this lack of access to proprietary information, a full quantitative analysis of green chemistry metrics is not possible, but it is worthwhile to highlight what is known and where the science is going in sustainability. The next 12 sections follow the 12 principles of green chemistry (see Chapter 10: Circularity: Environmental, Chemical, Agricultural), and provide illustrative examples of the application of green chemistry to the design, synthesis, and use of commercial CPCs.

12.3.1 Prevent waste

Waste prevention is the first and perhaps most important principle related to the manufacturing footprint of an active ingredient. This principle is especially relevant to agrochemicals because effective minimization of waste results in a lower cost of manufacture due to more efficient conversion of raw materials to products and reduced waste treatment cost. This will translate to lower cost to farmers and thereby greater usage. Producers of CPCs therefore face significant pressure to develop low-cost processes. Compared with pharmaceutical processes, where patients may be less aware of drug manufacturing costs, farmers typically operate with small profit margins and fixed commodity prices. For this reason, minimizing manufacturing costs by waste prevention is one critical factor in the commercial success of a new agrochemical active ingredient. Process chemists and engineers are often faced with competing priorities when designing a new chemical process, and compromises are often necessary to minimize manufacturing costs. In this context, greenness and green chemistry processes must provide economic value in addition to societal value.

The compound glyphosate is an example of the way economic factors heavily weight the selection of the manufacturing process.[31] The cost of manufacturing this compound has been reduced by lowering waste generated in the overall process. A key feature that reduced waste in this process was the use of catalytic as opposed to stoichiometric transformations, another of the green principles. The amount of solvent used was also reduced, showing that solvent can account for a significant amount of waste produced by active ingredient manufacturing processes.[32]

As we have illustrated, crop wastes are also agricultural wastes generated during the production and processing of food materials. Their generation is the other side of improved sustainability activities. They are now becoming an alarming source of environmental pollution, leading to an unhealthy society. There is an urgent need to develop robust methods to utilize these types of wastes as feedstocks for beneficial compounds or materials. Much work has been successfully done in these areas, and several strategies have been developed to produce biochemicals from biological wastes. In other words, value addition has been done to the crop waste materials. The chemicals like carbohydrates, minerals, proteins, and other compounds have been isolated from various crop residues.[33] The chemical feedstock from waste forms raw material streams from the biorefinery, which is the topic of the next chapter.

The (re)utilization of silicon, silica and silicates is also becoming highly important as there is a growing global need for renewable bioenergy, in the form of controlled biomass burning such as agricultural straws, wood pellets, grasses, etc. The anticipated growth in controlled biomass burning will generate significant quantities of ash often rich in silica and silicates. The valorization of these waste materials is vital to ensure recovery and reuse of the inorganic species, in line with an elemental sustainability and biorefinery vision, and to add economic value in the form of biobased adhesives and coatings. The use of silicon and silicates can lead to a potential role as replacements for traditional petroleum-based binders in particleboards.[34]

12.3.2 MAXIMIZE ATOM ECONOMY

Atom economy (AE) or atom utilization was one of the first defining terms in the sustainable chemistry movement. In contrast to the 12 (qualitative) principles of green chemistry, AE represents a metric for quantification purposes.[35] AE (atom efficiency/percentage) is the conversion efficiency of a chemical process in terms of all atoms involved and the desired products produced. The simplest definition was introduced by Barry Trost in 1991 and is equal to the ratio between the mass of desired product to the total mass of products, expressed as a percentage. A strong increase in agricultural productivity and sustainable food production has become indispensable due to the constantly growing world population. Crop yields suffer greatly from weeds, harmful insects, and fungal or other pathogens, therefore effective crop protection strategies are needed to protect crops and ensure higher yields. Changing regulatory requirements and resistance developments lead to a constant demand for new crop protection products with good sustainability profiles and novel modes of action (MOA). AE helps to make this possible through efficient and economical production routes.[36] Importantly, these compounds must be synthesied efficiently, as measured by AE.

AE in agricultural chemistry is important, and it is often found in the mechanistic explanations of successful chemical reactions. For example, 2,5-Dimethylpyrazine (2,5-DMP) is an important active intermediate and an important essence. Conventional chemical synthesis methods are often accompanied by toxic substances as by-products, and the biosynthesis efficiency of 2,5-DMP is insufficient for industrial applications. A genetically engineered Escherichia coli strain with the highest carbon recovery rate (30.18%) and the highest yield reported to date was successfully constructed, and 9.21 g·L-1 threonine was able to produce 1682 mg·L-1 2,5-DMP after 24 h. At the same time, an expression regulation strategy and whole-cell biocatalysts helped to eliminate the damage to cells caused by 2,5-DMP, aminoacetone, and reactive oxygen species generated by aminoacetone oxidase from *S. oligofermentans*, and the negative effect of 2-amino-3-ketobutyrate CoA ligase on the yield of 2,5-DMP in *E. coli* was also demonstrated.[37]

As an example, the methods for the preparation of nanomaterials including magnetic nanoparticles and their catalytic activities are examples of greener and atom-economic multicomponent reactions; the high-value utilization of abandoned yet abundant renewable resources from biowaste is beneficial for the environment. [38]

12.3.3 DESIGN LESS HAZARDOUS SYNTHESIS OF CPCS

In order to be effective, CPCs must be effective in their activity against pests. During the manufacturing process, the structures of crop protection active ingredients exhibit a wide variety of functional groups that must be installed using inexpensive raw materials because of the cost constraints of agriculture. Hazardous substances, such as chlorine, phosgene, isocyanates, mercaptans, sodium azide, sodium cyanide, and HF are used in numerous manufacturing processes for older active ingredients. These toxic synthetic intermediates are a problem, and the newer syntheses are designed to avoid their use. The toxicity of synthetic intermediates in a chemical process should be evaluated as part of development activities since workers could potentially be exposed to these materials during production.[39]

Green synthetic protocols refer to the development of processes for the sustainable production of chemicals and materials. For the synthesis of various biologically active compounds such as agrochemicals, energy-efficient and environmentally benign processes are important, such as microwave irradiation technology, ultrasound-mediated synthesis, photo-catalysis (ultraviolet, visible and infrared irradiation), molecular sieving, grinding and milling techniques, etc. These techniques are assets for sustainability and are valuable green tools to aid in synthesizing new drug molecules as they provide numerous benefits over conventional synthetic methods. They directly lead to less hazardous syntheses. For instance, using these tools, oxadiazole derivatives are synthesized under microwave irradiation conditions to reduce the formation of byproducts, thereby increasing the product yield and doing so in less time. Oxadiazole is drawing considerable interest for the development of new active candidates with potential for applications as a CPC. This shows that the synthesis of active molecules under microwave irradiation can become a green chemistry approach that aligns with the 12 principles to minimize or remove the production of hazardous toxic materials during the design, manufacture and application of chemical substances. This approach plays a major role in controlling environmental pollution by utilizing safer solvents, catalysts, suitable reaction conditions and thereby increases the atom economy and energy efficiency. Oxadiazole is a five-membered heterocyclic compound that possesses one oxygen and two nitrogen atoms in the ring system. [40] This provides one example of how modifying one step in a chemical synthesis can lead to multiple benefits toward sustainability.

12.3.4 DESIGN SAFER CHEMICALS AND PRODUCTS

Due to their intended function, crop protection products are highly regulated and registered for use in each country where they are sold. In addition, regulatory agencies require periodic re-registration of active ingredients. For example, in the United States agrochemical active ingredients must undergo review by the EPA every 10 years. These regulations are the baseline requirements for sale of crop protection products, and manufacturers often perform additional studies triggered by internal risk assessment procedures. Acute toxicological effects of CPCs often occur on those who come in contact with them from usage, causing skin sensitization, inhalation, and eye irritation, or during product development. Predictive *in vitro* assays have been developed that allow screening of acute toxicological effects without the need for animal testing.[41] Chronic toxicological effects are also investigated and provide key data to support registration. An overview of the regulatory requirements for the introduction of new crop protection active ingredients has been provided by Gehen.[42]

Various methods have been developed to estimate the environmental impact of agrochemical products to enable growers to make more informed decisions. The environmental impact quotient (EIQ) was developed as a composite qualitative score that incorporates use rate and data related to potential toxic effects on farm workers and aquatic and terrestrial nontarget organisms.[43] In addition, a variable in EIQ was included to incorporate the potential consumer exposure to an agrochemical related to its half-life in soil and plant matter. Although widely used, the EIQ has been criticized because of its use of weighted risk factors to generate a simple score that may not adequately relay the complexities of agrochemical environmental effects. As discussed earlier, the use of nano-scale and material dosage techniques are part of the approaches used in SMART farming.

12.3.5 USE SAFER SOLVENTS AND REACTION CONDITIONS

Solvent selection plays an important role in the design of an active ingredient manufacturing process, as we mentioned above with microwaves. Several useful solvent selection guides have been published by process chemists from the pharmaceutical industry.[44] The flammability, volatility, reactivity, and toxicity of solvents must be evaluated during process development. In addition, the ability to telescope sequential reactions using a single, common solvent is desirable. Special attention must be paid to understand hazards due to potentially incompatible raw materials that are often not recognized or evaluated by researchers on a small scale. One notable example is the incompatible combination of NaH with dimethylformamide, dimethyl sulfoxide, or dimethylacetamide that is often unwittingly used on a small scale.[45] Reactive chemical testing and hazard evaluation are critically important to minimize the potential for solvent-related safety issues. Continuous flow chemistry provides the ability to perform certain highly exothermic reactions more safely.

As an alternative to synthetic pesticides, natural chemistries from living organisms are not harmful to nontarget organisms and the environment. These can be used as biopesticides, nontarget. However, to reduce the reactivity of active ingredients, avoid undesired reactions, protect from physical stress, and control or lower the release rate, encapsulation processes can also be applied to biopesticides. The use of supercritical fluid technology (SFT), mainly carbon dioxide (CO_2), to encapsulate biopesticides is currently used. It reduces the use of organic solvents, has simpler separation processes, and achieves high-purity products. There is a lack of application of SFT for biopesticides in the published literature which is necessary to evaluate its potential and prospects.[46]

Ionic liquids are also being used. These can contain biologically active anions and cations, their potential application in the field of agrochemistry and agriculture have been studied. Ionic liquids can be used as herbicides, fungicides, antimicrobial agents, deterrents, and plant growth stimulants. There are advantages and disadvantages of using ionic liquids, such as their multitasking, toxicity, thermal stability and solubility in water in comparison with commercial chemicals. New results show encouraging prospects for the use of ionic liquids in agriculture, as well as the high value of using ILs as multifunctional biologically active substances.[47]

12.3.6 Design for energy efficiency and production of biofuel

Processes that efficiently produce active ingredients reduce the amount of energy used in manufacture and transport of raw aterials and disposal of waste. Lignocellulosic biomass is the most abundant renewable resource on earth and currently most of this biomass is considered a low-value waste. Specifically, lignin is a bioresource that is mostly burned for energy production but also has a few value-added products that can be derived from it. Since the agro-food industry produces large amounts of wastes that can be potential sources of high-quality lignin, this is now a potentially rich field. The extraction of lignin from various agro-food sources and together with applications of lignin in the agro-food chain means it is a critical area for product development. The extraction process efficiency (yield) and lignin purity are used as indicators of the raw material potential. Overall, it is notable that research interest on agro-food lignin has increased exponentially over the years, both as source and application. Wheat, sugarcane, and maize are the most studied sources and are the ones that render the highest lignin yields. As for the extraction methods used, alkaline and organosolvent methods are the most employed. The main reported applications are related to lignin incorporation in polymers and as antioxidant. Studies on agro-food system-derived lignin is highly important since there are numerous possible sources that are yet to be fully valorized and many promising applications that need to be further developed.[48]

As a second example, the increase in energy demand across the world has put immense pressure on utilization of widely available lignocellulosic agricultural waste biomass and forest residues. Physico-chemical characteristics of leaf litter from largely grown tree species such as *Mangifera indica*, *Populus deltoides* and *Polyalthia longifolia* have been evaluated for possible use in biorefinery. The calorific values of all biomasses were in the range of 18.37 to 19.32MJ/kg. Delineation of these physico-chemical characteristics together per se shows that leaf litter biomass can also act as a potential feedstock for biofuel production.[49]

The challenges in reducing the world's dependence on crude oil and the greenhouse gas accumulation in the atmosphere, while simultaneously improving engine performance through better fuel efficiency and reduced exhaust emissions, have led to the emergence of new fuels. The new fuel formulations blend petrodiesel, biodiesel, bioethanol and water in various proportions. The sustainability of new biofuel industries also requires a high level of bioresources and new methods. Simply, the industry must use a variety of resources that do not compete with edible crops (nor by using arable land for energy crops or food crops for energy production) and flexible conversion technologies satisfying the eco-design, eco-energy and eco-materials criteria. For example, the supercritical ethanolysis of lipid resources to produce ethyl biodiesel is a simple but efficient route that can reach sustainability. The production of ethyl biodiesel via triglyceride supercritical ethanolysis is a potentially viable pathway toward sustainability. The scientific and technical bottlenecks requiring further development include

- the kinetic and thermodynamic aspects (experiments and modeling) required for the process simulation,
- need for securing the life cycle assessment, to improve the supercritical process performance in terms of eco-material and eco-energy,
- the impacts of ethyl vs. methyl biodiesel fuels and of biodiesel-ethanol-petrodiesel blends (with or without water) on the diesel engine emissions and performance, and
- the technological flexibility of the supercritical process allowing its conversion toward production of other key products.

Is it possible to combine supercritical ethanolysis of lipids with the addition of CO_2, glycerol recovery, and cogeneration, according to the biorefinery concept?[50]

The land area used for bioenergy feedstock production is increasing because substitution of fossil fuels by bioenergy is promoted as an option to reduce greenhouse gas (GHG) emissions. However, agriculture itself contributes to rising atmospheric nitrous oxide (N_2O) and methane (CH_4) concentrations. The net exchanges of N_2O and CH_4 between soil and atmosphere differ between annual fertilized and perennial unfertilized bioenergy crops. This can be seen in N_2O and CH_4 soil fluxes from poplar short rotation coppice (SRC), perennial grass-clover and annual bioenergy crops (silage maize, oilseed rape, winter wheat). Using changes in farming methods of energy crops can be shown to improve sustainability.[51]

12.3.7 USE RENEWABLE FEEDSTOCKS

This is the area where agriculture is most important. Globally agricultural production generates a huge amount of solid waste. Improper agri-waste management causes environmental pollution which results in economic losses and human health-related problems. Hence, there is an urgent need to design and develop sustainable, cost-effective, and socially acceptable agri-waste management technologies. Agri-waste has high energy conversion efficiency as compared to fossil fuel-based energy materials. Agri-waste can also potentially be exploited for the production of second-generation biofuels, as shown above. Additionally, composted agri-waste can be an alternative to energy-intensive chemical fertilizers in organic production systems. Furthermore, value-added agri-waste can be a potential feedstock for livestock and industrial products. But comprehensive information concerning agri-waste management is lacking in the literature.[52]

12.3.7.1 Biorefinery

As will be seen in the following chapter in the biorefinery, there are examples of commercial processes that utilize microbes and enzymes to convert renewable agricultural feedstocks into complex natural products with agrochemical activity. These natural products are used in organic farming or as building blocks for semisynthetic, or naturally derived, crop protection products.[53] They use renewable material as their feedstock and thereby offer viable alternative to petroleum-based products.

12.3.7.2 CPC

CPCs can also come from renewable resources. The continuing demand for agrochemical insecticides that can meet increasing grower, environmental, consumer and regulatory requirements creates the need for the development of new solutions for managing crop pest insects. The development of resistance to the currently available insecticidal products adds another critical driver for new insecticidal active ingredients (AIs). An alternative path to meeting these challenges is the creation of new classes of insecticidal molecules to act as starting points, stimulating further efficacy and environmental impact refinements. A new class of insecticides is offering one measure of innovation within the agrochemical industry coming from natural products. Most insecticides owe their discovery to competitor-inspired (i.e. competitor patents/products) or next-generation (follow-on to a company's pre-existing product) strategies. In contrast, new insecticides primarily emerge from a bioactive hypothesis approach, with the largest segment resulting from the exploration of new areas of chemistry/heterocycles and underexploited motifs. Natural products also play an important role in the discovery of new insecticides. Understanding the origins of these active compounds and the approaches used in their discovery can provide insights into successful strategies for future insecticides.[54]

12.3.7.3 Bioplastics

The availability of fossil resources for production of various goods like plastics also causes increasing moral/ethical dilemmas for numerous industrial fields. Destabilizing political developments in several petrol exporting countries and the unpredictably fluctuating price of petrol create immense uncertainty, especially for the highly petrol-dependent polymer industry. Current efforts to switch from fossil to renewable resources as starting materials for polymer production is generally considered a promising strategy to overcome these problems, especially in combination with valorizing agri-waste. To make biobased and biodegradable polymers like polyhydroxyalkanoates (PHAs) economically more competitive with common resistant plastics from fossil resources, production costs have to be reduced considerably. The selection of suitable renewable resources as carbon feedstock for PHA production affects the cost in the PHA production chain. Sucrose from sugarcane, together with its high significance for nutrition purposes, constitutes a valuable renewable feedstock for the biomediated production of polymers like PHAs and other goods required by societies, especially as biofuels.

The selection of a suitable carbon source allows identification of potential microbial production strains that might guarantee fast growth, high product formation rates and robustness. Such organisms are found among Gram-negative and Gram-positive bacteria and even among extremophilic representatives of the archaea. The integration of PHA production into a sugar and ethanol factory starting from the raw material sugarcane can be done on a semi-industrial scale. This integration makes it possible to achieve an economically competitive production price for the biopolyester when compared to other PHA production processes on a larger scale. This cost reduction is possible due to an efficient utilization of by-products of the sugarcane plant, especially the lignocellulosic waste bagasse, and by the utilization of additional in-house waste streams for the biopolymer isolation from bacterial cells. In this process, bagasse is burned for generation of steam and electrical energy required for several process steps in PHA production. Further, the price advantage improves the availability of the substrate sucrose in high quantities.

Together with the combustion of the sucrose yielding ethanol as a "first generation biofuel", CO_2 emissions from the production plant return to the sugarcane fields via photosynthetic fixation by sugarcane resulting in a carbon balance of nearly zero. Major drawbacks in profitability and environmental embedding of PHA production are solved by a future-oriented, integrated process. In addition, the production of biobased polyethylene starting from sugar cane is discussed in the work. It is demonstrated that the application of Life Cycle Analysis and the strategies of Cleaner Production provide precious tools for quantifying the ecological footprint of sugarcane-based polymer production. Potential improvements of the process by a number of recently investigated microbial production strains are currently possible.[55]

12.3.7.4 Biopolycarbonates

As one of the main plastic resin types, polycarbonate-based plastics play an important role in manufacturing and increased global applications. However, with increasing awareness of environmental challenges related to petroleum-based polycarbonates, demand for production of bio-based polycarbonates from carbon dioxide (CO_2) and renewable feedstocks has attracted significant attention. Green chemistry and sustainable development are major influences. Recent advances in the efficient conversion of CO_2 and bio-based feedstocks to value-added bio-based polycarbonates with attractive properties can be done with significant energy savings. Specifically, renewable bio-based feedstocks that provide bio-based epoxides with long chains, resulting in "soft" bio-polycarbonate materials via innovative and efficient synthetic pathways illustrate major advances in this area. These bio-based feedstocks, including plant oils, industrial byproducts (crude glycerol) and pure fatty acids, offer new opportunities to fully take advantage of CO_2 and agricultural or industrial byproducts to bio-degradable polycarbonate plastics. Some challenges regarding comparable mechanical properties and scale-up remain.[56]

12.3.7.5 Oils

Oils and fats of vegetable origin have been the most important renewable feedstock of the chemical industry. A tremendous geographical and feedstock shift of oleochemical production has taken place from North America and Europe to southeast Asia. It will be important to introduce and to cultivate more and new oil plants containing fatty acids with interesting and desired properties for chemical utilization while simultaneously increasing the agricultural biodiversity. The problem of the industrial utilization of food plant oils has become more urgent with the development of the global biodiesel production. The remarkable advances made during the last decade in organic synthesis, catalysis, and biotechnology using plant oils and the basic oleochemicals derived from them will play a role in energy. This will include ω-functionalization of fatty acids containing internal double bonds, application of the olefin metathesis reaction, and de novo synthesis of fatty acids from abundantly available renewable carbon sources.[57]

12.3.8 AVOID DERIVATIVES IN SYNTHESIS STEPS

The use of protecting groups during active ingredient manufacturing increases the number of reaction steps and raw material usage and may increase waste generation. Protecting groups should be avoided wherever possible.

One approach would be to identify bioresources that already contain desired activity. Biopesticides obtained from renewable resources and associated with biodegradability have the potential to address resource limitations and environmental pollution, often caused by many conventional pesticides. This results from the facility of natural products to integrate into natural nutrient cycles. As an example, flavonoids are considered benign substitutes for pesticides, however, little comprehensive information of their pesticidal activities and critical evaluation of their associated advantages is available. Sources, structures, activities and the environmental fate of flavonoids is a valuable new area. Many flavonoids have shown pesticidal activity as either a pure compound or a flavonoid-containing extract, with quercetin, kaempferol, apigenin, luteolin and their glycosides as the most studied compounds. Agricultural or food waste, a potential sustainable source for flavonoids, represent significant sources of flavonoids, showing the currently underutilization of these preferable feedstocks. Analysis of pesticidal activities and target organisms also showed a broad target spectrum for the class of flavonoids, including fungi, insects, plants, bacteria, algae, nematodes, mollusks, and barnacles. Little information is available on the environmental fate and biodegradation of flavonoids, and a connection to studies investigating pesticidal activities is largely missing. This shows there is the need for a comprehensive understanding of flavonoids pesticidal activities with emphasis on structural features that influence activity and target specificity to avoid risks for non-target organisms. Only if the target spectrum and environmental fate of a potential biopesticide are known can they serve as a benign substitute. Flavonoids might be integrated in a valorization process of agricultural and food waste aiding the move to a more circular economy.[58]

12.3.9 USE CATALYSIS, NOT STOICHIOMETRIC REAGENTS

12.3.9.1 Chemical reactions

Catalysis and the use of catalysts are techniques to improve the efficiency of chemical reactions and to minimize the quantities of actual active reagents used. Catalytic reaction steps have been employed in agrochemical manufacturing processes for decades, as shown in earlier sections for prosulfuron, (S)-metolachlor, and glyphosate. Homogeneous catalysis is widely practiced by discovery chemists throughout industry, and these routes are often used for early kilo-lab synthesis. Pd-catalyzed cross couplings have become especially prevalent in agrochemical production routes.[1]

Pd-catalyzed cross-coupling reactions have become essential tools for the construction of carbon-carbon and carbon-heteroatom bonds. Over the last three decades, great efforts have been made with cross-coupling chemistry in the discovery, development, and commercialization of innovative agrochemicals (mainly herbicides, fungicides, and insecticides). Pd-catalyzed carbon-coupling reactions have been implemented as key steps in the synthesis of agrochemicals (on R&D and pilot-plant scales) such as the Heck, Suzuki, Sonogashira, Stille, and Negishi reactions, as well as decarboxylative, carbonylative, α-arylative, and carbon-nitrogen bond bond-forming cross-coupling reactions. There still remain further opportunities for these catalytic coupling processes in the discovery of agrochemicals. Cross-coupling chemistry approaches open-up new, low-cost, and more efficient industrial routes to existing agrochemicals, and such methods also have the capability to lead the new generation of pesticides with novel modes of action for sustainable crop protection.[59]

12.3.9.2 Catalysts from waste

The use of catalytic reactions can help control generation of waste. A wide variety of wastes are generated all around the world from fruits, marines, and plants, among several other sources. The remains of animal skeletal waste possess many precious compounds, which undergo decomposition in the environment and are incorporated into environmental matrices. Biowaste contains renewable chemical materials, which are useful in various applications including the treatment of wastewater, synthesis of high-value nanomaterials, and useful catalysts. Biowaste materials have excellent chemical properties, which are not only attractive in terms of price but are environment friendly and can be deployed as effective catalysts in the synthesis of organic compounds. Strategies to recover and reuse catalysts exemplify the main principles of green chemistry that aim to minimize the production of toxic materials in assorted reaction conditions.[38]

12.3.9.3 Asymmetric synthesis

Catalytic asymmetric synthesis has become an essential tool for the enantioselective synthesis of natural products and agrochemicals (mainly fungicides, herbicides, insecticides, and pheromones). With continuous growing interest in both modern agricultural chemistry and catalytic asymmetric synthesis chemistry, recent successful applications of various catalytic asymmetric syntheses methodologies show that this can be applied to synthesis of agriculture protection chemicals. Examples of these reactions include

- enantioselective hydroformylation,
- enantioselective hydrogenation,
- asymmetric Sharpless epoxidation,
- dihydroxylation, asymmetric cyclopropanation or isomerization, and
- organocatalyzed asymmetric synthesis.

All of these have been used as key steps in the preparation of chiral agrochemicals (on R&D, piloting, and commercial scales). Chiral agrochemicals can also lead to new generation of active ingredients having specific and novel modes of action for achieving sustainable crop protection and production. Use of catalytic asymmetric systems for the synthesis of chiral agrochemicals among the agrochemists will open new fields for agricultural chemistry in the twenty-first century.[60]

12.3.9.4 Enzymes

Enzymes are biocatalysts that are finding application in agriculture. Enzymes as industrial biocatalysts in chemical synthesis offer numerous advantages over traditional chemical processes with respect to sustainability and process efficiency. Enzyme catalysis has been scaled up for commercial processes in the food and beverage industries. Further enhancements in stability and biocatalyst functionality are required for optimal biocatalytic processes in the energy sector for biofuel production and in natural gas conversion applications in agrochemical industry. The technical barriers associated with the implementation of immobilized enzymes suggest that a multidisciplinary approach is necessary for the development of immobilized biocatalysts applicable in such industrial-scale processes. Specifically, the overlap of technical expertise in enzyme immobilization, protein use, and process engineering will define the next generation of immobilized biocatalysts and the successful scale-up of their applications. Biocatalysis has been successfully deployed and enzyme immobilization can improve industrial processes, with multiscale implementation for increased product yield at maximum market profitability and less burden on the environment and user.[61]

12.3.10 DESIGN PRODUCTS TO DEGRADE AFTER USE

It is important to plan the environmental lifetimes of crop protection chemicals. To minimize pesticide residues in food and the environment, crop protection active ingredients must not persist for long periods of time. Regulatory requirements dictate specific studies to understand the environmental fate of an agrochemical in water, soil, and nontarget organisms. Metabolites must be identified, and their fates and toxicities must also be evaluated. Regulatory agencies have cutoff criteria for acceptable rates of degradation.

The design and synthesis of the broad-spectrum fungicide, florylpicoxamid, embodies multiple green chemistry principles. The active ingredient was strategically designed to deliver maximum biological activity and rapidly degrade after application to minimize its environmental impact. Unlike many chiral crop protection chemicals, florylpicoxamid was purposely developed as a single stereoisomer to minimize the dose rate of this product in field applications and avoid any potential environmental burdens due to its less active stereoisomers. Fortunately, the most efficacious stereoisomer of the natural product-inspired active ingredient can be derived from the natural antipodes of lactic acid and alanine. The principles of green chemistry were also used to improve the synthetic route originally developed for structure-activity relationship studies, resulting in a convergent stereoselective synthesis that is more sustainable and cost-effective. The streamlined route decreases the use of protecting groups and utilizes safer reaction solvents. Development of a novel synthetic sequence to the 2,3,4-trisubstituted pyridine motif from furfural further increased the renewable carbon content of the active ingredient to nearly 50%. Through process development, improved reaction conditions, and industrially preferred procedures, alternative hazardous reagents for key transformations were identified. Following the green chemistry principles in the design and synthesis of florylpicoxamid provides new chemicals for farmers with a highly sustainable product for crop disease management can be designed and synthesized.[62]

12.3.11 ANALYZE IN REAL TIME

12.3.11.1 Manufacturing

In the manufacture of crop protection chemicals, chemical analyses are critical. In situ analysis of reaction mixtures using techniques such as infrared or NMR spectroscopy is commonly used in industry. These methods allow continuous monitoring of starting materials, intermediates, and products to develop mechanistic understandings. In addition, real-time analysis can provide information to reduce safety concerns about unstable reactive intermediates and aid in reaction optimization.

Since crop protection is such an important component of modern farming to maintain food production, considerable changes have occurred in the regulation of pesticides in the last decade. The aim has been to reduce their impact on people and the environment. This has resulted in a major reduction in the number of chemicals approved for application on crops. In parts of the world, a continuing expansion in the growing of genetically modified crops has also changed the pattern of pesticide use. Ways to monitor pesticides and the techniques used to mitigate their effects in the environment are needed. Real-time analyses are used to monitor their manufacture and their presence in the environment. Information on operator safety, protection of workers handling crops treated with pesticides and spray drift affecting those who live in farming areas also benefit from these analyses. By bringing together the most recent research on pesticides and using current analytical techniques, a vital resource is available for agricultural scientists, agronomists, plant scientists, plant pathologists, entomologists, environmental scientists, public health personnel, toxicologists and others working in the agrochemical industry. From this should come improvements in harmonizing regulation of pesticides in countries with limited resources for registration of pesticides.[63]

12.3.11.2 Pesticide analyses

Organophosphates, carbamates, and pyrethroids are among the most commonly used pesticides worldwide. However, these pesticides are toxic not only to insects but also to other non-targets such as animals, including humans. The increasing public concern in recent years about possible health risks due to pesticide residues in the diet has influenced the strategies for crop protection. Over the years, researchers have relied on several analytical methods. The importance of enzyme-linked immune sorbent assays (ELISA) for pesticide analysis has increased over the past decades. ELISA is a rapid, economical, and safe analytical procedure and is an alternative to chromatographic techniques for monitoring the residues of these target pesticides in vegetables. A colorimetric ELISA test kit can be used to detect organophosphates and carbamates directly, while the analysis of pyrethroid can be performed using paramagnetic particles attached to antibodies specifically to detect pyrethroids. To confirm the positive results, the samples can also be analyzed by chromatography. With the use of the ELISA kits, it is possible to rapidly determine the presence of organophosphates, carbamates, and pyrethroid residues in collected samples. The ELISA kits show quantification capacity at values below the detection limit of the chromatographic techniques. Linear relationships between the quantified values obtained by the chromatographic technique and results obtained through the ELISA test kits have been shown. Developed ELISA exhibit high accuracy and is ideally suited as a fast, high-throughput, and low-cost screening test for organophosphates, carbamates, and pyrethroid residues to monitor and control these residue levels in the local agriculture settings.[64]

Paraquat (1,10-dimethyl-4,40-dipyridinium chloride) (PQ), also known as methyl viologen, is widely used as a quaternary ammonium herbicide (broadleaf weed killer) all over the world owing to its excellent effect in plant cells for crop protection and horticultural use. However, it is dangerous because of its high acute toxicity even at low concentrations. Its detection in the environment is therefore necessary. As a consequence of its widespread usage, it causes genotoxic, teratogenic as well as other environmental and ecological adverse impacts. Exposure to PQ leads to a high mortality rate because no specific drug is effective for treatment. Excessive consumption of PQ can cause cellular damage and necrosis in the brain, heart, lungs, liver, and kidneys. The diversity and sensitivity of the analyses currently required to test for PQ necessitates that the experimenter use more advanced and efficient techniques, that can provide qualitative and quantitative results in complex environments. Electrochemical methods generally meet these criteria while offering other advantages which achieve excellent accuracy and fast handling. The determination of PQ using electrochemical methods combined with several modified electrodes in food samples, including milk, apple juice, tomato juice, and potato juice is both accurate and reliable. Comprehensively, it is strongly convincing that the synergy between the sensor substrate and the modifier architecture gives the electrodes a high capacity to detect PQ in complex matrices such as food.[65]

12.3.11.3   Field analyses/Smart Agriculture

Agriculture is the ability to monitor fields continuously. Smart technologies are combining advanced technologies like Big data analytics, Robotics, Artificial Intelligence (AI), and Internet of Things (IoT). This allows the stream controlled environment monitoring using different types of sensors and implementation of precision delivery technologies. Future research ideas will focus on "real time monitoring of nutrition solution management and pest management" for the plants growing in controlled environment to maximize the production.(See Chapter 11)

12.3.12   Minimize potential for accidents

Mitigation of reactive chemical hazards is necessary when less hazardous chemical syntheses (principle #3) are not practical. Reactions that involve unstable intermediates, such as diazonium salts, azides, hydrazines, etc., are practiced in the agrochemical industry. The identification of conditions to safely perform these types of chemistries requires an appropriate reactive chemical hazard evaluation. The testing strategy should be based on the scale of the process and should utilize appropriate calorimetric methods.[66] Shock and friction sensitivities should also be evaluated, especially for isolated solids. Calorimetric techniques also help to identify highly exothermic reactions, and appropriate control schemes (e.g., controlled addition of a limiting reagent) must be devised. Thorough reactive chemical testing should be performed to avoid incompatibilities between raw materials and solvents. In addition to chemical reaction hazards, additional operational hazards, such as the potential for dust explosion, must also be considered to minimize the potential for accidents.[67]

12.3.12.1 Personal Protection Equipment

Pesticides are an essential component of modern agriculture, playing an important role in crop productivity. However, many farmers in developing countries use pesticides without taking the necessary precautions, showing poor levels of personal safety in their use. There is an important role for technical efficiency and other factors in the safety behavior of farmers in the use of chemical pesticides in various parts of the world. When asked or when behavior is studied, more than half of the farmers showed unsafe or potentially unsafe safe behavior (i.e., poor use of personal protection items and poor implementation of hygiene practices) in the use of chemical pesticides. There is a strong correlation between the cultivation area and high yield of produce associated with improved safety levels of farmers in the use of chemical pesticides ($P < 0.05$). Farmers who efficiently produced crops showed better safety behaviors than inefficient farmers. Findings provide new evidence for effective interventions that could support farmers in the promotion of safety measures during pesticide handling. Developing and implementing suitable educational programs on pesticide safety must be a top priority for addressing gaps in farmers' knowledge of the hazards of pesticide exposure.[67]

12.3.12.2 Phytomanagement

The mitigation of potential health hazards and land scarcity due to land use change can be addressed by restoring functional and ecosystem services of contaminated land. Physico-chemical remediation options are costly and do not always provide environment-friendly solutions. The use of plants and associated microorganisms for remediation could be a sustainable, cost-effective option to reduce pollutant exposure. Phytomanagement aims at using valuable non-food crops to alleviate environmental and health risks induced by pollutants, and at restoring ecosystem stability. Suitable plant species must be tolerant to contaminants, reduce their transfer into the food chain, and efficiently produce marketable biomass. These plant species are favorable for phytostabilization and phytodegradation. The noninvasive hybrid *Miscanthus giganteus*, with a high lignocellulosic content, is a promising biomass crop for the bio-economy, notably the biorefinery and bioenergy industries. Planting this species on contaminated and marginal land is a promising option to reduce contamination in arable land and to offer a potential fuel feedstock. Key issues in promoting sustainable management of *Miscanthus sp.* on contaminated land are:

- crop suitability, integration, and sustainability in a region with a potential local market;
- site suitability in relation to the species' requirements and potential,

- biotic interactions in the landscape diversity; and
- increase in shoot yields in line with various stressors (e.g., pollutants, drought, cold temperatures), and with minimal inputs.[68]

12.4 CONCLUSION

CPCs illustrate specific sustainability challenges to agriculture and how agricultural green chemistry is providing solutions. The agronomic production of food, feed, fuel and fiber requires innovative solutions. The current and future challenges involve climate change, resistance issues and resistance management, increasing regulatory demands, renewable raw materials and requirements of food chain partnerships. The modern agrochemical industry has to support farmers to manage these diversified tasks in accordance with new understanding of the interrelationships between biological systems and new innovation. Currently, several modern active ingredients are already matching these expectations, and further ones will need to follow in order to fulfil these ambitious criteria.

The innovative agricultural chemicals launched in the first quarter of the twenty-first century meet these challenges, as reflected by novel fungicides, herbicides and safeners, insecticides and nematicides. Since 2007, a new generation of systemic, broad-spectrum fungicides (e.g. fluorine-substituted pyrazol-4-yl-carboxamide SDHIs) has been discovered, which can be used for seed treatment applications and as a perfect option for mixtures. With isotianil, a novel plant defense inducer could be developed to initiate systemic induction of the plant's own defense mechanism that controls rice diseases. Further optimised herbicide classes can be applied at much lower application rates. To overcome inherent selectivity of herbicides in special crops such as corn, safener technology has been further intensified (e.g. the combination of thiencarbazone-methyl with cyprosulfamide). New classes of insecticides with novel MoAs continue to be discovered. The pyridinyl-ethyl benzamide fungicide fluopyram, combining a good safety profile with a significant increase in yield and quality in a broad spectrum of crops, will be a benefit for nematode control. In the past decade, the side effects of agrochemicals that affect crop yield, virus vector control, plant health, resistance-breaking potential, new physicochemical properties such as phloem mobility, and quality have gained increasing importance. These beneficial effects are sometimes complemented by improved agricultural farming techniques. Modern agricultural chemistry is vital and will have the opportunity to shape the future of agriculture by continuing to deliver further innovative integrated solutions.

The crop protection industry manufactures hundreds of active ingredients on significantly large scales. (See Figure 12.3) Many of the commercial processes utilize hazardous raw materials, during production and manufacturing. Once manufactured, these active ingredients are formulated into products that are used by farmers to protect their crops and minimize food waste caused by pest damage. Since these products are key inputs into food production, their environmental fate and toxicological effects are closely regulated and subjected to periodic review.

FIGURE 12.3   Modern crop protection. (See www.shutterstock.com/image-illustration/
cultivation-drone-icon-isolated-on-white-2040030284)

Examples of chemical processes and products from the crop protection industry
have been used to illustrate specific green chemistry principles. The 12 principles
of green chemistry,[8] formulated by Anastas and Warner are especially important
for agrochemical process chemists and engineers to consider in designing
manufacturing processes and for any worker using agricultural technologies. These
principles provide a framework to reduce the environmental impact and improve the
safety of manufacturing processes that have been in operation for several decades.

REFERENCES

1. *The 12th IUPAC International Congress of Pesticide Chemistry in Conjunction with
    Royal Australian Chemical Institute's 13th National Convention 4th July-8th July
    2010 Melbourne Convention & Exhibition Centre.* Journal of Pesticide Science,
    2011. **36**(1): pp. 138–139.
2. Lamberth, C., et al., *Current challenges and trends in the discovery of agrochemicals.*
    Science, 2013. **341**(6147): pp. 742–746.
3. Dhillon, N.K., S.S. Gosal, and M.S. Kang, *Improving crop productivity under
    changing environment,* in *Improving crop productivity in sustainable agriculture.*
    2012. Wiley-VCH. pp. 23–48.
4. Rauzan, B.M. and B.A. Lorsbach, *Designing Sustainable Crop Protection Actives,*
    in *ACS Symposium Series,* B.M. Rauzan and B.A. Lorsbach, Editors. 2021.
    American Chemical Society. pp. 1–9.
5. Großkinsky, D.K., et al., *Phenotyping in the fields: Dissecting the genetics of
    quantitative traits and digital farming.* New Phytologist, 2015. **207**(4): pp. 950–952.
6. Krengel-Horney, S., et al., *Climate change and possible challenges for crop
    protection – yesterday, today, tomorrow.* Journal fur Kulturpflanzen, 2021. **73**(7–
    8): pp. 292–305.
7. de Marsily, G. and R. Abarca-del-Rio, *Water and food in the twenty-first century.*
    Surveys in Geophysics, 2016. **37**(2): pp. 503–527.
8. Anastas, P.T. and J.C. Warner, *Green Chemistry: Theory and practice.* 1998.
    Oxford: Oxford University Press.
9. O'Riordan, T.J.C., *UN Sustainable Development Goals: How can sustainable/green
    chemistry contribute? The view from the agrochemical industry.* Current Opinion in
    Green and Sustainable Chemistry, 2018. **13**: pp. 158–163.
10. Kharissova, O.V., et al., *Greener synthesis of chemical compounds and materials.*
    Royal Society Open Science, 2019. **6**(11): pp. 1–41.

11. Jeschke, P., *Progress of modern agricultural chemistry and future prospects*. Pest Management Science, 2016. **72**(3): pp. 433–455.

12. Jaggi, M., et al., *Mitogen-activated protein kinases in abiotic stress tolerance in crop plants: "-omics" approaches*, in *Improving Crop Productivity in Sustainable Agriculture*. 2012, Wiley-VCH. pp. 107–132.

13. Jeschke, P., *Manufacturing approaches of new halogenated agrochemicals*. European Journal of Organic Chemistry, 2022. **2022**(12): pp. 1–11.

14. Palmieri, D., et al., *Advances and perspectives in the use of biocontrol agents against fungal plant diseases*. Horticulturae, 2022. **8**(7): pp. 1–34.

15. Duke, S.O., *Why have no new herbicide modes of action appeared in recent years?* Pest Management Science, 2012. **68**(4): pp. 505–512.

16. Powles, S., *Global herbicide resistance challenge*. Pest Management Science, 2014. **70**(9): pp. 1305–1305.

17. El-Dabaa, M.A.T., G.A. Abo-Elwafa, and H. Abd-El-Khair, *Safe methods as alternative approaches tochemical herbicides for controlling parasitic weeds associated with nutritional crops: a review*. Egyptian Journal of Chemistry, 2022. **65**(4): pp. 53–65.

18. Abu-Qare, A.W. and H.J. Duncan, *Herbicide safeners: Uses, limitations, metabolism, and mechanisms of action*. Chemosphere, 2002. **48**(9): pp. 965–974.

19. Kraehmer, H., et al., *Herbicides as weed control agents: State of the art: I. weed control research and safener technology: the path to modern agriculture*. Plant Physiology, 2014. **166**(3): pp. 1119–1131.

20. Kraehmer, H., et al., *Herbicides as weed control agents: State of the Art: II. Recent achievements*. Plant Physiology, 2014. **166**(3): pp. 1132–1148.

21. Mužinić, V. and D. Želježić, *Non-target toxicity of novel insecticides*. Arhiv za Higijenu Rada i Toksikologiju, 2018. **69**(2): pp. 86–102.

22. Li, S., et al., *Molecularly imprinted electroluminescence switch sensor with a dual recognition effect for determination of ultra-trace levels of cobalt (II)*. Biosensors and Bioelectronics, 2019. **139**: pp. 1–6.

23. Sparks, T.C., et al., *Insecticide resistance management and industry: the origins and evolution of the Insecticide Resistance Action Committee (IRAC) and the mode of action classification scheme*. Pest Management Science, 2021. **77**(6): pp. 2609–2619.

24. Casida, J.E., *Pesticide interactions: mechanisms, benefits, and risks*. Journal of Agricultural and Food Chemistry, 2017. **65**(23): pp. 4553–4561.

25. Sparks, T.C. and R. Nauen, *IRAC: Mode of action classification and insecticide resistance management*. Pesticide Biochemistry and Physiology, 2015. **121**: pp. 122–128.

26. Tang, Z., et al., *A review on constructed treatment wetlands for removal of pollutants in the agricultural runoff*. Sustainability (Switzerland), 2021. **13**(24): pp. 1–28.

27. Chen, J., Q.X. Li, and B. Song, *Chemical nematicides: recent research progress and outlook*. Journal of Agricultural and Food Chemistry, 2020. **68**(44): pp. 12175–12188.

28. Kim, Y.H. and J.W. Yang, *Recent research on enhanced resistance to parasitic nematodes in sweetpotato*. Plant Biotechnology Reports, 2019. **13**(6): pp. 559–566.

29. Chen, J.X. and B.A. Song, *Natural nematicidal active compounds: Recent research progress and outlook*. Journal of Integrative Agriculture, 2021. **20**(8): pp. 2015–2031.

30. Chen, H., et al., *Eighteen-year farming management moderately shapes the soil microbial community structure but promotes habitat-specific taxa.* Frontiers in Microbiology, 2018. **9**(AUG): pp. 1–14.

31. Tian, J., et al., *Coupling mass balance analysis and multi-criteria ranking to assess the commercial-scale synthetic alternatives: A case study on glyphosate.* Green Chemistry, 2012. **14**(7): pp. 1990–2000.

32. Jimenez-Gonzalez, C., et al., *Using the right green yardstick: Why process mass intensity is used in the pharmaceutical industry to drive more sustainable processes.* Organic Process Research and Development, 2011. **15**(4): pp. 912–917.

33. Duenas, M. and I. Garciá-Estévez, *Agricultural and food waste: Analysis, characterization and extraction of bioactive compounds and their possible utilization.* Foods, 2020. **9**(6).

34. Petrič, M. and A.S. Matharu, *CHAPTER 8: Biosilicate Binders*, in *RSC Green Chemistry*, R. Hofer, A.S. Matharu, and Z. Zhang, Editors. 2019. Royal Society of Chemistry. pp. 183–204.

35. Eissen, M., et al., *Atom Economy and Yield of Synthesis Sequences.* Helvetica Chimica Acta, 2004. **87**(2): pp. 524–535.

36. Pirzada, T., et al., *Recent advances in biodegradable matrices for active ingredient release in crop protection: Towards attaining sustainability in agriculture.* Current Opinion in Colloid and Interface Science, 2020. **48**: pp. 121–136.

37. Yang, C., et al., *Redistribution of intracellular metabolic flow in E. coli improves carbon atom economy for high-yield 2,5-dimethylpyrazine production.* Journal of Agricultural and Food Chemistry, 2021. **69**(8): pp. 2512–2521.

38. Kheilkordi, Z., et al., *Waste-to-wealth transition: application of natural waste materials as sustainable catalysts in multicomponent reactions.* Green Chemistry, 2022. **24**(11): pp. 4304–4327.

39. Ita-Nagy, D., et al., *Reviewing environmental life cycle impacts of biobased polymers: current trends and methodological challenges.* International Journal of Life Cycle Assessment, 2020. **25**(11): pp. 2169–2189.

40. Banik, B.K., et al., *Green synthetic approach: An efficient eco-friendly tool for synthesis of biologically active oxadiazole derivatives.* Molecules, 2021. **26**(4): pp. 1–23.

41. Clippinger, A.J., et al., *Pathway-based predictive approaches for non-animal assessment of acute inhalation toxicity.* Toxicology in Vitro, 2018. **52**: pp. 131–145.

42. Gehen, S., et al., *Challenges and opportunities in the global regulation of crop protection products.* Organic Process Research and Development, 2019. **23**(10): pp. 2225–2233.

43. Ottinger, M.A., et al., *Assessing the consequences of the pesticide methoxychlor: Neuroendocrine and behavioral measures as indicators of biological impact of an estrogenic environmental chemical.* Brain Research Bulletin, 2005. **65**(3): pp. 199–209.

44. Prat, D., J. Hayler, and A. Wells, *A survey of solvent selection guides.* Green Chemistry, 2014. **16**(10): pp. 4546–4551.

45. McKusick, B.C., *Prudent practices for handling hazardous chemicals in laboratories.* Science, 1981. **211**(4484): pp. 777–780.

46. Do Nascimento Junior, D.R., et al., *Biopesticide encapsulation using supercritical CO2: A comprehensive review and potential applications.* Molecules, 2021. **26**(13): pp. 1–19.

47. Ten, A., et al., *Ionic liquids in agrochemistry.* Current Organic Chemistry, 2020. **24**(11): pp. 1181–1195.
48. Cassoni, A.C., et al., *Systematic review on lignin valorization in the agro-food system: From sources to applications.* Journal of Environmental Management, 2022. **317**: pp. 1–13.
49. Akhtar, N., D. Goyal, and A. Goyal, *Physico-chemical characteristics of leaf litter biomass to delineate the chemistries involved in biofuel production.* Journal of the Taiwan Institute of Chemical Engineers, 2016. **62**: pp. 239–246.
50. Coniglio, L., et al., *Biodiesel via supercritical ethanolysis within a global analysis "feedstocks-conversion-engine" for a sustainable fuel alternative.* Progress in Energy and Combustion Science, 2014. **43**: pp. 1–35.
51. Walter, K., A. Don, and H. Flessa, *Net N2O and CH4 soil fluxes of annual and perennial bioenergy crops in two central German regions.* Biomass and Bioenergy, 2015. **81**: pp. 556–567.
52. Babu, S., et al., *Exploring agricultural waste biomass for energy, food and feed production and pollution mitigation: A review.* Bioresource Technology, 2022. **360**: pp. 1–12.
53. Dayan, F.E., C.L. Cantrell, and S.O. Duke, *Natural products in crop protection.* Bioorganic and Medicinal Chemistry, 2009. **17**(12): pp. 4022–4034.
54. Sparks, T.C. and R.J. Bryant, *Innovation in insecticide discovery: Approaches to the discovery of new classes of insecticides.* Pest Management Science, 2022. **78**(8): pp. 3226–3247.
55. Koller, M., et al., *Sugarcane as feedstock for biomediated polymer production*, in *Sugarcane: Production, Cultivation and Uses.* 2012. Nova Science Publishers, Inc. pp. 105–136.
56. Cui, S., et al., *Recent advances of "soft" bio-polycarbonate plastics from carbon dioxide and renewable bio-feedstocks via straightforward and innovative routes.* Journal of CO2 Utilization, 2019. **34**: pp. 40–52.
57. Biermann, U., et al., *Oils and fats as renewable raw materials in chemistry.* Angewandte Chemie – International Edition, 2011. **50**(17): pp. 3854–3871.
58. Schnarr, L., et al., *Flavonoids as biopesticides – Systematic assessment of sources, structures, activities and environmental fate.* Science of the Total Environment, 2022. **824**: pp. 1–13.
59. Devendar, P., et al., *Palladium-catalyzed cross-coupling reactions: a powerful tool for the synthesis of agrochemicals.* Journal of Agricultural and Food Chemistry, 2018. **66**(34): pp. 8914–8934.
60. Sharma, S.K., A.S.R. Paniraj, and Y.B. Tambe, *Developments in the catalytic asymmetric synthesis of agrochemicals and their synthetic importance.* Journal of Agricultural and Food Chemistry, 2021. **69**(49): pp. 14761–14780.
61. Chapman, J., A.E. Ismail, and C.Z. Dinu, *Industrial applications of enzymes: Recent advances, techniques, and outlooks.* Catalysts, 2018. **8**(6): pp. 1–26.
62. Babij, N.R., et al., *Design and synthesis of florylpicoxamid, a fungicide derived from renewable raw materials.* Green Chemistry, 2020. **22**(18): pp. 6047–6054.
63. Matthews, G.A., *Pesticides: Health, Safety and the Environment: Second Edition.* Pesticides: Health, Safety and the Environment: Second Edition. 2015. Wiley. 1–277.
64. López Dávila, E., et al., *ELISA, a feasible technique to monitor organophosphate, carbamate, and pyrethroid residues in local vegetables. Cuban case study.* SN Applied Sciences, 2020. **2**(9), pp. 1–12.

65. Laghrib, F., et al., *Electrochemical sensors for improved detection of paraquat in food samples: A review.* Materials Science and Engineering C, 2020. **107**: pp. 1–9.

66. Frurip, D.J., *Selection of the proper calorimetric test strategy in reactive chemicals hazard evaluation.* Organic Process Research and Development, 2008. **12**(6): pp. 1287–1292.

67. Levin, D., *Managing hazards for scale up of chemical manufacturing processes*, in *ACS Symposium Series.* 2014. American Chemical Society. pp. 3–71.

68. Nsanganwimana, F., et al., *Suitability of Miscanthus species for managing inorganic and organic contaminated land and restoring ecosystem services. A review.* Journal of Environmental Management, 2014. **143**: pp. 123–134.

13 Sustainable agricultural chemistry
The biorefinery

If there is one development in the twenty-first century that captures the essence of sustainability, while epitomizing agricultural chemistry (AC), it is the biorefinery (BR). While not originating from AC, it does draw upon agricultural chemistry and advances sustainable agriculture.

13.1 BIOREFINERY OVERVIEW

It has been shown earlier that the energy requirements of the world are largely met by fossil fuels. The limited deposits of these fossil fuels coupled with environmental problems, such as greenhouse gases, have prompted the search for sustainable resources as alternatives to meet the increasing energy and chemical composites demand. Biomass is one of the few resources that has the potential to meet the challenges of sustainable and green energy systems. Biomass is plant matter of recent (non-geologic) origin and could be used to produce various useful chemicals and fuels. A system similar to a petroleum refinery ("biorefinery") can produce useful chemicals and fuels from biomass.[2]

The BR refers to a series of chemical processes which converts biomass to energy, chemicals, and other beneficial byproducts. The major advantage (from the processes and starting resources) is the reduction of environmental impacts, such as lower emission of pollutants as well as the reduction in the emission of hazardous products. The BR is defined as "the sustainable processing of biomass into a spectrum of marketable products and energy".[3] Energy and non-energy products are produced in an ideal biorefinery process, and economic and environmental benefits result from its use, as depicted in Figure 13.2.

The concept of producing products from agricultural commodities (i.e., biomass) is not new, but its applications are expanding in the demand for sustainability. Using biomass as an input to produce multiple products from complex processing methods can produce an approach similar to a petroleum refinery where fossil fuels are used as input. Biomass consists of carbohydrates, lignin, proteins, fats, and to a lesser extent, various other chemicals, such as vitamins, dyes, and flavors.[5] The goal of a biorefinery is to transform such plentiful biological materials into useful products using a combination of technologies and processes. Figure 13.2 describes

DOI: 10.1201/9781003157991-13

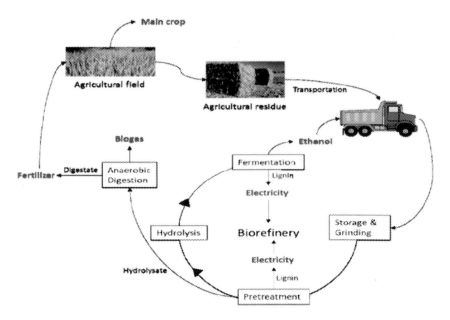

FIGURE 13.1   Systematic process to produce diverse products involving the biorefinery.
[1]

the elements of a biorefinery in which biomass feedstocks are used to produce various useful products such as fuel, power, and chemicals using biological and chemical conversion processes.

The main goal of a BR is to produce products from agricultural products, produce biomass, and food waste, using a series of unit operations. The operations are designed to maximize the valued compounds while minimizing the waste streams by efficiently converting intermediates into valuable materials. Guided by sustainability, the high-value products enhance the economy, while the fuels help to meet the global energy demand. The power produced from a biorefinery also helps to reduce the overall cost of the process. In contrast to a petroleum refinery, a biorefinery uses renewable resources and produces fuels and chemicals that contribute less to environmental pollution.[6]

13.1.1   PRINCIPLES OF A SUSTAINABLE BIOREFINERY

Biorefining can be defined as the sustainable processing of biomass into a spectrum of marketable products and energy.[7] Although only a few general categories of biorefineries have been clearly identified and the number of different types of biorefineries is still growing, it is agreed that all kinds of biorefineries should be designed and operated in a sustainable way from the social, economic, and environmental perspectives. In this regard, a biorefinery system should obey the following general sustainability principles:[8] (See Table 13.1)

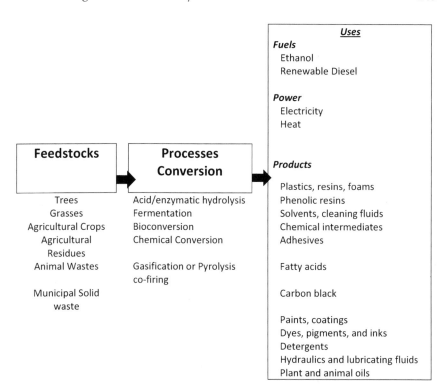

FIGURE 13.2 Simple three-step biomass-process-products procedure.[4]

TABLE 13.1
Principles of a Biorefinery

Principle	Explanation
1	A biorefinery should ensure the maximum economic benefits through minimizing the use of raw material and energy resources as well as the overall supply chain cost in operations intrinsic to biorefining.[9]
2	A biorefinery should ensure minimum environmental impacts through minimizing the generation of hazardous wastes and reduced use of nonrenewable energy resources.
3	A biorefinery should provide maximum social benefits through creating employment, supporting regional developments, improving public image, etc.
4	A biorefinery should meet the market demands on renewable energy products and the conformity of end-products to end-user requirements
5	A biorefinery should not result in competition with food supply.
6	A biorefinery should have affordable biomass-to-products value.
7	A biorefinery system should have good conversion efficiency

13.1.2 GLOBAL DRIVERS

The need for renewable energy sources is being driven by the increase of fossil fuel (crude oil) prices, depletion of natural resources, public awareness regarding energy security, as well as environmental pressures. As a renewable energy source, biomass has the potential to fulfill future energy and chemical demands, especially in the areas of biofuels and biochemicals derived from it are considered to have inherently lower carbon footprints than those derived from fossil sources. Life cycle assessment shows that the potentials to reduce greenhouse gas (GHG) by corn ethanol and cellulosic ethanol are 52% and 86% over their gasoline counterpart, respectively.[10] Additionally, the nation's reliance on foreign sources of oil has left the United States and other nations vulnerable to a shortage of petroleum-based fuel and products due to natural disasters, political disruptions, and price volatility. Therefore, complementing the nation's energy supply with renewable, sustainable, and domestically produced sources of energy, such as bioenergy, can help ensure future energy security of any nation.[8]

13.1.3 TYPES OF BIOREFINERIES

Biorefineries can be classified into different categories:

- Conventional Biorefineries (CBR),
- Green Biorefineries (GBR),
- Whole Crop Biorefineries (WCBR),
- Lignocellulosic Feedstock Biorefineries (LCFBR),
- Marine Biorefineries (MBR),
- Two-Platform Concept Biorefineries (TPCBR), and
- Thermo-Chemical Biorefineries (TCBR).[1]

CBR are based on existing biomass feedstock like sugar, starch, and food waste. GBR are based on wet biomass (e.g. green grasses and crops) [11,12]. WCBR use either dry or wet milling of biomass such as cereals. Phase I biorefineries use single feedstock, single process and single product. Most of the Phase I biorefinery are currently in practice and are economically viable. Phase II biorefinery was, for example, the conversion of corn starch to different products. Phase III biorefinery use multiple feed stocks, multiple processes and multiple products. This is the most advanced biorefinery. Lignocellulosic biomass used in Phase III and is composed of cellulose, hemicelluloses and lignin, which can be used for the production of a wide variety of energy and chemical products

Biorefineries are processing facilities that convert biomass into value-added products such as biofuels, biochemicals, bioenergy/biopower, and other biomaterials. Various types of biorefineries have been presented in the literature, and most of them are mainly based on the individual feedstock they process, such as corn-based biorefinery, wood-based biorefinery, forest-based biorefinery, palm-based biorefinery, algae-based biorefinery, etc. On the other hand, some biorefineries are based on the generation or complexity of feedstock, which are first-generation biorefinery (energy crop, edible oil seeds, food crops, animal fats, etc.), second-generation biorefinery (lignocellulosic biomass), and third- or fourth-generation biorefinery (algae and other microbes). The integrated biorefinery, which focuses on the integration of various biomass conversion technologies, combines the above. Within integrated biorefinery, multiple products can be used to generate various types of products. As the characteristic and properties of the feedback are significantly different from one and other; therefore, it is a huge challenge to design an integrated biorefinery. Various conversion platforms (thermochemical, biological, catalytic, and physical) have been studied to convert the feedstock into products. In order to design a sustainable biorefinery/integrated biorefinery, various process design and analysis tools have also been developed. Process integration in the design stage further improves the economic and environmental performances of the biorefinery/integrated biorefinery.[13]

13.1.4 PRETREATMENT PROCESSES

Pretreatment processes for biorefineries are necessary for efficient and effective product production. They are divided into two biomass sources: vegetal and animal. Vegetal biomass is the material produced by plants on land or in water (algae). The plants consume sunlight, CO_2, water, and soil nutrients. This range of nutrients comes from residues or main products from, for example, intensive grass crops, forestry, and industrial and agricultural activities. Animal biomass is the residual biomass generated from the production of food from animals (e.g., manure and whey). The pretreatment includes physical pretreatments, microwave-assisted extraction, and water treatments for vegetal biomass. There are also chemical and biological treatments.[14]

A second pretreatment technology used is chemical extraction. In particular, deep eutectic solvents (DESs) are used. These are green solvents that are used in many types of applications as well as fundamental investigations. The physicochemical properties of DESs are one of the most important factors which led to their increased interest in science and technology. DESs are thermally and chemically stable, non-flammable and have a negligible vapor pressure. Furthermore, most of the newly formulated DESs are liquids at room temperature. This use of green solvents is both economical and sustainable. DESs are more economical and less expensive compared to ionic liquids. They are frequently prepared from renewable and non-toxic precursors, in addition, there are wide selections of biocompatible and biodegradable DESs. Hence, DESs have been used in many applications and processes such as biorefinery, lignocellulose dissolution, bioactive compound extraction and electrochemical applications.[15]

13.1.5 SEPARATION AND RECOVERY TECHNOLOGIES

The biorefinery is the most strategically important and developing industry. Many other technologies of transforming and utilizing other feedstocks and alternative resources are still immature. With multiple products being delivered in a complicated way, advanced separation technologies have been developed so as to maximize the biorefinery efficacy and biomass potential. One of the most important technologies is the membrane-based separation in biorefinery. While chemically simple, it provides separation technologies that are biofriendly and energy efficient. Membrane separation efficiency is always a key aspect of membrane-based separation. For gas separation and pervaporation, membrane intrinsic performance is closely tied to the material intrinsic physiochemical properties. With advent of a large number of materials, membranes with recommendable performance have already been commercialized.[16]

An example of valued products from membrane filtration is Brewers' spent grain (BSG). BSG is the main by-product derived from the brewing industry, where it accounts for 85% of the total waste generated. The total annual production worldwide of this waste is 39 million tons. This lignocellulosic material is traditionally used as cattle feed and sold at a low retail price (~USD 45.00 per ton). However, efforts for the revalorization of this by-product are emerging since research has established that it can be used as a low-cost source of bioactive molecules and commodity chemicals that can bring value to integral biorefinery ventures. Among commodities produced, phenolic compounds have attracted attention as added-value products due to their antioxidant properties with applications in the food, cosmetic, and pharmaceutical industries. These phytochemicals have been associated with antiaging and anticancer activities that have potential applications in cosmetic products. Compounds from BSG together with bioactive extracts in the cosmetic industry and their reported beneficial effects are a good example of how the biorefinery can bring added value.[17]

A second example is the use of Ionic liquids (ILs). ILs can play multiple roles in lignocellulose biorefineries, including utilization as agents for the separation of selected compounds or as reaction media for processing lignocellulosic materials (LCM). Imidazolium-based ILs have been utilized for separating target components from LCM biorefinery streams, during the dehydration of ethanol-water mixtures or during the extractive separation of biofuels (ethanol, butanol) or lactic acid from the respective fermentation broths. As in other industries, ILs are potentially suitable for removing volatile organic compounds or CO_2 from gaseous biorefinery effluents. On the other hand, cellulose dissolution in ILs allows homogeneous derivatization reactions to be carried out, opening new ways for product design or for improving the quality of the products. Imidazolium-based ILs are also suitable for processing native LCM, allowing the integral benefit of the feedstocks via separation of polysaccharides and lignin. Even strongly lignified materials can yield celullose-enriched substrates highly susceptible to enzymatic hydrolysis upon ILs processing. Recent developments in enzymatic hydrolysis include the identification of ILs causing limited enzyme inhibition and the utilization of enzymes with improved performance in the presence of ILs.[18]

13.1.6 ROLE OF ENZYMES

Enzymes are useful in the production of ethanol. The efficient generation of cellulosic ethanol depends on the optimal use of biomass within the biorefinery, and this requires the integration of unit operations that are involved in the production of fuel and chemicals. Enzymes are important tools to improve the efficiency and sustainability of a BR process. Therefore, a comprehensive approach and full understanding of the structure and function relationships that are involved in the enzymatic hydrolysis of lignocellulosic materials is a fundamental step toward the optimization of these bioconversion processes.[19]

The increase of pollutants in wastewater, expensive cultivation of microalgae, and difficulties in industrial scale production within the biorefinery are the main challenges for successful coupling of microalgae with wastewater. Nitrogen, carbon, and phosphorus in wastewater are feedstocks consumed by microalgae and cyanobacteria for their growth. This could act as a green technology for wastewater treatment. The role and mechanistic approaches of microalgae and cyanobacteria for removal of various (in)organic compounds from wastewater is becoming clear. Distinct pathways have been reported for improving wastewater treatment technologies through large-scale cultivation of microalgal. The techno-economic feasibility and major commercial production challenges along with genetic engineering research have been addressed. A BR approach with integrated biology, ecology, and engineering would lead to a feasible microalgal-based technology for various applications.[20]

Lastly, lignocellulosic biomass is the most abundant renewable resource in nature and has received considerable attention as one of the most promising alternatives to oil resources for the production of energy and certain raw materials. The phenolic polymer lignin is the second most abundant constituent of this biomass resource and has been shown to have the potential to be converted into industrially important aromatic chemicals after degradation. However, due to its chemical and structural nature, it exhibits high resistance toward mechanical, chemical, and biological degradation, and this causes a major obstacle for achieving efficient conversion of lignocellulosic biomass. In nature, lignin-degrading microorganisms have evolved unique extracellular enzyme systems to decompose lignin using radical mediated oxidative reactions. These microorganisms produce a set of different combinations of enzymes with multiple isozymes and isoforms by responding to various environmental stimuli such as nutrient availability, oxygen concentration and temperature, which are thought to enable effective decomposition of the lignin in lignocellulosic biomass. Use of these enzymes could provide a valuable way forward.[21]

13.1.7 MICROORGANISMS

The biorefinery process for production of biofuels and chemicals relies upon the breakdown of biomass. This represents an attractive and sustainable alternative to fossil fuel-based refineries, but necessitates the development of novel technologies. The inherent recalcitrance of lignocellulosic biomass requires a physical-chemical pretreatment prior to the biocatalytic conversion of monomeric constituents of plant cell walls into bioproducts. However, the pretreatment step often generates fermentation inhibitors, such as furfural, 5-hydroxymethylfurfural, aliphatic acids, and phenolic acids and aldehydes. Thus, a detoxification step prior, or simultaneous to microbial fermentation is valuable to improve product yields. Among detoxification strategies, biological detoxification, also known as bioabatement, as an eco-friendlier, milder, and cheaper alternative to physical-chemical detoxification methods. Recent valuable advances in the field of microbial detoxification of lignocellulosic hydrolysates are critical. The generation of inhibitory compounds from lignocellulosic biomass pretreatment and their effects on enzymatic hydrolysis and cell physiology provide biological detoxification strategies for the following microbial groups: filamentous fungi, non-Saccharomyces yeasts, bacteria, and Saccharomyces cerevisiae. Furthermore, there are advantages and disadvantages of microbial detoxification, and the main challenges currently faced by researchers in the field are important for long term usage.[22]

13.1.8 BIOREFINERY AND CIRCULAR ECONOMY

Within the circular economy, the biorefinery provides vast opportunities. The huge energy demand due to global economics exacerbates the non-renewable resources depletion and ecological-social challenges. For this reason, renewable energy has become a crucial element in sustainable strategy. The BR emerged as a sustainable approach and utilized promising transformation platforms for products from biomass to achieve circular bioeconomy. The biowaste biorefinery has provided a sustainable approach for integrated bioproducts and further applied this technology in industrial, commercial, agricultural and energy sectors. Biowaste is a renewable resource and utilizes resource utilization technologies from the perspective of energy, nutrient and material recovery in the action of a biorefinery. The BR can support biowaste management as a way for conversion of biowaste into value-added materials and thus contribute as a driving force to provide needed resources, reduce climate change, and meet the huge material demand in circular bioeconomy. In practice, the optimal use of biorefinery technologies will depend on environmentally friendly, economic and technical feasibility, social and policy acceptance. (See Table 13.1. Principles of a Biorefinery). Additionally, policy interventions will be necessary to increase use of biowaste biorefinery for circular bioeconomy and contribute to low-carbon and cleaner environment.[23]

The example of Sugarcane straw (SCS) illustrates the biorefinery and the circular economy. SCS represents about one-third of the total primary sugarcane energy. Since its burning in pre-harvesting periods has been banned in several countries, SCS available amount has grown, and the development of valuable applications for this lignocellulosic material became crucial for the development of a circular economy. It has been used in combustion processes for energy generation. However, in the biorefinery concept, the SCS processing is used as a potential feedstock for obtaining bioproducts and biomaterials, such as ethanol, xylitol, biogas, enzymes and oligosaccharides. This has also attracted a great deal of attention. SCS processing with its valorization through the production of high-value products and the development of environmental-friendly and cost-effective processes is both probable and highly desirable.[24]

13.1.9 GREEN BIOREFINERIES

The use of biorefineries for the production of chemicals as well as materials and energy products is key to ensuring a sustainable future for the chemical and allied industries. Through the integration of green chemistry into biorefineries, and the use of low environmental impact technologies, genuinely green and sustainable chemical products are produced.[25] Human health is mainly associated with suitable nutrition based on the intake of non-toxic food products with bioaccessible nutrients rich in bioactive compounds. In this sense, the application of the biorefinery in the processing of vegetable matrices, the production of new functional products and high value-added co-products is a potential alternative for safe food production. New BR models, including the integration and intensification of processes, using non-thermal and clean emerging technologies such as supercritical carbon dioxide, pressurized liquid, high-intensity ultrasound, and pulsed electric field show that this is a vibrant field.[26]

There is a growing need for protein for both feed and food in order to meet future demands. It is imperative to explore and utilize novel protein sources such as protein from leafy plant material, which contains high amounts of the enzyme ribulose-1,5-biphosphate carboxylase/oxygenase (RuBisCo). Leafy crops such as grasses and legumes can produce in humid climates high protein in a sustainable way when compared with many traditional seed protein crops. Despite this, very little RuBisCo is utilized for foods because proteins in the leaf material have a low accessibility to traditional methods. In order to utilize the leaf protein for food purposes, the protein needs to be extracted from the fiber rich leaf matrix. This conversion of green biomass to valuable products is an example of the green biorefinery. The GBR may be tailored to produce different products. The existing knowledge on the extraction, purification, and concentration of protein from green biomass can be used as a biorefinery starting point. Additionally, the quality and potential application of the leaf protein in food products and side streams from the BGR and possible uses of side streams from the protein production are illustrative of the benefits to sustainable agriculture in the twenty-first century.[27]

13.2 CONNECTION TO AGRICULTURE

In a more modern approach, the U.S. Department of Energy/NREL has described conversion technologies for expanded biomass utility. (See Figure 13.3) This basic technology could generate base or platform chemicals from which industry could make a wide range of fuels, chemicals, materials, and power. In essence five platforms that show the breadth of the biorefinery have been suggested:

- sugar platform biorefineries (SPBs),
- thermochemical or syngas platform,
- biogas platform,
- carbon-rich chains platform, and
- plant products platform.

FIGURE 13.3 Linking agriculture and the biorefinery. (See www.shutterstock.com/image-vector/biofuel-biomass-ethanol-life-cycle-earth-393421855)

The SPB, for example, focuses on the fermentation of sugars extracted from biomass feedstocks. The objective is to biologically process the sugars to produce fuel, such as ethanol, or other building block chemicals. SPBs are closely related to LCF biorefineries.[2] The other platforms indicate the desired products being produced.

13.2.1 Lignocellulosic biorefinery (LBR)

This is a type of biorefinery that captures a large range of feedstocks. An increasing demand for energy and depleting petroleum sources has elevated the need for producing alternative renewable resources. Owing to the prominence of lignocellulosic biomass as bio-renewable and the most abundant resource on earth, the potential of lignocellulosic biomass for production of second generation (2G) ethanol and value added products in a BR manner is quite high. The efficient utilization of all three components of lignocellulosic biomass (i.e., cellulose, hemicellulose, and lignin) would play a significant role in the economic viability of cellulosic ethanol and other products. The pretreatment of these is key to the success of bioconversion processes and greatly influences the economics of BR process. Within the BR, biotechnology tools and process engineering play pivotal roles in development of integrated processes for production of biofuels, biochemicals and biomaterials from lignocellulosic biomass. LBR has ample technologies, but commercial production of biofuels and chemicals is still challenging. Since there is such a pervasive resource, the efforts to utilize it are critical.[28]

13.2.2 Food waste

Food waste provides an immense and valuable resource for the biorefinery. In a society where the environmental conscience is gaining attention, it is necessary to evaluate the potential valorization options for agricultural biomass to create a change in the perception of the waste agricultural biomass from waste to resource. In that sense, the BR technologies have been proposed as the roadway to increase profit in the agricultural sector and, at the same time, ensure environmental sustainability. The BR processes integrate biomass conversion processes to produce fuels, power, and chemicals from biomass found in waste. The extraction of value-added compounds, anaerobic digestion, and composting of agricultural waste in the biorefinery is essential. This BR approach is an application of configurations found in other biorefinery configurations, such as bioethanol production or white biotechnology. However, any of these processes can now be applied to a single operation unit for agricultural waste valorization. The potential valorization of agricultural waste biomass, focusing on valuable compound extraction, anaerobic digestion, and composting of agricultural waste, whether they are not, partially, or fully integrated is a reachable goal.[29]

Potential feedstocks for future waste biorefineries include the food content of municipal waste and the industrial waste of the food processing industry. Potential products include energy and bioproducts (chemicals, materials, value-added ingredients) that add economic value to the simultaneous environmental benefits (reduction of fossil fuels and GHG emissions, waste disposal and treatment). Food waste is rich in organic materials (carbohydrates, proteins, oils and fats) and other organic species (high-market-value products such as antioxidants). Organic components offer multiple paths to the production of chemicals but will require design studies to improve raw material and energy efficiencies. Compared to biomass-based biorefineries (e.g., biorefineries using wood chips and cereal straw), waste-based biorefineries are simpler and require fewer stages (e.g., whey and molasses could be used directly as fermentation media).[30]

The global agriculture sector is one of the major sources of bio-wastes, and around the world the amount and type of agricultural waste vary. Cereal crops are among the major crops responsible for the agricultural waste in the world. Some of these crops are used as fodder for domesticated animals, composting and energy production. Composting is a technique, where aerobic fermentation changes agri-waste into soil conditioner. Another strategy for valorization of agri-residue is its bioconversion to second generation biofuels. Agri-waste can also be converted into briquetting or biogas.[31] Research activities are going on across the globe for bioconversion of agri-waste to value added products.[1]

13.2.3 Biomass

Liquid biomass is a valuable renewable resource with high energy density and can be processed into green fuels that are similar to petroleum fuels using hydroprocessing technology. Hydroprocessing is a classic petroleum operation for eliminating heteroatoms from crude oil such as S, N, and O before splitting the higher carbon-containing molecules into lower ones. Traditional technology has recently been employed to convert first-generation (1-G) and second-generation (2-G) liquid biomass into a wider range of hydrocarbons, such as diesel and jet fuels. The transformation of 1-G and 2-G liquid biomass into green fuels using hydroprocessing technology is a valuable step toward sustainability. Hydroprocessing of 1-G and 2-G liquid biomass to produce green fuels can be accomplished using mono, bi, tri and renewable catalysts under varying operating parameters. Commercial green fuels are possible. The future perspective and economic analysis of hydroprocessing technology is an important component to make these fuels. Hydroprocessing is a promising technology for future biorefinery success.[32]

Lignocellulose biomass is considered to be the prevalent and economic substrate for biofuel generation. The presence of certain refractory components in biomass causes major obstacles in enzymatic hydrolysis and thus they must be improved by the advancement of pretreatment technology. The pretreatment focuses on the enhancement of hydrolysis by converting the polymeric substance into component monomers. A suitable pretreatment would convert the biomass into easily accessible components for enzymes and thus enhance fermentation during biofuel production. The research and development focused on the goal of fuels together with the integrated pretreatment for biofuel production are needed to be optimized in terms of both economics and environmental concerns. The problems associated with the economic and environmental result from the enormous research covering several decades in order to replace fossil fuels with the lignocellulosic feedstock. The various new approaches in lignocellulose based biofuel production consider the net cost and the energy demand by upgrading the process design. The energy demand on pretreatment and the techno-economic and environmental aspects remain key issues to resolve.[33]

13.2.4 FRUIT

Fruit processing industries contribute more than 0.5 billion tons of waste worldwide. The global availability of this feedstock and its untapped potential has encouraged researchers to perform detailed studies on value-addition potential of fruit processing waste (FPW). Compared to general food or other biomass derived waste, FPW is minimally developed. The peels, pomace and seed fractions of FPW could potentially be a good feedstock for recovery of bioactive compounds such as pectin, lipids, flavonoids, and dietary fibers. A novel biorefinery approach would aim to produce a wider range of valuable chemicals from FPW. The wastes from majority of the extraction processes may further be used as renewable sources for production of biofuels. The path to value addition from fruit derived waste is diverse. Fruit waste derived bioactives may be a beneficial field to pursue, but the financial challenges encountered in existing methods must be addressed.[34]

The growing interest in the use of by-products from the agri-food industry with the aim of achieving zero waste has shown that pectin and FPW may be more of a problem rather than a solution.[35] The composition of fruit processing waste also presents opportunities for its utilization in the biorefinery. FPW feedstock can potentially contribute to the production of renewable energy, biofuels, and high-value compounds. However, commercialization of fruit waste biorefineries is limited, due mainly to the unavailability of information on feedstock availability, process design, and scale-up information. Consequently, capital and operating cost requirements, associated resource use, barriers to mobilizing available feedstocks as well as the anticipated social and environmental impacts of such systems are poorly understood. There is a need for current fruit processing trends and methods to utilize fruit processing waste while developing potential integrated multi-feed BR schemes that can utilize the waste for co-production of bioenergy and biofuels with low volume high-value compounds (essential oils, pectin, and polyphenols). Multi-feedstock operations are possible to enable all-year-round operation of the biorefineries while co-production of bioenergy and bioactives is anticipated to boost the economic and environmental viability of these waste streams. Limitations in technology and process development, coupled with limited availability of information on economic viability and sustainability at industrial scale, are challenges to adoption of integrated biorefineries. Hence, future work aimed at evaluating the applicability in terms of economic viability and sustainability of integrated multi-feed fruit waste-based biorefineries is a worthwhile effort in sustainable agriculture.[36]

13.2.5 RICE

The global volume of rice production warrants further exploration of its uses. The increasing interest in reusing coproducts from rice crops, such as straw, husks, and bran are a few potential topics. The major challenges for further processing of rice coproducts within the biorefinery are important. Current studies and applications mainly include the use of such coproducts on energy and biofuel generations, production of building blocks, adsorption of recalcitrant substances, animal feeding and fertilization crops, extraction of bioactive compounds, and production of carbon-based and silica-based materials. Furthermore, the processing steps (chemical, biochemical, and thermochemical routes), conventional and novel technologies and value-added products/derivatives with interest in several industrial fields all indicate the expansion in rice is both necessary and desirable. One new green area involves the activity of some companies and government projects that are trying to use sub/supercritical water hydrolysis as a promising technology with high potential to decompose rice lignocellulosic biomass into small-chain sugars and bioproducts. Overall, as the future outlook for making most of the processing routes of rice coproducts becomes more feasible, more infrastructure and scientific research is needed to overcome some barriers and drawbacks that still exist.[37]

13.2.6 CORN

Corn provides a significant opportunity for agriculture and agricultural chemistry to have an impact in sustainability. Corn is the second most abundant crop produced worldwide annually. Its residue, corn stover, is among the three most abundant agricultural residues globally and the first in the United States. The available corn stover in the world could be used to produce more than 45 million cubic meters of bioethanol (60% of the world production) instead of being mainly burnt on the fields. This practice could potentially reduce Green House Gases (GHGs) emissions by more than 150 million tons of CO_2 equivalent per year. The results from studies show that pretreatment is the most crucial step that hinders bioethanol production from corn stover, whereas newly developed methods like ionic liquids show promising results of 90% lignin degradation. The integrated biorefinery approach can be employed as an efficient strategy to cut the ethanol minimum selling price to half with a lower average capital cost.[38]

13.3 ROLE OF AGRICULTURAL CHEMISTRY IN THE BIOREFINERY

Since the agricultural business produces enormous amounts of waste, this becomes a good starting point. Improper disposal of the waste in open dumpsites and non-sanitary landfills or by burning has ill effects on human and environment health. This waste is coming primarily from agriculture and agri-food processing sectors, which includes mainly lignocellulosic waste. It is estimated that approximately 30% of the global food production is lost in food or agri-food sector.[39] For sustainable growth of agriculture and agricultural chemistry, a policy shift towards "reduce, reuse, recycle, regenerate" is required. Agriculture waste must be viewed as a sustainable resource. Sustainable integrated biorefineries using agricultural waste will need to develop efficient and economically feasible conversion technologies.[29]

13.3.1 FOOD WASTE BIOREFINERIES (FWB)

The goals of the biorefinery include the production of higher-value chemicals with potential for wide dissemination and an untapped marketability. The FWB is an example of this effort. The initial step is to construct a clearer picture of rural waste treatment system and to regularly utilize agricultural waste management technologies from their state of the art. The present challenges for designing the biorefinery and implementing a system of circular economy with self-efficient business model will be major accomplishments, if successful. The drivers that can make the biorefinery model appropriate to waste management and the conceivable outcomes for its improvement to full scale are critical. Technological, strategic and market imperatives influence the effective usage of these frameworks. The state-of-the-art biorefinery offers opportunities beyond conventional strategies as an economically viable solution to overcome numerous current challenges such as waste minimization and the biosynthesis of different high-value bioproducts.[1]

13.3.2 MODELING BIOWASTE BIOREFINERIES

The biorefining of biowaste is a logical strategy to manage the challenge of sustainability but is mostly still in its conceptual phase. Biowaste biorefineries allow (rural) communities to convert their biowaste into value-added biofuels, biochemical compounds, and fertilizers that could be economically beneficial. Several different types of biowaste biorefineries have already been developed, but these designs have not found commercial application. Their further development and commercial implementation are hampered by the high investment costs and risks. There is not sufficient trust in these novel technologies, the projected yields and profits, and its operating reliability. Modeling these integrated processes, together with their supply chains, would optimize the considered biorefinery designs, speeding up the R&D-process. The optimized biorefinery designs and supply chains would additionally result in an increased amount of trust in potential investors in terms of the economic sustainability of the considered processes. Therefore, the complete picture of biorefinery models and supply chain network models is needed. The biorefinery models can also be categorized according to the conversion platform they use, being thermochemical, biological, or hybrid ones.[40]

13.3.3 INTEGRATED BIOREFINERIES OF AGRICULTURAL WASTE

Agricultural wastes can serve as a potential source for integrated biorefinery for multiple product production. The major issues associated with agricultural wastes include increased greenhouse gas emission and leaching of the wastes cause contamination of water supplies. Agricultural wastes have unique components which can be used to produce an array of value-added products.[41] Ecological as well as economic impacts of agricultural waste management are difficult to assess due to their complex nature and seasonality.[3] The main challenge of agricultural waste management is the lack of prediction tools to provide a clear pathway to the policy makers and end users.

Integrated biorefineries currently have large investment risks in novel technologies. Adopting integrated technologies can yield multiple products which in turn improve the overall process economics. For example, production of chemicals from orange peel is not economically viable. But the overall process economics can be improved by adopting an integrated strategy to produce multiple products like limonene, pectin and cellulose by microwave technology. This integrated process reduces environmental foot print as well as increase the number of more renewable compounds.[42] Anaerobic digestion of agricultural wastes for the production of biogas is considered a mature technology but it has several limitations like low conversion yield as well as low economic value.[43] Operation conditions play a significant role in product yield. Since the agricultural wastes contains considerable amount of lignin, some pretreatments to be carried out for its removal. A cost effective and eco-friendly strategy for the production of biogas is by adopting microbial electrolysis cell technology.[44]

Another alternative strategy is the implementation of two-stage anaerobic digestion process comprised of a fermentation stage for the production of biohydrogen followed by a methanization for the production of biogas enriched with hydrogen.[45] The two-phase anaerobic digestion process improves the process efficiency in comparison to conventional strategies adopted for biogas production.[1]

13.3.4 CHALLENGES IN INTEGRATED BIOREFINERY OF AGRICULTURAL WASTE

Agricultural waste is predominantly lignocellulosic biomass, which is a recalcitrant structure that can be broken to obtain sugar chains and lignin. Lignocellulosic wastes can be processed through multiple avenues, both thermochemical and biochemical, such as combustion, gasification, pyrolysis, pretreatment followed by saccharification, and fermentation, to produce multiple products in a biorefinery, allowing for a circular economy. Finding the best and more economic route is not easy. The most often used route for utilization of agricultural residue is biochemical route to produce ethanol in a biorefinery setting (Figure 13.1). Even with a vast potential in biorefinery setting, agricultural biomass faces some challenges for successful commercialization. The main challenge in employing integrated strategies is the compatibility between the thermo-chemical and biological processes.

Agricultural waste is one of the most abundant and inexpensive feedstocks for biorefinery. For sustainable year-round availability, some of the ideal properties of this is as a feedstock and supply chain would be its low cost and ease of transportation, smaller collection radius, easily densified feedstock, consistent composition, and price stability. The current model of large cellulosic biorefineries with a daily processing capacity of 2000–5000 dry tons feedstock is the centralized facilities with all unit operations at one location. It has been shown that from a logistics standpoint, these biorefineries are impractical.[46]

Biorefinery studies have shown that lower processing capacities lead to higher operating cost per unit of main product, favoring the larger biorefinery in the long run.[47] To overcome some of the challenges of larger biorefineries, regional, decentralized depots can be created, with facilities acting as transitory storage locations. Farmers can use these facilities to store biomass in the form of round or square bales, which can be accessible for year-round availability for biorefineries. Some pre-processing such as densification and pretreatment can also be performed at these facilities, which will reduce the transportation cost of the densified biomass to biorefineries.[48]

The role of agricultural chemistry in this area becomes clear. The biggest challenge in the success of sustainable integrated biorefinery using agricultural waste is the conversion technologies. The main challenge is understanding the chemistry of the heterogenous nature of the lignocellulosic biomass. New strategies must be adopted for the production of value added products from lignocellulosic biomass. Fine tuning for each unit operations should be done to make the process economically viable. The technologies should be simple, robust, and at the same time, economically feasible and environmentally friendly. The most used technology for the conversion of agricultural waste to biofuel and added value-products involves pretreatment, followed by hydrolysis and fermentation. The biomass pretreatment is one of the most cost-intensive steps in lignocellulosic biomass processing.[49] The recalcitrant structure of the biomass needs to be disrupted in order for the enzymes or microbes to access the carbohydrates. Agricultural chemistry can help make this more efficient and profitable.

All the pretreatments have different challenges such as physical and physicochemical methods being energy intensive, chemical pretreatments being expensive with a need to remove the chemical and neutralize the slurry before downstream processes, and biological pretreatments being time consuming. The pretreatment stages need to be optimized to balance the sugar yields during hydrolysis and then energy and cost involved during the process. The challenge in these biorefinery settings is optimization of individual processes. The sustainability of a biorefinery is highly dependent on the optimization of the inter-relation of all processes.

Hydrolysis technological challenge faced by agricultural waste biorefineries. It must be technologically feasible and economically viable. Currently it is expensive due to the addition of high-value enzymes.[50] The pentose sugar production rate and yields are currently much lower than the optimum required for a commercially feasible process. To obtain high yields of lignin, the front-end conversion processes need to be selected in order to not affect the quality of lignin and to ensure minimum losses. In order to achieve this, a trade-off with low sugar yields might be necessary. Overall, optimization is required to maximize the sugar yields as well as high-quality lignin yields.[11] However, the feasibility of a biorefinery, energetically and economically, for successful commercialization requires performing techno-economic analyses, which can provide a better understanding of processes that need improvement.

13.4   SUSTAINABILITY CONTRIBUTIONS BY BIOREFINERIES

The following list includes top twelve building blocks identified by a NREL and PNNL study.

- 1,4-succinic, -fumaric, and -malic acids
- 2,5-furan dicarboxylic acid
- 3-hydroxy propionic acid
- aspartic acid
- glucaric acid
- glutamic acid
- itaconic acid
- levulinic acid

- 3-hydroxybutyrolactone
- glycerol
- sorbitol
- xylitol/arabinitol

The NREL and PNNL study analyzed the synthesis for each of the top building blocks and their derivatives as a two-part pathway, where the first part is the transformation of the sugars into the building blocks and the second part is the conversion of the building blocks to secondary chemicals or families of derivatives. Biological transformations account for the majority of the routes from plant feedstocks to building blocks, but chemical transformations predominate in the conversion of building blocks to molecular derivatives and intermediates. The challenges and complexity of these pathways, as briefly examined by the NREL and PNNL study highlight R&D needs that could help improve the economics of producing these building blocks and derivatives.[2]

Biorefineries built on lactic acid production from renewable materials have gained enormous attention due to the several functional properties it may offer in different fields. The high yield productivity of lactic acid by utilizing economical and easily available substrates has received immense attraction in the petrochemical industry. Extraction of lactic acid from biomass or waste materials from various sources is highly desirable because they cause severe environmental pollution if disposed of improperly. Hence, the employment of an integrated biorefinery platform for waste materials is an ideal option to produce high-value bio-products while remediating the waste. Optical pure lactic acid production through fermentation has gained interest due to its high potential applications in food, pharmaceutical, textiles, and cosmetic industries as well as highly promising packaging materials. The manufacturing of biodegradable bioplastic from polylactic acid materials is a green alternative to that derived from petrochemicals. However, high manufacturing costs have impeded the widespread application of polylactic acid due to the high cost of lactic acid production. Pretreatment and enzymes hydrolysis and fermentative technologies of biowastes can help meet this need.[51]

13.4.1 Wastewater

The world is facing a dwindling supply of fresh water. The freshwater microalgae *Haematococcus pluvialis* and *Chlorella zofingiensis* are attractive biorefinery enzymes in view of their ability to simultaneously synthesize astaxanthin and other valuable metabolites. Nonetheless, there are concerns regarding the sustainability of such biorefineries due to their high freshwater footprint for microalgae cultivation. The integration of wastewater as an alternative growth media is a potential approach to reduce freshwater demand. Wastewater-based cultivation enables the recovery of essential nutrients required for microalgae growth and might result in phytoremediation of wastewater, thus promoting the feasibility of a circular economy and further enhancing the sustainability of the process. Recent developments in wastewater-integrated cultivation of *H. pluvialis* and *C. zofingiensis* for astaxanthin production provide examples of new technologies for overcoming the inherent challenges of wastewater-based cultivation. Moreover, the biorefinery potential of wastewater-grown *H. pluvialis* and *C. zofingiensis* increases the potential of wastewater-based biorefineries.[12]

13.4.2 ENERGY

With conventional fossil energy reserves diminishing as well as atmospheric pollution increasing through excessive use of these non-renewable resources, cleaner, renewable replacement sources of energy for a sustainable future are needed. Bioenergy from microalgae is a promising alternative among the different renewable energy sources, and it has desirable properties. Microalgae can yield a very high lipid content, which can be processed to biodiesel. They can capture carbon dioxide and can be grown even on degraded land with wastewater. Additionally, it is possible to generate a variety of products such as ethanol, biogas and pharmaceuticals from microalgae. In spite of these promising attributes, its commercial scale implementation has been limited. The major challenge is the relatively high cost associated with the entire process. Additionally, each of the processing steps presents technological challenges. There is a great potential for development of an integrated microalgal biorefinery.[52]

Green technologies are in demand worldwide to meet the fuel needs of the increasing population together with combating global warming. Agriculture and agricultural chemistry are in a key position to help here. While biofuels from conventional energy crops and lignocellulosic biomass can substitute for fossil fuels, there remain questions regarding impacts of using food for fuel on food supply and security and their cost effectiveness. Given this background, algae have a high photosynthetic efficiency and production rate. This makes algae attractive for third generation biofuels. Algae (both microalgae and macroalgae) represent an economically and environmentally sustainable renewable source of biomass for the production of biofuels. Algal biofuels are attractive in that they do not compromise food supplies and arable land. Microalgae have received more attention than macroalgae. Nevertheless, the latter can also serve as a viable option for biofuel generation. Carbohydrates obtained from macroalgal biomass can be used for fermentative production of bioethanol. Macroalgal biomass can store large amounts of oil which can be exploited for the production of biodiesel. Methane and hydrogen can readily be produced from macroalgal biomass through biologically mediated degradation. The biorefinery approach, integrating bioprocessing and low-environmental-impact chemical technologies, can efficiently utilize macroalgal biomass, generating diverse products from a single biomass feedstock. As the use of biorefineries grow, algae will become a major feedstock.[53]

Commercial-scale production of microalgae biomass for biofuel and biochemicals requires a substantial amount of water and nutrients. One of the challenges will be to recover the water for recycling. Studies have been conducted to explore the potential of various aqueous streams originating from algae harvesting stage and different energy recovery steps as an alternative for water and nutrient supply. The presence of toxic organic compounds, unassimilated ions, particulate matter, and high alkalinity in post-harvest water limit its recyclability. Nutrient recovery from biomass via anaerobic digestion (AD) and various hydrothermal processes is being explored. So, there is a need to understand the impact of harvesting methods, nature and impact of organic compounds, buildup of algogenic organic matter (OM), amount of unused nutrients, and salinity on water recycling. Optimum conditions for maximum nutrient recovery from AD and hydrothermal processes are needed for effective nutrient recycling. A possible way forward could include the recycling of aqueous streams for water and nutrient requirement for sustainable microalgae cultivation. The effectiveness of a recycling stream is defined here as "biomass ratio." Possible growth inhibiting factors and their solutions are important potential directions for future research. Large-scale sustainable cultivation of microalgae through recycling of different streams depends on better understanding of biological activity of algal OM through detailed characterization and in-depth understanding of physiochemical properties.[54]

Studies on the techno-economic and life cycle assessment of microalgal biodiesel production suggest that the standalone production of biodiesel is currently unviable. The production of microalgal biomass using the currently available technologies has costs that are unacceptably high for biodiesel production. The challenges lie in high biomass production cost with unfavorable energetic balance. Significant process and engineering advancements are desirable before mass-scale production of algal oil and biodiesel at a cost-competitive price is realized. On the other hand, various high-value products sourced from microalgae are already commercialized. The biomass production cost is more than acceptable if such products are also derived and sold. It is expected that with process modifications and engineering advancements, the biomass production cost can be brought down to levels that will improve profitability. Moreover, coupling phytoremediation of pollution loads in waste streams to microalgal biomass production offers economic and environmental gains (90% reduction in water footprint, improved GHG balance, and a substantial reduction in external input of fertilizers). Such approaches are more likely to translate into an economically appealing and environmentally desirable business model.[55]

There are different sources of biomass that can be used as raw material for the production of biogas. The focus is mainly on the use of plants that do not compete with the food supply. Biogas obtained from edible plants entails a developed technology and good yield of methane production; however, its use may not be sustainable. Biomass from agricultural waste is a cheap option, but in general, with lower methane yields than those obtained from edible plants. On the other hand, the use of algae or aquatic plants promises to be an efficient and sustainable option with high yields of methane produced, but it necessary to overcome the existing technological barriers. Moreover, these last raw materials have the additional advantage that they can be obtained from wastewater treatment and, therefore, they could be applied in the biorefinery. Different strategies to improve the yield of biogas, such as physical, chemical, and biological pretreatments, are needed. Other alternatives for enhanced the biogas production such as bioaugmentation will also be helpful.[56]

13.4.3 BIOWASTE

Valorization of biowaste, if achievable, is a significant contribution to sustainability. This is a topic deserving closer scrutiny for its role in transition of fossil-based economy to bioeconomy. Increasing waste generation and climate change are among the issues that can be alleviated if biowaste can be fully utilized. Various conversion methods have been used and/or developed as the most effective methods to valorize biowaste into value-added products such as biofuel, biochar and other biomaterials. Recognizing the need to identify areas with higher demand for scientific effort will increase the ways this may be improved. There are rheological limitations that hinder the upscaling feasibility of biowaste valorization. Process optimization and development of new products with competitive edge from biowaste are the key avenues to close the loop of circular bioeconomy.[57]

13.4.4 CHEMICAL PRODUCTION

The current climate emergency and the risks to biodiversity that the planet is facing, have made the management of food resources increasingly complex and scientifically challenging. One-third of the edible parts of food produced throughout the whole food supply chain gets lost or wasted globally every year. At the same time, demographic growth makes it necessary to adjust in order to achieve sustainable economic development while satisfying market demands. The European Union supported the idea of a Circular Economy from 2015 and arranged annual Action Plans toward a greener, climate-neutral economy. Utilizing the biorefinery, food waste becomes byproducts that can be recovered and exploited as high added-value materials for industrial applications. The use of sustainable extraction processes to manage food byproducts is a task that research has to support through the development of low environmental impact strategies. Following eco-pharmacogenetic approaches inspired by green chemistry guidelines, agricultural chemistry can help achieve sustainability. In particular, the use of innovative hybrid techniques to maximize yields and minimize the environmental impact of processes is a necessary path forward.[58]

Cellulosic ethanol production can be used as transportation fuels by blending with gasoline, which combines the virtue of carbon benefits and decarbonization. However, due to changing feedstock composition, natural resistance, and a lack of cost-effective pretreatment and downstream processing, current cellulosic ethanol biorefineries are not able to be sustainable. There are technical and non-technical barriers, numerous R&D advancements in biomass pretreatment, and enzymatic hydrolysis which need to be addressed. New research has discovered fermentation strategies for low-cost sustainable fuel ethanol. Moreover, selection of a low-cost efficient pretreatment method, process simulation, unit integration, state-of-the-art in one-pot saccharification and fermentation, system microbiology/ genetic engineering for robust strain development, and comprehensive techno-economic analysis are all major bottlenecks that must be considered for long-term ethanol production for use in the transportation sector.[59]

During the production of chemicals, the technology of two-stage anaerobic digestion (AD) is gaining popularity because its stability and opportunity of recovering multiple-resources such as biohydrogen and organic acids from the first stage dark fermentation (DF) and methane in the AD as the second stage while treating the organic waste. As the performance of two-stage processes is influenced by the type of substrate and operational conditions, work has been done to determine the optimum conditions. An updated overview of advancements in biohythane and organic acids production from food waste (FW) in the two-stage DF-AD process is very encouraging. The improved economics and future prospectives of utilizing organic acids for different biochemicals such as polyhydroxyalkanoates, polylactate, and microalgal biomass production render this process even more appealing. The integration of optimum operational parameters, pretreatment methods and types of bioreactors is essential for optimum combined DF-AD processes. The parameters and reactor configuration have to be optimized depending upon the targeted end-products. More research into the techno-economic analysis of different bioreactor configurations for long-term operations in an integrated DF-AD process with FW as a feedstock is needed to realize its viability for commercial application.[60]

Lignocellulosic biomass and hemicellulose have been recognized as some of the most important renewable bioresources for production of alternative biofuels and biochemicals. The successful utilization of biomass derived from agricultural feedstocks to replace petroleum and petrochemical products are necessary to support the sustainable bioeconomy and biorefinery industry around the world. C_5 and C_6 sugars produced by the deconstruction of lignocellulosic materials through hydrolysis processes can be further converted to key intermediate chemicals, including furfural, 5-hydroxymethylfurfural, furan and organic acids. Among these, furfural is considered a promising biomass-derived platform which can be a key intermediate for producing a variety of C_4 and C_5 compounds, such as, furan, tetrahydrofuran, pentanediol, lactic acid, and levulinic acid. (See Figure 13.4) The catalytic processes, especially, heterogeneous catalysis, for converting furfural into value-added biochemicals and biofuels are key for success here.[61]

Identification and production of desired food chemicals is a valuable target of the biorefinery. The sustainable one-pot conversion methods of cellulose into very important platform chemicals such as 5-hydroxymethylfurfural and isosorbide finds applications in many fields. Various new techniques based on such as ionic liquids, acid functionalized mesoporous materials, organic acids, functionalized nanoparticles, and catalytic depolymerization causes direct conversion of cellulose to 5-hydroxymethylfurfural. There is great need for a comparative analysis of

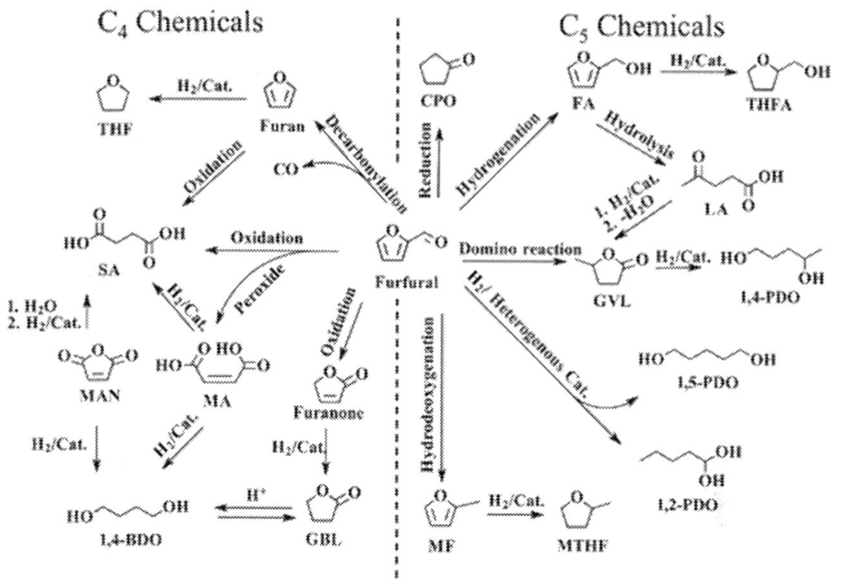

FIGURE 13.4   The pathways for furfural conversion into C_4 and C_5 fine chemicals.[61]

recently developed successful methods for 5-hydroxymethylfurfural production from cellulose in terms of efficiency, selectivity, and cost-effectiveness. This green chemistry example employs promising extraction methods for the 5-hydroxymethylfurfural using special solvents. Another very interesting platform chemical is isosorbide. Several factors of cellulose to isosorbide transformation including metal nanoparticle size, crystallinity order of the cellulose, and extraction medium which controls the rate of conversion and product distillation are technologies found in the biorefinery. Green chemistry has provided potential discoveries in one-pot conversion of cellulose into biofuels. The strategies of cellulose and lignocellulose conversions to compounds such as liquid fuels focus on the example of γ-valerolactone as an important intermediate to access liquid hydrocarbons and valeric esters. Increasing the cellulose value-chain using the direct conversions to liquid fuels (e.g. cellulose to levulinic acid platform to obtain valeric biofuels) is accomplished using nanostructured metal catalysts. Chemical production in the biorefinery still faces technical barriers, commercial promise, and environmental issues.[58]

A significant area of valorization is in the isolation of polyhydroxyalkanoates (PHAs). PHAs are microbial biopolymers (polyesters) that have a wide range of functions and applications. They serve in nature mainly as carbon and energy storage materials for a variety of microorganisms. Once isolated, they can be utilized ways ranging from in commodities and degradable plastics to specialty performance materials in medicine. PHA biosynthesis is well understood, and it is now possible to design bacterial strains to produce PHAs with desired properties.

There are many substrates for the fermentative production of PHAs. They can be derived from food-based carbon sources (e.g., fats and oils (triglycerides)), which could challenge the sustainability of their productions in terms of crop area and food. However, hemicellulose hydrolysates, crude glycerol, and methanol are very promising carbon sources for the sustainable production of PHAs. Therefore, the integration of PHA production within a modern biorefinery is an important issue and can result in simultaneous production of biofuels and bioplastics. Furthermore, many chemical-synthetic procedures by means of efficient catalysts can give access to a variety of PHAs. From these examples, the biorefinery can foster a sustainable PHA-based industry.[62]

Work by agricultural chemists, among others, help achieve technological developments to enable the production of new value-added products from lignocellulosic residues such as lignin. This effort has allowed the forest industry to diversify its product portfolio and increase the economic returns from fruit waste feedstock, while simultaneously working towards sustainable alternatives to petroleum-based products. Although previous research has explored industrial-scale production opportunities, many challenges persist, including the cost of woody biomass and its supply chain reliability. While numerous studies have addressed these issues, their emphasis has traditionally been on bioenergy, with little focus on biochemical, biomaterials, and bioproducts. Significant work has been done in the United States and Canada with an emphasis on bioenergy production: part of the work is focused on biomass to materials and chemicals. Between 2012 and 2015, published work highlighted biomass to materials and chemicals and both biomass to energy and biomass to materials and chemicals. This reflects a greater interest in diversified biorefinery portfolios. However, further work concerning forest biomass supply chain optimization and new high-value bio-based materials and chemicals is necessary.[63]

13.4.5 CIRCULAR ECONOMY

A circular economy approach in the agriculture sector will improve resource efficiency and waste management. The circular economy will also have the potential to reduce greenhouse gas emissions, waste, pressure on land, and preservation of the biodiversity.[64] The continuous and balanced use of agricultural resources for production and consumption needs proper planning. This must include reuse of the material after disposal. United States and Europe are two prominent regions of the world where the concept of the biorefinery and bio-based product is in policies, but in other regions of the world still no comprehensive policies on these important topics exist. Lack of specific policies limits the overall acceptance of bio-based economy.[65] The three following examples illustrate how the circular economy can be actualized.

The need for alternative source of fuel has led to the use of non-edible foods in fuel production. Now there is a 3rd generation feedstock which includes microalgae, seaweed and cyanobacteria. These phototrophic organisms are unique in a sense that they utilize natural sources like sunlight, water and CO_2 for their growth and metabolism thereby producing diverse products that can be processed to produce biofuel, biochemical, nutraceuticals, feed, biofertilizer and other value-added products. But due to low biomass productivity and high harvesting cost, microalgae-based production has not received much attention. In order to move this into a sustainable matrix, the microalgae based biorefinery approach needs to discover an economical and sustainable process. The three major segments that need to be considered for economic microalgae biorefinery is low cost nutrient source, efficient harvesting methods and production of by-products with high market value. Various wastewaters can also be used as nutrient source for simultaneous biomass production and bioremediation. Harvesting methods used for microalgae can yield various products from both raw biomass and delipidified microalgae residues in order to establish a sustainable, economical microalgae biorefinery with a touch of circular bioeconomy. In order for it to be fully realized it must manage various challenges followed by a techno-economic analysis of the microalgae based biorefinery model.[66]

Bio-based plastics are a new area of biorefinery development. It currently only represents 1% of world production, while 98% are plastics of fossil origin. This shows how far we are from an ecologically correct scenario, and how much more needs to be done to develop bioplastic processes. It can be shown that biopolymer production from monomers generated by fermentation, such as some organic acids, is a sustainable alternative to petrochemical sources. One of the main challenges in the use of organic acids in the production of biopolymers is economic competitiveness, since oleochemicals are cheaper than carbon sources used in the production of biomonomers. Thus, the use of agro-industrial residues as substrates in organic acid production can lead to economically viable biopolymers. The production of biomonomers using hydrolyzed biomass has the potential to be competitive with petrochemical-based plastics, in which acetic, citric, fumaric, gluconic, lactic itaconic, and succinic acids are the main candidates to be employed in a biorefinery concept.[67] This is a clear example of the circular bioeconomy.

Finally, the projection of world population growth with concurrent generation of large volumes of agro-industrial waste that negatively affect the environment is of great concern. Therefore, the nexus between concepts of Circular Bioeconomy, Zero Waste Technology, Sustainable Development, Biorefineries, and Green Chemistry need to coalesce to form processes that generate less environmental impact. It is important to have emerging technologies and innovation in order to promote the replacement of fossil-derived raw materials with renewable raw materials and develop more environmentally friendly processes and industries. The current state of biomass research, together with current pretreatment research on biomass to obtain bioproducts and biofuels in a biorefinery that promotes clean production for the extraction of phytochemicals using green solvents and technologies to recover high-value added materials with enhanced properties are one example of a circular economy solution. The need to develop technologies and markets to commercialize high value-added products coming from biorefineries will increase the income globally and will strengthen the productivity and profitability of a sustainable agroindustry. The goal is to improve environmental sustainability, while also contributing to the circular bioeconomy that promotes sustainable development.[68]

REFERENCES

1. Awasthi, M.K., et al., *Agricultural waste biorefinery development towards circular bioeconomy.* Renewable and Sustainable Energy Reviews, 2022. **158**: pp. 1–17.
2. Fernando, S., et al., *Biorefineries: Current status, challenges, and future direction.* Energy and Fuels, 2006. **20**(4): pp. 1727–1737.
3. Nandi, R. and S. Sengupta, *Microbial production of hydrogen: an overview.* Critical Reviews in Microbiology, 1998. **24**(1): pp. 61–84.
4. Morey, R.V., et al. *A biomass supply logistics system.* in *American Society of Agricultural and Biological Engineers Annual International Meeting 2009.* 2009. Reno, NV.
5. Parajuli, R., et al., *Biorefining in the prevailing energy and materials crisis: A review of sustainable pathways for biorefinery value chains and sustainability assessment methodologies.* Renewable and Sustainable Energy Reviews, 2015. **43**: pp. 244–263.
6. Silvestri, C., et al., *Green chemistry contribution towards more equitable global sustainability and greater circular economy: A systematic literature review.* Journal of Cleaner Production, 2021. **294**: pp. 1–24.
7. Cherubini, F., *The biorefinery concept: Using biomass instead of oil for producing energy and chemicals.* Energy Conversion and Management, 2010. **51**(7): pp. 1412–1421.
8. Liu, Z. and M.R. Eden, *Biorefinery principles, analysis, and design,* in *Sustainable Bioenergy Production.* 2014. CRC Press. pp. 447–476.
9. Batsy, D.R., et al., *Product Portfolio Selection and Process Design for the Forest Biorefinery,* in *Integrated Biorefineries: Design, Analysis, and Optimization.* 2012. CRC Press. pp. 3–35.
10. Liu, M., J. Cao, and C. Wang, *Bioremediation by earthworms on soil microbial diversity and partial nitrification processes in oxytetracycline-contaminated soil.* Ecotoxicology and Environmental Safety, 2020. **189**: pp. 1–10.
11. Rigamonti, L. and E. Mancini, *Life cycle assessment and circularity indicators.* International Journal of Life Cycle Assessment, 2021. **26**(10): pp. 1937–1942.
12. Nishshanka, G.K.S.H., et al., *Wastewater-based microalgal biorefineries for the production of astaxanthin and co-products: Current status, challenges and future perspectives.* Bioresource Technology, 2021. **342**: pp. 1–14.
13. Ng, D.K.S., K.S. Ng, and R.T.L. Ng, *Integrated Biorefineries,* in *Encyclopedia of Sustainable Technologies.* 2017. Elsevier. pp. 299–314.
14. Cantero, D., et al., *Pretreatment processes of biomass for biorefineries: Current status and prospects.* Annual Review of Chemical and Biomolecular Engineering, 2019. **10**: pp. 289–310.
15. Elgharbawy, A.A.M., et al., *A grand avenue to integrate deep eutectic solvents into biomass processing.* Biomass and Bioenergy, 2020. **137**: pp. 1–21.
16. Jiang, L.Y. and J.M. Zhu, *Separation Technologies for Current and Future Biorefineries-Status and Potential of Membrane-Based Separation,* in *Advances in Bioenergy: The Sustainability Challenge.* 2015. Wiley Blackwell. pp. 193–208.

17. Macias-Garbett, R., et al., *Phenolic compounds from brewer's spent grains: toward green recovery methods and applications in the cosmetic industry.* Frontiers in Sustainable Food Systems, 2021. **5**: pp. 1–10.

18. Peleteiro, S., et al., *Utilization of ionic liquids in lignocellulose biorefineries as agents for separation, derivatization, fractionation, or pretreatment.* Journal of Agricultural and Food Chemistry, 2015. **63**(37): pp. 8093–8102.

19. Silveira, M.H.L., et al., *The essential role of plant cell wall degrading enzymes in the success of biorefineries: Current status and future challenges*, in *Biofuels in Brazil: Fundamental Aspects, Recent Developments, and Future Perspectives.* 2014. Springer International Publishing. pp. 151–172.

20. Singh, J.S., et al., *Cyanobacteria: A precious bio-resource in agriculture, ecosystem, and environmental sustainability.* Frontiers in Microbiology, 2016. **7**(APR): pp. 1–19.

21. Furukawa, T., F.O. Bello, and L. Horsfall, *Microbial enzyme systems for lignin degradation and their transcriptional regulation.* Frontiers in Biology, 2014. **9**(6): pp. 448–471.

22. Sodré, V., et al., *Microorganisms as bioabatement agents in biomass to bioproducts applications.* Biomass and Bioenergy, 2021. **151**: 1–15.

23. Duan, Y., et al., *Sustainable biorefinery approaches towards circular economy for conversion of biowaste to value added materials and future perspectives.* Fuel, 2022. **325**: pp. 1–10.

24. Aguiar, A., et al., *Sugarcane straw as a potential second generation feedstock for biorefinery and white biotechnology applications.* Biomass and Bioenergy, 2021. **144**: pp. 1–16.

25. Clark, J.H., F.E.I. Deswarte, and T.J. Farmer, *The integration of green chemistry into future biorefineries.* Biofuels, Bioproducts and Biorefining, 2009. **3**(1): pp. 72–90.

26. Saldaña, M.D.A., et al., *Green Processes in Foodomics: Biorefineries in the Food Industry*, in *Comprehensive Foodomics*. 2020. Elsevier. pp. 808–824.

27. Møller, A.H., et al., *Biorefinery of green biomass-how to extract and evaluate high quality leaf protein for food?* Journal of Agricultural and Food Chemistry, 2021. **69**(48): pp. 14341–14357.

28. Patel, A. and A.R. Shah, *Integrated lignocellulosic biorefinery: Gateway for production of second generation ethanol and value added products.* Journal of Bioresources and Bioproducts, 2021. **6**(2): pp. 108–128.

29. Fermoso, F.G., et al., *Valuable compound extraction, anaerobic digestion, and composting: a leading biorefinery approach for agricultural wastes.* Journal of Agricultural and Food Chemistry, 2018. **66**(32): pp. 8451–8468.

30. Kokossis, A.C. and A.A. Koutinas, *Food Waste as a Renewable Raw Material for the Development of Integrated Biorefineries: Current Status and Future Potential*, in *Integrated Biorefineries: Design, Analysis, and Optimization*. 2012. CRC Press. pp. 469–488.

31. Xiao, R., et al., *Recent developments in biochar utilization as an additive in organic solid waste composting: A review.* Bioresource Technology, 2017. **246**: pp. 203–213.

32. Aslam, M., *Transformation of 1-G and 2-G liquid biomass to green fuels using hydroprocessing technology: A promising technology for biorefinery development.* Biomass and Bioenergy, 2022. **163**: pp. 1–25.

33. Preethi, et al., *Lignocellulosic biomass as an optimistic feedstock for the production of biofuels as valuable energy source: Techno-economic analysis, Environmental Impact Analysis, Breakthrough and Perspectives.* Environmental Technology and Innovation, 2021. **24**: pp. 1–22.

34. Banerjee, J., et al., *Bioactives from fruit processing wastes: Green approaches to valuable chemicals.* Food Chemistry, 2017. **225**: pp. 10–22.

35. Sabater, C., M. Villamiel, and A. Montilla, *Integral use of pectin-rich by-products in a biorefinery context: A holistic approach.* Food Hydrocolloids, 2022. **128**: pp. 1–21.

36. Manhongo, T.T., et al., *Current status and opportunities for fruit processing waste biorefineries.* Renewable and Sustainable Energy Reviews, 2022. **155**: pp. 1–19.

37. Abaide, E.R., et al., *Reasons for processing of rice coproducts: Reality and expectations.* Biomass and Bioenergy, 2019. **120**: pp. 240–256.

38. Aghaei, S., et al., *A comprehensive review on bioethanol production from corn stover: Worldwide potential, environmental importance, and perspectives.* Biomass and Bioenergy, 2022. **161**: pp. 1–13.

39. Regalado, R.E.H., et al., *Optimization and analysis of liquid anaerobic co-digestion of agro-industrial wastes via mixture design.* Processes, 2021. **9**(5), pp. 1–16.

40. De Buck, V., M. Polanska, and J. Van Impe, *Modeling Biowaste Biorefineries: A Review.* Frontiers in Sustainable Food Systems, 2020. **4**: pp. 1–19.

41. Zema, D.A., et al., *Valorisation of citrus processing waste: A review.* Waste Management, 2018. **80**: pp. 252–273.

42. Clark, J.H., D.J. Macquarrie, and J. Sherwood, *A quantitative comparison between conventional and bio-derived solvents from citrus waste in esterification and amidation kinetic studies.* Green Chemistry, 2012. **14**(1): pp. 90–93.

43. Budde, J., et al., *Energy balance, greenhouse gas emissions, and profitability of thermobarical pretreatment of cattle waste in anaerobic digestion.* Waste Management, 2016. **49**: pp. 390–410.

44. Premier, G.C., et al., *Integration of biohydrogen, biomethane and bioelectrochemical systems.* Renewable Energy, 2013. **49**: pp. 188–192.

45. Ferrari, G., et al., *Environmental assessment of a two-stage high pressure anaerobic digestion process and biological upgrading as alternative processes for biomethane production.* Bioresource Technology, 2022. **360**: pp. 1–10.

46. Juneja, A., et al., *Bioprocessing and technoeconomic feasibility analysis of simultaneous production of D-psicose and ethanol using engineered yeast strain KAM-2GD.* Bioresource Technology, 2019. **275**: pp. 27–34.

47. Maity, S.K., *Opportunities, recent trends and challenges of integrated biorefinery: Part i.* Renewable and Sustainable Energy Reviews, 2015. **43**: pp. 1427–1445.

48. Kongjan, P., et al., *Biohydrogen production from wheat straw hydrolysate by dark fermentation using extreme thermophilic mixed culture.* Biotechnology and Bioengineering, 2010. **105**(5): pp. 899–908.

49. Croce, S., et al., *Anaerobic digestion of straw and corn stover: The effect of biological process optimization and pre-treatment on total bio-methane yield and energy performance.* Biotechnology Advances, 2016. **34**(8): pp. 1289–1304.

50. Kazi, F.K., et al., *Techno-economic comparison of process technologies for biochemical ethanol production from corn stover.* Fuel, 2010. **89**(SUPPL. 1): pp. S20–S28.

51. Ahmad, A., F. Banat, and H. Taher, *A review on the lactic acid fermentation from low-cost renewable materials: Recent developments and challenges*. Environmental Technology and Innovation, 2020. **20**: pp. 1–21.

52. Gupta, S.S., S. Bhartiya, and Y. Shastri, *The practical implementation of microalgal biodiesel: Challenges and potential solutions*. CAB Reviews: Perspectives in Agriculture, Veterinary Science, Nutrition and Natural Resources, 2014. **9**: pp. 1–12.

53. Pjohn, R. and G.S. Anisha, *Macroalgae and their potential for biofuel*. CAB Reviews: Perspectives in Agriculture, Veterinary Science, Nutrition and Natural Resources, 2011. **6**: pp. 1–15.

54. Farooq, W., *Sustainable production of microalgae biomass for biofuel and chemicals through recycling of water and nutrient within the biorefinery context: A review*. GCB Bioenergy, 2021. **13**(6): pp. 914–940.

55. Kumar, D. and B. Singh, *Algal biorefinery: An integrated approach for sustainable biodiesel production*. Biomass and Bioenergy, 2019. **131**: pp. 1–16.

56. Martínez-Gutiérrez, E., *Biogas production from different lignocellulosic biomass sources: advances and perspectives*. 3 Biotech, 2018. **8**(5): pp. 1–18.

57. Cheng, S.Y., et al., *Incorporating biowaste into circular bioeconomy: A critical review of current trend and scaling up feasibility*. Environmental Technology and Innovation, 2020. **19**: pp. 1–16.

58. Dutta, S. and S. Pal, *Promises in direct conversion of cellulose and lignocellulosic biomass to chemicals and fuels: Combined solvent-nanocatalysis approach for biorefinary*. Biomass and Bioenergy, 2014. **62**: pp. 182–197.

59. Raj, T., et al., *Recent advances in commercial biorefineries for lignocellulosic ethanol production: Current status, challenges and future perspectives*. Bioresource Technology, 2022. **344**: pp. 1–12.

60. Dangol, S., et al., *Biohythane and organic acid production from food waste by two-stage anaerobic digestion: a review within biorefinery framework*. International Journal of Environmental Science and Technology, 2022. **19**(12): pp. 12791–12824.

61. Khemthong, P., et al., *Advances in catalytic production of value-added biochemicals and biofuels via furfural platform derived lignocellulosic biomass*. Biomass and Bioenergy, 2021. **148**: pp. 1–12.

62. Winnacker, M., *Polyhydroxyalkanoates: Recent Advances in Their Synthesis and Applications*. European Journal of Lipid Science and Technology, 2019. **121**(11): pp. 1–9.

63. Dessbesell, L., et al., *Forest biomass supply chain optimization for a biorefinery aiming to produce high-value bio-based materials and chemicals from lignin and forestry residues: A review of literature*. Canadian Journal of Forest Research, 2017. **47**(3): pp. 277–288.

64. Velis, C.A., *Circular economy and global secondary material supply chains*. Waste Management and Research, 2015. **33**(5): pp. 389–391.

65. Proto, A.R., et al., *Assessment of wood chip combustion and emission behavior of different agricultural biomasses*. Fuel, 2021. **289**: pp. 1–19.

66. Sarma, S., et al., *Valorization of microalgae biomass into bioproducts promoting circular bioeconomy: a holistic approach of bioremediation and biorefinery*. 3 Biotech, 2021. **11**(8).

67. Magalhães Júnior, A.I., et al., *Challenges in the production of second-generation organic acids (potential monomers for application in biopolymers).* Biomass and Bioenergy, 2021. **149**: pp. 1–14.

68. Orejuela-Escobar, L.M., A.C. Landázuri, and B. Goodell, *Second generation biorefining in Ecuador: Circular bioeconomy, zero waste technology, environment and sustainable development: The nexus.* Journal of Bioresources and Bioproducts, 2021. **6**(2): pp. 83–107.

14 Epilogue
Building a sustainable agricultural chemistry

14.1 CHALLENGE OF BUILDING SUSTAINABLE AGRICULTURAL CHEMISTRY

The tangled ball of yarn can be unraveled, then knitted into a beautifully patterned piece of fabric (See Figure 14.1). It takes patience, skill, and a vision of the final product. The current tangled, unsustainable practice of agriculture and agricultural chemistry can also be systematically unraveled and woven into an integrated sustainable solution.

14.1.1 AGRICULTURAL CHEMISTRY AS A TANGLED BALL OF YARN

The practice of agriculture (AG) and agricultural chemistry (AC) in the twenty-first century has as its goal to become sustainable. This goal is identical to the needs of our planet for survival. As this book has shown, there are many facets to the emerging dimensions of both AG and AC that will direct us toward sustainability. Currently, there are many practices and challenges ("threads") that are problems or obstacles for sustainability. To untangle and separate them is the first step. Let's begin to compile a list of these, knowing full well that more will be added over time. But they will also show the problems that are needed to be identified and that solutions are possible.

The "problems" are the result of the current environment, the previous social and scientific decisions, and the awareness of new imperatives. To set the stage for the work that lies ahead, the following "threads" are listed:

- To increase agricultural production with limited resources, technological advancements have been implemented throughout much of human history. These have not always been sustainable, environmentally benign, nor safe.
- A growing population along with climate change increases tension between food/non-food supply and demand. Food security is also threatened by the increased use of food crops for biofuel production, bioenergy, and other industrial uses.

DOI: 10.1201/9781003157991-14

FIGURE 14.1 Tangled problems to integrated solutions

- The world's population is estimated to reach 9 billion by 2050, which is about a twenty-five percent increase over the current population.[1] The population growth will be larger, mostly in emerging countries such as Mexico, India, China, and others.[2]
- The living standard is expected to increase in the future, it will further increase food demand and shift land use, particularly in emerging nations.
- Due to the continuous increase in the global population, the nutritional values and food quality will be a concern.[3]
- The existing farming land is being reduced because of economic and political factors, including population growth, climate, and land-use patterns, while the rapid urbanization process is continuously putting pressure on the availability of arable land.
- Crop production is already reduced due to a lack of arable land over the last few decades.[4] The total agricultural land area of the world and is expected to decrease in the future.[5]
- The quality and quantity of crops rely on climate, topography, and soil characteristics. Major features such as available nutrients, soil types, soil health, insect resistance, and quality and quantity of irrigation determine its adaptability and the quality of certain crops.
- Farmers/growers must monitor their croplands at a near to real-time scale to apply necessary measures to enhance crop production. For this reason, there is a need for smart agriculture.
- To achieve sustainable agriculture, the effects of climatic and environmental conditions should be minimal.

14.1.2 NECESSARY METHODOLOGY

The way to begin to address these issues is to start with the problems affecting sustainability. Since the agricultural sector must adapt and grow (no pun intended) under these constraints, the issues will be presented. As Figure 14.1 showed, the unraveling of these problems can lead to their solution. For example, agricultural residues provide multiple ways to provide chemicals, agricultural products, food processing products, and pharmaceuticals for the development of novel goods from the biorefinery. Future possibilities exist on developing more effective and efficient bioconversion technologies for converting agricultural leftovers into high-value goods.[6]

By producing greater quantities of food at lower costs, AG and AC can significantly improve sustainability through all applications. New chemicals and agricultural technologies will require terrestrial risk assessment and the estimation of the environmental concentrations of them remains difficult. The effects of various technologies on soil life (microbes and fauna), which play important roles in agricultural soil functioning have high environmental significance. In addition to this, adsorption/desorption, dissolution to ions, volatility in the soil column, and translocation to groundwater are some important knowledge gaps. The physicochemical degradation of any new technology through hydrolysis (water) and photodegradation (light) is important for understanding their fate in soil media. Also, it is important to provide a comprehensive understanding of risk to organisms in co-contaminated soil.[7]

14.1.3 IMPORTANCE OF AGRICULTURAL CHEMISTRY

The role of agricultural chemistry is enormous. There will be many opportunities and important obligations to society that will make the road to sustainability challenging and rewarding. The issue is certainly not that there are not problems for agricultural chemistry to solve. In fact, AC may now be the most important of the sciences in its potential to impact sustainable agriculture.[8]

Chemistry is no longer just about measurements and molecules, but about what it, as a field with unique capabilities in manipulating systems and matter, can do to understand, manipulate, and control complex systems composed (in part) of atoms and molecules. In agriculture, AC work extends from plant protection chemicals to large farms, and from improving plant growth to improving valorizing waste. To deal efficiently with these problems, AC will need to integrate "solving problems" and "generating understanding" better.[8]

There is a wide range of problems (e.g., environmental maintenance, sustainable practices, reduction in the costs of agriculture, food security, education, raising the standard of agricultural workers), which are critical to sustainable agriculture.

14.2 PROBLEMS AFFECTING SUSTAINABILITY

As we untangle the tightly intertwined threads of the industry of AG and AC in our world of the twenty-first century, many challenges/problems emerge. Analyzing them carefully, the very problem contains the seeds of potential solutions. As is the case with knitting, or any art, envisioning the future form of the final product will help guide its realization. In this section we will identify specific problems that will lead to the solutions agriculture and agricultural chemistry will integrate into a sustainable enterprise.

14.2.1 INTENSIFICATION

Though clearly successful at boosting food production in developing countries (where it is badly needed), agricultural intensification has imposed heavy environmental costs. In irrigated areas, the problems include waterlogging and salinization, soil degradation, waterborne pollutants, loss of genetic diversity, and long-term changes in soil structure. Intensive irrigation in areas with poor drainage can lead to a rise in the water table, which in turn can cause water to flush out accumulated salts. Waterlogging, on the other hand, is a common problem in the region. It lowers productivity by slowing the decomposition of organic matter, reducing nitrogen availability, and contributing to a buildup of soil toxins.[9]

Further, the influences of climate on soil microbial communities are shifted after intensive agriculture, and this affects the total and dominant microbial groups present in the soil. These results can also be used to project changes in microbial diversity in other land uses under the global climate changes.[10]

During intensification activities, chemical (synthetic) fertilizers used indiscriminately for improved production pose a major threat to long-term soil fertility, the soil environment, and its components. While the use of synthetic fertilizers has a profound impact on plant growth, it also significantly alters the makeup of the microbial community towards a detrimental low, especially N and P fertilizers.[11]

14.2.2 DIETARY CHALLENGES

The increasing use of nanotechnology and engineered nanomaterials in food and agriculture (nano-agrifoods) may provide numerous benefits to society. At the same time, there is also a chance that nano-agrifood innovations may pose new or unknown risks to human or environmental health and safety. To understand these issues and be more responsive to public concerns, researchers must discuss and adopt an emerging best practice in science and technology communities known as "responsible innovation" (RI). While key findings from this work are focused on the need for RI of nano-agrifoods in the U.S., implementation of these best practices could have positive benefits for other emerging technologies and in other national contexts as well.[12]

14.2.3 LAND USE

The larger effects of intensification of crop production as a strategy have caused near and long term effects in land use.[13] A growing global population increases the need for food, fuel and shelter, is resulting in changing food consumption patterns towards commodities that are more land intensive to supply. The location of food production is changing, due to the globalization of food supply and increasing international trade in agricultural commodities. Demands for land unrelated to food production are also increasing, for example from bioenergy feedstock supply. Greater demands for agricultural commodities can be met by intensification (improved yield by greater inputs, such as fertilizer, pesticides, or water, and or changes to management practices), agricultural expansion, or both. Improvements in agricultural yield have helped to mitigate the impact on these demands, but land use changes have still occurred. Negative environmental impacts can result from land use change or agricultural intensification, including greenhouse gas emissions, deteriorating soil quality, use of scarce water resources, and biodiversity loss.[14]

Crop production worldwide is under pressure from reductions in available arable land and sources of water.[15] This is being exacerbated by changes in human diet, reflecting a rise in meat consumption. It is likely that this will supersede population change as the greatest driver for increases in agricultural land use. Assuming the rate of population growth continues to slow, a moderation in diet, coupled with continued yield improvements may be needed to attain global food security.[14]

14.2.4 CLIMATE CHANGE

The effects of changing climate harshen the biotic stresses including increasing the population of insect/pest and disease, enhancing weed growth, reducing soil beneficial microbes, and heightening abiotic stresses including extreme drought/ flooding, large fluctuations in temperature, and salinity/alkalinity changes. All of these affect the agricultural crops in multiple ways. These effects from climate change severely challenge world sustainability.[16]

14.2.5 SMART AGRICULTURE

Smart Agriculture (SA) is a solution, but it brings additional challenges. The agricultural industry is becoming more data-centric and requiring precise, more advanced data and technologies than before. This is creating a large knowledge gap and fostering a disparity between wealthy and poor farmers, despite both being familiar with agricultural processes. The agriculture industry is being advanced by various information and advanced communication technologies, such as the Internet of Things (IoT). The rapid emergence of these advanced technologies has restructured almost all other industries, as well as advanced agriculture, which has shifted the industry from a statistical approach to a quantitative one. This radical change has shaken existing farming techniques and produced the latest prospects in a series of challenges.[1]

14.2.6 DISAPPEARING WATER

The increasing global water scarcity is affecting agriculture in many ways. Broadly, it affects whether crops can be grown and how healthy/plentiful they will be. The efficient assessment methods for crop water stress assessment will be necessary, as will ways to precisely irrigate crops. As a future perspective, the potential use of deep learning approaches will be needed to highlight the future trends. [13] Furthermore, the use of groundwater is increasing worldwide, particularly in agriculture. This can lead to environmental disasters and the degradation of groundwater quality. Questions must be raised about the intensive agricultural systems driving groundwater demand and failure to recognize the multifunctional nature of groundwater.[17]

14.2.7 SOCIETAL EDUCATION

A balanced working environment for those employed in agriculture must include working conditions and sustainable wages. It is an unquestionable objective. This must include the global agricultural workers, especially where agriculture still plays a major role in employment and the economy. While modern agricultural development is based on technical progress and the centrality of the productive optimum, it has resulted in major productivity increases through farm differentiation and concentration followed by a massive exit of farm workers. The result today is major environmental and social sustainability challenges prevent agricultural modernization from reaching individual workers. Environmental challenges result from "smart-modernization" while alternative approaches like agro-ecology remain limited to local experiences. Mainstream policies prevent addressing the continuing structural issues faced by many developing countries, as well as global sustainability issues. Improving working conditions in agriculture is part of a necessary global approach for the development of the agricultural sector, where the multiple roles that agriculture plays beyond the production of food and feed must be acknowledged.[18]

14.3 INTEGRATION OF SOLUTIONS

The problems presented above are being evaluated and solutions are beginning to emerge. The art will be found in integrating them into a sustainable enterprise. Sustainable agriculture is extremely critical for the sustenance of all the life forms across the globe. Essentially, sustainable agriculture must involve holistic management of livestock, crops, and fisheries, to make the farming process self-sustaining for a longer period. The food and agriculture sector contributes heavily in meeting the human demands and thus sustainable agricultural practices would result in positive and long-lasting benefits. Conventional farming methods failed to fully utilize the available resources. To correct this situation, modern developments can be effectively used to sustainably enhance the crop quality and productivity. [19]

14.3.1 IMPROVING POLICY ENVIRONMENT/EDUCATION

Human health will benefit from new sustainable agricultural policies. Minerals are the key factor determining human beings' optimum growth and development. The deficiencies of minerals and vitamins hinder the human normal growth and development and economic status. In the twenty-first century, macro and micronutrient deficiencies are significant challenges to improving the nutritional value of foods at the socio-economic level. Bio-fortification is a simple strategy to improve the nutritional value of the human diet. Several bio-fortification strategies, including traditional breeding, transgenic, agronomic, and modernized agriculture practices, can be employed to biofortify crops to meet nutritional needs.[20]

Practical knowledge will always be necessary to promote and maintain agricultural practices. The need for fundamental changes in the way humans interact with nature is now widely acknowledged in order to achieve sustainable development. Agriculture figures prominently in this quest, being both a major driver and a major threat to global sustainability. There is definite potential for small-scale changes that might be transferable to other regions and higher levels of governance, but more work is need to develop guidance that would lead to national/global policies that will mandate this effort.[21]

14.3.2 NEW PRACTICES AND TECHNOLOGIES

New technologies are emerging daily. Their evaluation and implementation is critical, if they can also be shown to economically and environmentally sustainable. Intensification of agriculture and changing climatic conditions, while nourishing a growing global population, requires optimizing environmental sustainability and reducing ecosystem impacts of food production. For example, the use of microbiological systems to ameliorate agricultural production in a sustainable and eco-friendly way is widely accepted as a future key-technology. However, the multitude of interaction possibilities between the numerous beneficial microbes and plants in their habitat calls for systematic analysis and management of the microbiome. These innovations will open a new avenue for designing and implementing intensive farming microbiome management approaches to maximize resource productivity and stress tolerance of agro-ecosystems, which in return will create value to the increasing worldwide population, for both food production and consumption.[13]

A second example is found in the area of water availability. It is commonly held that soil salinity adversely affects plant growth and has become a major limiting factor for agricultural development worldwide. A valuable innovation is found in saline agriculture. With the aim of better utilizing symbiotic microorganisms in saline agriculture, there is the possibility of developing plants that will thrive in saline conditions by taking advantage of newly developed genome editing technology. This will open a new avenue for capitalizing on symbiotic microorganisms to enhance plant saline tolerance for increased sustainability and yields in saline agriculture.[22]

14.3.3 FERTILIZERS/CROP PROTECTION

New work in fertilizers and crop protection is finding immediate application. Sustainable farming practices can reduce the depletion of natural resources and maintain both productivity and soil fertility through some innovative chemistry. The use of minerals that contain fertilizer nutrients in their native state is a very promising approach to reducing emissions associated with the processing chemical industries. Organic material from natural sources (food waste, manure from livestock, agricultural biomass, etc.) acts as a source of microbial culture and encourages the release of nutrients into the soil during mineral weathering.[11] This can be augmented through the use of nanomaterials in agriculture as growth regulators, pesticides, fertilizers, antimicrobial agents, and transfer of target genes. Nanotechnology can also be a potential strategy for the mitigation of abiotic stress in plant to achieve sustainable agriculture.[23]

Understandably, over the last century, the demand for food resources has been continuously increasing with the rapid population growth. Adoption of sustainable farming practices that can enhance crop production without the excessive use of fertilizers is desirable. Nanomaterials can improve plant nutrition as an alternative to traditional chemical or mineral fertilizers. Using this technology, the efficiency of micro- and macro-nutrients in plants can increase. Various nanomaterials have been successfully applied in agricultural production and have been compared to conventional fertilizers. Among the major plant nutrients, phosphorus (P) is the least accessible since most farmlands are frequently P deficient. Hence, P use efficiency should be maximized to conserve the resource base and maintain agricultural productivity.[24] The use of metal nanoparticles is shown to be a good alternative to control phytopathogenic fungi in agriculture. Numerous metal nanoparticles (e.g., Ag, Cu, Se, Ni, Mg, and Fe) have been incorporated into nanoparticles and used as potential antifungal agents.[25]

Advanced nano fertilizers with high carrier efficiency and slow and controlled release are now considered the gold standard for promoting agricultural sustainability while protecting the environment. Graphene's attractive properties include large surface area, chemical stability, mechanical stability, tunable surface chemistry and low toxicity making it a promising material on which to base agricultural delivery systems. Recent research has demonstrated considerable success in the use of graphene for agricultural applications, including its utilization as a delivery vehicle for plant nutrients and crop protection agents, as well as in post-harvest management of crops.[26]

14.3.4 IMPROVE SEED GROWTH

Global food systems are under significant pressure to provide enough food, particularly protein-rich foods according to FAO.[27] Crop production worldwide is under pressure from multiple factors, including reductions in available arable land and sources of water, along with the emergence of new pathogens and development of resistance in pre-existing pathogens. To meet these needs, different techniques have been employed to produce new crops with novel heritable mutations.[13]

Nano-priming is an innovative seed priming technology that helps to improve seed germination, seed growth, and yield by providing resistance to various stresses in plants. Nano-priming is a considerably more effective method compared to all other seed priming methods. Nano-priming induces the formation of nanopores in shoots and helps in the uptake of water absorption, activates reactive oxygen species (ROS)/antioxidant mechanisms in seeds, and forms hydroxyl radicals to loosen the walls of the cells and acts as an inducer for rapid hydrolysis of starch. The use of nano-based fertilizers and pesticides as effective materials in nano-priming and plant growth development complements the nano-priming of seeds.[28]

The application of plant growth-promoting rhizobacteria (PGPRs) can be an excellent and eco-friendly alternative to the use of chemical fertilizers. Despite the plethora of research that has been performed to date, there remains a huge knowledge gap that needs to be addressed to facilitate the commercialization of PGPRs for sustainable soilless agriculture.[29]

14.3.5 VALORIZING WASTE

Waste in the agricultural food industry exists and offers tremendous opportunity. The investigation of the potential and environmental impacts of different biomass valorization processes is focused on bioenergy. Bioenergy is a clean renewable energy to mitigate anthropogenic greenhouse gas (GHG) emissions. It has been studied through life cycle assessment (LCA) to evaluate the impact of the energy used, release of toxic substances, natural resources involved in all life cycle stages of a product or process.[30] The depletion of conventional fuel resources and environmental concerns is a driving force to explore the harvesting of biofuels from biomass. The ready availability of huge quantities of agriculture biomass can be transformed to biofuels by utilizing various procedures. However, issues such as environmental damage and competing uses of agriculture biomass need to be investigated, as well as its effect on the soil and conversion to biofuels. Techno-economic analysis performed shows the feasibility of utilizing agriculture biomass as a competitive energy source.[31]

The rising demand for food and feed results in increased agricultural waste. Simultaneously, plastic pollution increases due to hostile human activities. Bioplastics, from agricultural waste, are a sustainable way to promote the bioeconomy concept. Increasing research in bioplastics focus on developing trends in sustainable bioplastic production, agriculture waste management, biopolymer, and biological processes. Further research may create a sustainable agricultural sector and produce higher added-value products. The agro-biopolymer area needs more focus on sustainable development considering the economic, social, and environmental dimensions.[32]

14.3.6 SMART AGRICULTURE THROUGH IoT

The high demand for sufficient and safe food, and continuous damage of environment by conventional agriculture are major challenges facing the globe. The necessity of smart alternatives and more sustainable practices in food production is crucial to accommodate the steady increase in human population and careless depletion of global resources. Nanotechnology implementation in agriculture offers smart delivery systems of nutrients, pesticides, and genetic materials for enhanced soil fertility and protection, along with improved traits for better stress tolerance. Additionally, nano-based sensors are the ideal approach towards precision farming for monitoring all factors that impact agricultural productivity. Nanotechnology can also play a significant role in post-harvest food processing and packaging to reduce food contamination and wastage.[33]

The utility of sensors in agriculture is undeniable. Advanced technologies with sensors in advanced agricultural applications can lead to sustainability. Numerous sensors that can be implemented for specific agricultural practices require best management practices (e.g., land preparation, irrigation systems, insect, and disease management). The integration of all suitable techniques, from sowing to harvesting, packaging, transportation, and advanced technologies available for farmers throughout the cropping system can be improved through sensors. The continued utilization of other tools such as unmanned aerial vehicles (UAVs) for crop monitoring and other beneficiary measures help to optimize crop yields. In summary, advanced programs based on the IoT are essential tools for sustainable agriculture.[1]

The Agriculture 4.0, also called Smart Agriculture or Smart Farming, is at the origin of the production of a huge amount of data that must be collected, stored, and processed in a very short time. The agricultural sector must adjust to the realities of global warming, environmental pollution, climate change, and weather disasters. Solving the interconnections between food-water-energy-climate nexus, and achieving agricultural transformation from traditional to digital, agricultural digital transformation will play role in sustainable development. Digital agriculture combined under precision agriculture and Agriculture 4.0 require monitoring, control, prediction, and logistics. The IoT-based irrigation and fertilization systems can help enhance the efficiency of irrigation processes and minimize water and fertilizer losses in agricultural fields and greenhouses, and solutions based on drones and robotics that reduce herbicide and pesticide use. As a result, for a sustainable future, technological innovations that increase crop productivity and improve crop quality, protect the environment, provide efficient resource use and decrease input costs can help agriculture of today successfully meet many of the economic, social, and environmental challenges.[34]

14.3.7 BIOREFINERY

Achieving sustainable socio-economic development requires approaches that enhance resource use efficiencies and can address current cross-sectoral challenges in an integrated manner. Existing evidence suggests an urgent need for polycentric and transformative approaches, as global and local systems have come under strain. Transformative approaches have been identified to provide pathways towards global climate targets and protection of the environment from further degradation.[35]

The biorefinery offers an example. Fruit and vegetable production have increased significantly in response to rising demand for fresh produce resulting from population growth and a shift toward better eating habits. The biorefinery offers a way to valorize the 40% lost in food waste. Based upon biochemical and electrochemical pathways a variety of conversion pathways can be used to generate a diverse range of bio-products, including liquid products (bioethanol, volatile fatty acids), gaseous products (methane and hydrogen), and solid products (digestate and hydro-char). Because anaerobic digestion (AD) is capable of producing a varied range of products (methane, hydrogen, volatile fatty acids (VFAs), and digestate), it is an appealing option to function as the core of a biorefining operation for fruit and vegetable wastes.[36]

14.4   FROM THIS DAY FORWARD...

In section 14.2 we began to unravel the ball of problems that are obstacles to sustainability. There is more, so more work is to be done. In section 14.3 we began to list solutions. This too is incomplete, as more have been reported and more have yet to be discovered. The final fabric is beginning to take shape. The art of creating this sustainable enterprise knit of component discoveries/technologies is found in maintaining the vision of sustainability:

> *meeting the needs of the present without compromising the ability of future generations to meet their own needs.*[37]

Ultimately, if we keep our promise to future generations in mind and so align our actions, we will accomplish sustainability. (See Figure 14.2)

FIGURE 14.2   From many promising threads, we can weave a sustainable future... (See www.shutterstock.com/image-photo/weaving-shuttle-on-color-warp-1135787750)

REFERENCES

1. Khan, N., et al., *Current progress and future prospects of agriculture technology: Gateway to sustainable agriculture.* Sustainability (Switzerland), 2021. **13**(9): pp. 1–31.

2. McNabb, D.E., *Global Pathways to Water Sustainability.* Global Pathways to Water Sustainability. 2019. Springer International Publishing. 1–303.

3. Ayaz, M., et al., *Internet-of-Things (IoT)-based smart agriculture: Toward making the fields talk.* IEEE Access, 2019. **7**: pp. 129551–129583.

4. Dai, K., et al., *Trade-off relationship of arable and ecological land in urban growth when altering urban form: A case study of shenzhen, china.* Sustainability (Switzerland), 2020. **12**(23): pp. 1–20.

5. Navulur, S., A.S.C.S. Sastry, and M.N. Giri Prasad, *Agricultural management through wireless sensors and internet of things.* International Journal of Electrical and Computer Engineering, 2017. **7**(6): pp. 3492–3499.

6. Dey, M.D., et al., *Utilization of two agrowastes for adsorption and removal of methylene blue: Kinetics and isotherm studies.* Water Science and Technology, 2017. **75**(5): pp. 1138–1147.

7. Singh, D. and B.R. Gurjar, *Nanotechnology for agricultural applications: Facts, issues, knowledge gaps, and challenges in environmental risk assessment.* Journal of Environmental Management, 2022. **322**: pp. 1–16.

8. Whitesides, G.M., *Reinventing chemistry.* Angewandte Chemie – International Edition, 2015. **54**(11): pp. 3196–3209.

9. Pingali, P.L. and M. Shah, *Policy re-directions for sustainable resource use: The rice-wheat cropping system of the indo-gangetic plains.* Journal of Crop Production, 2001. **3**(2): pp. 103–118.

10. Liu, J., et al., *The effects of climate on soil microbial diversity shift after intensive agriculture in arid and semiarid regions.* Science of the Total Environment, 2022. **821**: pp. 1–8.

11. Syed, S., et al., *Bio-organic mineral fertilizer for sustainable agriculture: Current trends and future perspectives.* Minerals, 2021. **11**(12): pp. 1–11.

12. Merck, A.W., K.D. Grieger, and J. Kuzma, *How can we promote the responsible innovation of nano-agrifood research?* Environmental Science and Policy, 2022. **137**: pp. 185–190.

13. Rosegrant, M.W. and R. Livernash, *Growing more food, doing less damage.* Environment, 1996. **38**(7): pp. 67–11,28.

14. Alexander, P., et al., *Drivers for global agricultural land use change: The nexus of diet, population, yield and bioenergy.* Global Environmental Change, 2015. **35**: pp. 138–147.

15. Ali, Q., et al., *Genome engineering technology for durable disease resistance: recent progress and future outlooks for sustainable agriculture.* Frontiers in Plant Science, 2022. **13**: pp. 1–19.

16. Shahzad, A., et al., *Nexus on climate change: agriculture and possible solution to cope future climate change stresses.* Environmental Science and Pollution Research, 2021. **28**(12): pp. 14211–14232.

17. Petit, O., et al., *Learning from the past to build the future governance of groundwater use in agriculture.* Water International, 2021. **46**(7–8): pp. 1037–1059.

18. Losch, B., *Decent employment and the future of agriculture. how dominant narratives prevent addressing structural issues.* Frontiers in Sustainable Food Systems, 2022. **6**: pp. 1–16.

19. Sahoo, A., et al., *Nanotechnology for precision and sustainable agriculture: recent advances, challenges and future implications.* Nanotechnology for Environmental Engineering, 2022.

20. Jaiswal, D.K., et al., *Bio-fortification of minerals in crops: current scenario and future prospects for sustainable agriculture and human health.* Plant Growth Regulation, 2022. **98**(1): pp. 5–22.

21. Melchior, I.C. and J. Newig, *Governing transitions towards sustainable agriculture— taking stock of an emerging field of research.* Sustainability (Switzerland), 2021. **13**(2): pp. 1–27.

22. Ren, C.G., et al., *A perspective on developing a plant 'holobiont' for future saline agriculture.* Frontiers in Microbiology, 2022. **13**: pp. 1–13.

23. Farooq, M.A., et al., *The potential of nanomaterials for sustainable modern agriculture: present findings and future perspectives.* Environmental Science: Nano, 2022. **9**(6): pp. 1926–1951.

24. Basavegowda, N. and K.H. Baek, *Current and future perspectives on the use of nanofertilizers for sustainable agriculture: the case of phosphorus nanofertilizer.* 3 Biotech, 2021. **11**(7).

25. Cruz-Luna, A.R., et al., *Metal nanoparticles as novel antifungal agents for sustainable agriculture: Current advances and future directions.* Journal of Fungi, 2021. **7**(12): pp. 1–20.

26. Bhattacharya, N., et al., *Graphene as a nano-delivery vehicle in agriculture– current knowledge and future prospects.* Critical Reviews in Biotechnology, 2022. doi: 10.1080/07388551.2022.2090315.

27. Knežić, T., et al., *Using vertebrate stem and progenitor cells for cellular agriculture-state-of-the-art, challenges, and future perspectives.* Biomolecules, 2022. **12**(5): pp. 1–39.

28. Nile, S.H., et al., *Nano-priming as emerging seed priming technology for sustainable agriculture—recent developments and future perspectives.* Journal of Nanobiotechnology, 2022. **20**(1): pp. 1–31.

29. Azizoglu, U., et al., *The fate of plant growth-promoting rhizobacteria in soilless agriculture: future perspectives.* 3 Biotech, 2021. **11**(8). doi: 10.1007/ s13205-021-02941-2.

30. Alhazmi, H. and A.C.M. Loy, *A review on environmental assessment of conversion of agriculture waste to bio-energy via different thermochemical routes: Current and future trends.* Bioresource Technology Reports, 2021. **14**: pp. 1–15.

31. Saleem, M., *Possibility of utilizing agriculture biomass as a renewable and sustainable future energy source.* Heliyon, 2022. **8**(2): pp. 1–11.

32. Patel, N., M. Feofilovs, and D. Blumberga, *Agro biopolymer: a sustainable future of agriculture-state of art review.* Environmental and Climate Technologies, 2022. **26**(1): pp. 499–511.

33. Ali, S.S., et al., *Nanobiotechnological advancements in agriculture and food industry: Applications, nanotoxicity, and future perspectives.* Science of the Total Environment, 2021. **792**: pp. 1–31.

34. Dayioğlu, M.A. and U. Türker, *Digital transformation for sustainable future- agriculture 4.0: A review.* Tarim Bilimleri Dergisi, 2021. **27**(4): pp. 373–399.

35. Naidoo, D., et al., *Transitional pathways towards achieving a circular economy in the water, energy, and food sectors.* Sustainability (Switzerland), 2021. **13**(17): pp. 1–15.

36. Magama, P., I. Chiyanzu, and J. Mulopo, *A systematic review of sustainable fruit and vegetable waste recycling alternatives and possibilities for anaerobic biorefinery.* Bioresource Technology Reports, 2022. **18**: pp. 1–12.
37. Keeble, B.R., *The Brundtland Commission: environment and development to the year 2000.* Medicine and War, 1987. **3**(4): pp. 207–210.

Index

Printed in the United States
by Baker & Taylor Publisher Services